1997
NATIONAL PLUMBING & HVAC ESTIMATOR

Includes inside the back cover:
A 3½" high density disk with all the cost estimates in the book plus an estimating program for *Windows*™.

by

Ray E. Prescher

Edited by

Martin D. Kiley
and
Marques Allyn

Craftsman Book Company
6058 Corte del Cedro / P.O. Box 6500 / Carlsbad, CA 92018

Acknowledgments

The sample "Standard Form Subcontract" and "Subcontract Change Order" forms used in the final section of this book are reprinted with the permission of the publisher, the Associated General Contractors of America (National Office), 1957 E Street NW, Washington, District of Columbia 20006.

Cover photos by Pete Rintye, Hi-Country Photography
Plumbing: General Contractor, Van Dosen Construction, Lake Arrowhead, CA
            Plumbing Contractor, Crestline Plumbing, Crestline, CA

HVAC: General Contractor, Fenn-Robbins Homes, Lake Arrowhead, CA
        HVAC Contractor, Cantrell's Heating & Air Conditioning, Crestline, CA

© 1996 Craftsman Book Company
ISBN 1-57218-034-X
Published October 1996 for the year 1997.

Contents

The National Estimator (the disk in the envelope inside the back cover) is an electronic version of this manual. Use the disk to transfer the National Estimator program and all the information in this book to the hard disk of your computer. Instructions for using National Estimator begin on page 331.

How to Use This Book

This *1997 National Plumbing & HVAC Estimator* is a guide to estimating labor and material costs for plumbing, heating, ventilating and air conditioning systems in residential, commercial and industrial buildings.

 Inside the back cover of this book you'll find an envelope with a 3-1/2" floppy disk. This disk has the entire *1997 National Plumbing & HVAC Estimator* in a form that can be installed on the hard disk of a computer. Once installed, it's easy to use costs in this book to compile estimates for your jobs. Instructions for using National Estimator begin on page 331.

Costs in This Manual will apply within a few percent on a wide variety of projects. Using the information given on the pages that follow will explain how to use these costs and suggest procedures to follow when compiling estimates. Reading the remainder of this section will help you produce more reliable estimates for plumbing and HVAC work.

 Manhour Estimates in This Book will be accurate for some jobs and inaccurate for others. No manhour estimate fits all jobs because every construction project is unique. Expect installation times to vary widely from job to job, from crew to crew, and even for the same crew from day to day.

There's no way to eliminate all errors when making manhour estimates. But you can minimize the risk of a major error by:

1. Understanding what's included in the manhour estimates in this book, and

2. Adjusting the manhour estimates in this book for unusual job conditions.

The Craft@Hrs Column. Manhour estimates in this book are listed in the column headed *Craft@Hrs*. For example, on page 15 you'll see an estimate for installing 6" solvent-welded ABS DWV pipe. In the *Craft@Hrs* column opposite 6" you'll see:

P2@.055

To the left of the @ symbol you see an abbreviation for the recommended work crew. In this book, all crews will be composed of either plumbers or sheet metal workers.

- P2 identifies a crew of one plumber and one plumber's helper.

- S2 identifies a crew of one sheet metal worker and one sheet metal worker's helper.

To the right of the @ symbol you see a number. The number is the estimated manhours (not crew hours) required to install each unit of material listed. In the case of 6" solvent-welded ABS DWV pipe, P2@.055 means that .055 manhours are required to install 1 linear foot of pipe. That's the same as about 18 linear feet per man per hour or 36 linear feet per crew hour for a crew of 2 men.

Costs in the Labor $ Column are based on manhour estimates in the Craft@Hrs column. Multiply the manhour estimate by the assumed hourly labor cost to find the installation cost in the Labor $ column. For example, .055 manhours times $24.64 (the average wage for crew P2) is $1.36 (rounded down to the nearest whole cent).

Manhour Estimates include all productive labor normally associated with installing the materials described. These estimates assume normal conditions: experienced craftsmen working on reasonably well planned and managed new construction with fair to good productivity. Labor estimates also assume that materials are standard grade, appropriate tools are on hand, work done by other crafts is adequate, layout and installation are relatively uncomplicated, and working conditions don't slow progress.

All manhour estimates include tasks such as:

- Unloading and storing construction materials, tools and equipment on site.

- Working no more than two floors above or below ground level.

- Working no more than 10 feet above an uncluttered floor.

- Normal time lost due to work breaks.

- Moving tools and equipment from a storage area or truck not more than 200 feet from the work area.

- Returning tools and equipment to the storage area or truck at the end of the day.

- Planning and discussing the work to be performed.

■ Normal handling, measuring, cutting and fitting.

■ Regular cleanup of construction debris.

■ Infrequent correction or repairs required because of faulty installation.

If the work you're estimating won't be done under these conditions, you need to apply a correction factor to adjust the manhour estimates in this book to fit your job.

Applying Correction Factors. Analyze your job carefully to determine whether a labor correction factor is needed. Failure to consider job conditions is probably the most common reason for inaccurate estimates.

Use one or more of the recommended correction factors in Table 1 to adjust for unusual job conditions. To make the adjustment, multiply the manhour estimate by the appropriate conversion factor. On some jobs, several correction factors may be needed. A correction factor less than 1.00 means that favorable working conditions will reduce the manhours required.

Supervision Expense to the installing contractor is not included in the labor cost. The cost of supervision and non-productive labor varies widely from job to job. Calculate the cost of supervision and non-productive labor and add this to the estimate.

Hourly Labor Costs also vary from job to job. This book assumes an average manhour labor cost of $24.64 for plumbers and $25.39 for sheet metal workers. If these hourly labor costs are not accurate for your jobs, adjust the labor costs up or down by an appropriate percentage. Instructions on the next page explains how to make these adjustments. If you're using the National Estimator disk, it's easy to set your own wage rates. See page 336.

Hourly labor costs in this book include the basic wage, fringe benefits, the employer's contribution to welfare, pension, vacation and apprentice funds, and all tax and insurance charges based on wages. Table 2 at the top of the next page shows how hourly labor costs in this book were calculated. It's important that you understand what's included in the figures in each of the six columns in Table 2. Here's an explanation:

Column 1, the base wage per hour, is the craftsman's hourly wage. These figures are representative of what many contractors are paying plumbers, sheet metal workers and helpers in 1997.

Condition	Correction Factor
Work in large open areas, no partitions	.85
Prefabrication under ideal conditions, bench work	.90
Large quantities of repetitive work	.90
Very capable tradesmen	.95
Work 300' from storage area	1.03
Work 400' from storage area	1.05
Work 500' from storage area	1.07
Work on 3rd through 5th floors	1.05
Work on 6th through 9th floors	1.10
Work on 10th through 13th floors	1.15
Work on 14th through 17th floors	1.20
Work on 18th through 21st floors	1.25
Work over 21 floors	1.35
Work in cramped shafts	1.30
Work in commercial kitchens	1.10
Work above a sloped floor	1.25
Work in attic space	1.50
Work in crawl space	1.20
Work in a congested equipment room	1.20
Work 15' above floor level	1.10
Work 20' above floor level	1.20
Work 25' above floor level	1.30
Work 30' above floor level	1.40
Work 35' to 40' above floor level	1.50

Table 1 Recommended Correction Factors

Column 2, taxable fringe benefits, includes vacation pay, sick leave and other taxable benefits. These fringe benefits average about 5.15% of the base wage for many plumbing and HVAC contractors. This benefit is in addition to the base wage

Column 3, insurance and employer-paid taxes in percent, shows the insurance and tax rate for plumbers and sheet metal workers. The cost of insurance in this column includes workers' compensation and contractor's casualty and liability coverage. Insurance rates vary widely from state to state and depend on a contractor's loss experience. Note that taxes and insurance increase the hourly labor cost by approximately 30%. There is no legal way to avoid these costs.

Column 4, insurance and employer taxes in dollars, shows the hourly cost of taxes and insurance. Insurance and taxes are paid on the costs in both columns 1 and 2.

Column Number	1	2	3	4	5	6
Craft	Base wage per hour	Taxable fringe benefits (at 5.15% of base wage)	Insurance and employer taxes (%)	Insurance and employer taxes ($)	Non-taxable fringe benefits (at 4.55% of base wage)	Total hourly cost used in this book
Plumber	20.50	1.06	29.20	6.29	.93	28.78
Plumber's Helper	14.60	.75	29.20	4.48	.66	20.49
Sheet Metal Worker	20.40	1.05	33.85	7.26	.93	29.64
Sheet Metal Helper	14.60	.75	33.35	5.12	.66	21.13
Average for a crew of one plumber and one plumber's helper, Crew P2	17.55	.90	29.20	5.39	.80	24.64
Average for a crew of one sheet metal worker and one sheet metal worker's helper Crew S2	17.50	.90	33.60	6.19	.80	25.39

Table 2 Labor Costs Used in This Book

Column 5, non-taxable fringe benefits, includes employer paid non-taxable benefits such as medical coverage and tax-deferred pension and profit sharing plans. These fringe benefits average 4.55% of the base wage for many plumbing and HVAC contractors. The employer pays no taxes or insurance on these benefits.

Column 6, the total hourly cost in dollars, is the sum of columns 1, 2, 4, and 5. The labor costs in Column 6 were used to compute costs in the Labor $ column of this book.

Adjusting Costs in the Labor $ Column. The hourly labor costs used in this book may apply within a few percent on many of your jobs. But wage rates may be much higher or lower in some areas. If the hourly costs shown in Column 6 of Table 2 are not accurate for your work, adjust labor costs to fit your jobs.

For example, suppose your hourly labor costs are as follows:

Plumber	$14.00
Plumber's helper	$10.00
Total hourly crew cost	$24.00

Your average cost per manhour would be $12.00 ($24.00 per crew hour divided by 2 because this is a crew of two).

A labor cost of $12.00 is about 50% of the $24.64 labor cost used for crew P2. Multiply costs in the Labor $ column by .50 to find your estimated cost.

For example, notice on page 15 that the labor cost for installing a 6" solvent-welded ABS DWV pipe is $1.36 per linear foot. If installed by your plumbing crew working at an average cost of $12.00 per manhour, your estimated cost would be 50% of $1.36 or $.68 per linear foot.

Adjusting the labor costs in this book will make your estimates much more accurate. Making adjustments to labor costs is both quick and easy if you use the National Estimator disk.

 Material Costs in this manual are intended to reflect what medium- to low-volume contractors will be paying in 1997 after applying normal discounts. These costs include charges for delivery to within 25 to 30 miles of the supplier.

Overhead and Profit for the installing contractor are not included in the costs in this manual unless specifically identified in the text. Markup can vary widely with local economic conditions, competition and the installing contractor's operating expenses. Add the markup that's appropriate for your company, the job and the competitive environment.

How Accurate Are These Figures? As accurate as possible considering that the editors don't know your material suppliers, haven't seen the plans or specifications, don't know what building code applies or where the job is, had to project material costs at least six months into the future, and had no record of how much work the crew that will be assigned to the job can handle.

You wouldn't bid a job under those conditions. And I don't claim that all plumbing and HVAC work is done at these prices.

Estimating Is an Art, not a science. There is no one price that applies on all jobs. On many jobs the range between high and low bid will be 10% or more. There's room for legitimate disagreement on what the correct costs are, even when complete plans and specifications are available, the date and site are established, and labor and material costs are identical for all bidders.

No estimate fits all jobs. Good estimates are custom made for a particular project and a single contractor through judgment, analysis and experience. This book is not intended as a substitute for judgment, analysis and sound estimating practice. It's an aid in developing an informed opinion of cost, not an answer book.

Additional Costs to Consider

Here's a checklist of additional costs to consider before submitting any bid.

1. Sales taxes
2. Mobilization costs
3. Payment and performance bond costs
4. Permits and fees
5. Storage container rental costs
6. Utility costs
7. Tool costs
8. Callback costs during warranty period
9. Demobilization costs

Exclusions and Clarifications

Neither the job specifications nor the contract may identify exactly what work should be included in the plumbing and HVAC bid. Obviously, you have to identify what work is included in the job.

The most efficient way to define the scope of the work is to prepare a list of tasks not normally performed by your company and attach that list to each bid submitted. Here's a good list of work that should be excluded from your bid:

Your Bid Should Exclude

Final cleaning of plumbing fixtures

Backings for plumbing fixtures

Toilet room accessories

Electrical work, including motor starters

Electrical wiring and conduit over 100 volts

Temporary utilities

Painting, priming and surface preparation

Structural cutting, patching or repairing

Fire protection and landscape sprinklers

Equipment supports

Surveying and layout of control lines

Removal or stockpiling of excess soil

Concrete work, including forming and rebar

Setting of equipment furnished by others

Equipment and personnel hoisting

Wall and floor blockouts

Pitch pockets

The costs of performance or payment bonds

Site utilities

Asbestos removal or disposal

Contaminated soil removal or disposal

Major increases in copper material prices

Fire dampers not shown on the plans

Your Bid Should Include

Trash sweep-up only. Others haul it away.

Site utilities from building to property line only.

Piping to 5 feet outside the building only.

Plumbing & HVAC permits for your work only.

Beware of Price Changes

There's no way to be sure what prices will be in three to six months. All labor, equipment, material and subcontract prices in a bid should be based on costs anticipated when the project is expected to be built, not when the estimate is compiled. That presents a problem. Except for the installation of underground utilities, most plumbing and HVAC work is done six months to a year after the bid is submitted. When possible, get price protection in writing from your suppliers and subcontractors. If your suppliers and subs won't guarantee prices, include an escalation allowance in your bid to cover anticipated price increases.

Material Pricing Conditions

All equipment and material prices quoted by your vendors will be conditional. They usually don't include sales tax and are subject to specific payment and shipping terms. Every estimator should understand the meaning of common shipping terms. They define who pays the freight and who has responsibility for processing freight-damage claims. Here's a summary of important conditions you should understand.

F.O.B. Factory (Free On Board at the Factory): Title passes to the buyer when the goods are delivered by the seller to the freight carrier. The buyer pays the freight and is responsible for freight-damage claims.

F.O.B. Factory F.F.A. (Free On Board at the Factory, Full Freight Allowed): The title passes to the buyer when the goods are delivered by the seller to the freight carrier. The seller pays the freight charges, but the buyer is responsible for freight-damage claims.

F.O.B. (city of destination) (Free On Board to your city): The title passes to the buyer when the goods are delivered by the seller to the freight terminal in the city, or nearest city, of destination. The seller pays the freight and is responsible for freight-damage claims to the terminal. The buyer pays the freight charge and is responsible for freight-damage claims from the terminal to the final destination.

F.O.B. Job Site (Free On Board at job site, or contractor's shop): The title passes to the buyer when the goods are delivered to the job site (or shop). The seller pays the freight and is responsible for freight-damage claims.

F.A.S. Port [of a specific city] (Free Alongside Ship at the nearest port): The title passes to the buyer when goods are delivered to the ship dock or port terminal. The seller pays the freight and is responsible for freight-damage claims to the ship dock or port termi-

nal only. The buyer pays the freight and is responsible for freight-damage claims from the ship dock or port terminal to the designated delivery point.

Obviously, it's to your advantage to instruct all vendors to quote costs F.O.B. the job site or your shop.

Reducing Costs

Most construction specifications allow the use of alternative equipment and materials. It's the estimator's responsibility to select the most cost-effective products. Research and compare your costs before making any decisions. Avoid selecting any material or equipment simply because that's what you've always done.

Don't recommend plastic products such as ABS, PVC, or polypropylene pipe or corrugated flexible ducts until you've checked local code requirements. Most building codes prohibit use of these materials inside public buildings such as schools, care centers and hospitals.

It's wise to select 100% factory-packaged equipment. Beware of equipment labeled "Some assembly required." Field labor costs for mounting loose coils, motors and similar equipment are very high.

Value Engineering

Let's suppose you've submitted a combined plumbing and HVAC bid for $233,000. Your cutthroat competitor put in a bid at $4,000 less, $229,000. Obviously there's no way you're going to get the job. Right?

Not so fast! Maybe value engineering can help you win that contract — while fattening your profit margin.

Suppose the proposal you submitted had two parts. Part I is the bid for $233,000, based entirely on job plans and specs, just the way they were written. But appended to your proposal is Part II, a list of suggestions for saving money without sacrificing any of the capacity or quality designed into the system. Here's an example of what might be in Part II:

1. Deduct for providing pipe hanger spacings per UPC in lieu of specified spacings:

    $1,750.00

2. Deduct for reducing heating hot water pipe sizes by using 40 degrees F Delta T in lieu of specified 20 degrees F Delta T: $4,600.00

3. Deduct for providing pressure/temperature taps at air handling units, pumps and chillers in lieu of specified thermometers and pressure gauges: $875.00

4. Deduct for eliminating water treatment in closed piping systems: $1,800.00

5. Deduct for piping chilled and heating hot water pumps in parallel in lieu of providing 100% standby pumps: $2,900.00

Total deductions: **$11,925.00**

Adopting these suggestions would make you low bidder by nearly $8,000. A saving like that will be tempting to most owners, especially if the owner understands that your suggestions result in a system that is every bit as good and maybe better than the system as originally designed.

You're not offering to undercut the competition. Far from it. You're using knowledge and experience to create better value for the owner. That's called value engineering and it's likely to win the respect of nearly all cost-conscious owners.

Notice that reducing costs is only part of what value engineering is all about. You don't cut costs at the expense of system quality, integrity, capacity or performance.

Don't waste your time, and your client's, by offering to substitute cheaper or lower-quality fixtures or equipment. Any cutthroat contractor with a price list can do that. Recommend the use of inferior materials and you'll be associated with the inferior goods you promote. Some owners consider even the suggestion to be insulting.

The recommendations you make (like most of those in the example) will require design changes. You can expect to be examined (or even challenged) on these points. Be ready to explain and defend each of your suggestions. Convince the client (or the design engineer) that your ideas are based on sound engineering principles and you're well on the way to winning the owner's confidence and the contract.

Now, let's go back to the list and see how we might justify the five value engineering recommendations.

1. **Pipe Hanger Spacing.** The pipe hanger spacings recommended in the Uniform Plumbing Code (UPC) are calculated by experienced, professional structural engineers. The safety factors used in these calculations are very conservative. They've been widely used for many years and have proved to be more than adequate. There's no need for more hangers than the UPC requires.

2. **Changing HHW Delta T.** In hydronic heating systems, heat measured in Btus is pumped to terminal units. The proposed change of the Delta T, from 20 degrees F to 40 degrees F, has no effect whatsoever on how many Btus the system delivers. You're not changing anything but the volume of water being pumped. At lower volume levels, the size of the pump, the pipe and the pipe insulation can all be reduced. Not one of these changes will affect the system's ability to transmit heat. Furthermore, operating costs will also drop, since less pump horsepower will be needed to run the smaller pump.

3. **Thermometers/Pressure Gauges.** Thermometers and pressure gauges installed on or near vibrating machinery have a very short life expectancy. Gauges quickly lose accuracy under harsh conditions. Readings will become less and less reliable. That's potentially dangerous. You can avoid this problem by using insertion-type pressure/temperature taps instead. Store these sensitive gauges in a desk drawer or a tool crib when not in use. Safely stored, they're protected from damage. They'll give accurate readings longer and won't need to be replaced as often. And they're simple to use. Just insert a gauge in one of the conveniently located taps. Make the reading, then remove the gauge and put it away.

4. **Water Treatment.** ITT Bell & Gossett has done studies on corrosion in closed hydronic systems that have a make-up water rate of no more than 5% per year. These studies show that corrosion virtually stops when entrained air is either removed or depleted. No water treatment is needed in this closed system.

5. **100% Standby Pumps.** Two pumps piped and operated in parallel are more economical. Even if one pump fails, the other pump can maintain delivery at 75 to 80% of the designed flow rate. That's usually adequate for emergency operation.

These cost-saving ideas are small, but could tip the balance in your favor. I hope they demonstrate the potential that value engineering has when bidding jobs. Any time you're compiling an estimate, keep an eye out for ways to save money or reduce the owner's cost. Jot a note to yourself about each potential saving you identify. Before submitting the bid, make a list of your alternate suggestions. Maybe best of all, markup on your value engineering suggestions can be higher than your normal markup. If value engineering can cut costs by $10,000, maybe as much as $4,000 of that should end up in your pocket!

The Estimating Procedure

Every plumbing and HVAC estimator works under deadline pressure. You'll seldom have the luxury of spending as much time as you would like on an estimate. Estimators who aren't organized waste valuable time and tend to make careless errors. Try to be well-organized and consistent in your approach to estimating. For most projects, I recommend that you follow the procedures listed below and in the order listed:

1. Get a second set of project drawings and specifications for use by your suppliers and subcontractors. Remember that your subs and suppliers need access to the plans and specs and time to prepare their quotes.

2. Study the plans and specs carefully. Highlight important items. Make a list of specific tasks that require labor unit correction factors. The estimate is never complete until you're totally familiar with the project and the applicable construction codes.

3. Get the general contractor or owner to identify the proposed construction schedule and subcontractor lay-down (storage) area. Work schedule and site conditions always affect your costs.

4. Contact all potential suppliers and subcontractors as early as possible. Set a time when each can come to your office to make their take-offs from the spare set of contract documents.

When this important preliminary work is done, or in progress, it's time to begin your detailed take-off.

Guidelines for Good Estimating

You can compile estimates on a legal pad, a printed estimating form or on a computer. Regardless of the method, these guidelines will apply:

List Each Cost Separately on your take-off sheet. Don't combine system estimates, even if the materials are the same type. A combined system estimate may have to be completely redone if materials for one system are changed at a later date. Use the Estimate Detail Sheet on the facing page if you don't already have a good material take-off form.

Use Engineer's Identification Numbers when listing equipment. The word pump without any other description is ambiguous when there are several pumps included in the project.

Don't Forget Labor Adjustment factors if your labor costs are significantly higher or lower than the costs used in this book. See instructions on page 7 for adjusting labor costs.

Use Colored Pencils or highlighters to mark the items you've taken off and listed. Use a different color for each piping or ducting system.

Log Telephone Quotes and other important phone conversations on a telephone quote form. See the sample on page 14.

Project Estimated Costs for labor, material and equipment to the time when the work is expected to be done, not when the job is being estimated.

The only good estimate is a complete estimate. You've probably heard this saying, "He who makes the most mistakes is likely to be low bidder, and live to regret it."

Preparing the Proposal

It's both common courtesy and good business practice to deliver an unpriced copy of your bid or proposal letter to the general contractor three or four days before the bid deadline date. This gives the contractor time to study your proposal and obtain alternate pricing for items you may have excluded. To avoid misunderstandings, make sure your proposals include, as a minimum, the following elements:

1. The complete name and address of the proposed project.

2. Specification title and issue date.

3. A complete listing of drawings and their issue or revision date.

4. A complete list of addenda and their dates of issue.

5. A list of specification section numbers covered by your proposal.

6. A list of exclusions, clarifications and assumptions.

Your final bid can be phoned in or sent by fax, but it should reach the general contractor or owner no more than five or ten minutes before the bid deadline. Prices submitted too early may have to be revised because of last-minute price changes by subcontractors or suppliers.

Estimate Detail Sheet

Data carried forward from Take-Off Quantity Survey Sheet(s)

Company/Department _____

Project _____

Address _____

Job description _____ Estimate # _____

CSI Division/ Account _____ Estimate due _____

Estimator _____ Date _____

Checked by _____ Date _____

Notes:

Item Description	Quantity	Unit	Crew @ MH/Unit	Manhours Ext.	Materials		Labor		Equipment		Subcontract		Total
					Unit $	Ext. $	Unit $	Ext. $	Unit $	Ext. $	Unit $	Ext. $	$
Totals This Sheet				Manhours	Material $		Labor $		Equipment $		Subcontract $		Total $

Carry totals forward to Estimate Summary Sheet

Estimate # _____ Estimate Details Sheet _____ of _____

Quotation Sheet

Job: _____

Supplier: _____

Salesperson: _____ **Phone No:** _____

Per Plans/Specs: _____ **Freight:** _____ **Terms:** _____

Description	Delivery Time	Price

By: _____

Record of Telephone Conversation

Date:_____ Time:_____ Project:_____

Telecon with:_____

Company:_____ Phone No:_____

Subject:_____

Details of Conversation:_____

_____

_____

_____

_____

_____

_____

_____

_____

_____

_____

_____

_____

_____

_____

_____

_____

_____

_____

_____

_____

By:_____

ABS (Acrylonitrile Butadiene Styrene) ASTM D 2661-73 is made from various grades of rubber-based resins and is primarily used for non-pressurized drain, waste and vent (DWV) systems.

ABS can also be used for low-temperature (100 degree F. maximum) water distribution systems, but consult the manufacturers for the proper ASTM type, maximum pressure limitations and recommended joint solvent.

The current Uniform Plumbing Code (UPC) states in Sections 401 and 503 that "ABS piping systems shall be limited to those structures where combustible construction is allowed." This does not preclude its use for underground DWV systems.

This section has been arranged to save the estimator's time by including all normally-used system components such as pipe, fittings, hanger assemblies and riser clamps under one heading. The cost estimates in this section are based on the conditions, limitations and wage rates described in the section "How to Use This Book" beginning on page 5.

Description	Craft@Hrs	Unit	Material $	Labor $	Total $

ABS DWV pipe with solvent-weld joints

1½"	P2@.020	LF	.37	.49	.86
2"	P2@.025	LF	.50	.62	1.12
3"	P2@.030	LF	1.03	.74	1.77
4"	P2@.040	LF	1.48	.99	2.47
6"	P2@.055	LF	3.61	1.36	4.97

ABS DWV 1/8 bend with solvent-weld joints

1½"	P2@.120	Ea	1.04	2.96	4.00
2"	P2@.125	Ea	1.14	3.08	4.22
3"	P2@.175	Ea	3.26	4.31	7.57
4"	P2@.250	Ea	6.46	6.16	12.62
6"	P2@.325	Ea	38.90	8.01	46.91

ABS DWV 1/8 street bend with solvent-weld joints

1½"	P2@.060	Ea	1.08	1.48	2.56
2"	P2@.065	Ea	1.47	1.60	3.07
3"	P2@.088	Ea	3.63	2.17	5.80
4"	P2@.125	Ea	8.21	3.08	11.29
6"	P2@.163	Ea	47.50	4.02	51.52

ABS DWV 1/4 bend with solvent-weld joints

1½"	P2@.120	Ea	.88	2.96	3.84
2"	P2@.125	Ea	1.25	3.08	4.33
3"	P2@.175	Ea	3.87	4.31	8.18
4"	P2@.250	Ea	7.58	6.16	13.74
6"	P2@.325	Ea	46.30	8.01	54.31

ABS DWV 1/4 street bend with solvent-weld joints

1½"	P2@.060	Ea	1.53	1.48	3.01
2"	P2@.070	Ea	2.16	1.72	3.88
3"	P2@.088	Ea	4.55	2.17	6.72
4"	P2@.125	Ea	12.90	3.08	15.98
6"	P2@.175	Ea	49.40	4.31	53.71

ABS, DWV with Solvent-Weld Joints

Description	Craft@Hrs	Unit	Material $	Labor $	Total $
ABS DWV 1/4 bend heel outlet with solvent-weld joints					
3 x 2"	P2@.235	Ea	7.58	5.79	13.37
4 x 2"	P2@.325	Ea	17.90	8.01	25.91
ABS DWV closet bend with solvent-weld joints					
3 x 4"	P2@.250	Ea	8.09	6.16	14.25
ABS DWV closet flange with solvent-weld joints					
4"	P2@.125	Ea	4.59	3.08	7.67
ABS DWV P-trap with solvent-weld joints					
1½"	P2@.150	Ea	3.29	3.70	6.99
2"	P2@.190	Ea	5.32	4.68	10.00
3"	P2@.260	Ea	24.40	6.41	30.81
4"	P2@.350	Ea	49.90	8.62	58.52
ABS DWV sanitary tee with solvent-weld joints					
1½"	P2@.150	Ea	1.50	3.70	5.20
2"	P2@.190	Ea	2.25	4.68	6.93
3"	P2@.260	Ea	5.73	6.41	12.14
4"	P2@.350	Ea	11.00	8.62	19.62
6"	P2@.450	Ea	68.60	11.10	79.70
ABS DWV reducing sanitary tee with solvent-weld joints					
2 x 1½"	P2@.160	Ea	2.16	3.94	6.10
3 x 1½"	P2@.210	Ea	4.38	5.17	9.55
3 x 2"	P2@.235	Ea	4.42	5.79	10.21
4 x 2"	P2@.300	Ea	14.80	7.39	22.19
4 x 3"	P2@.325	Ea	17.40	8.01	25.41
ABS DWV double sanitary tee with solvent-weld joints					
1½"	P2@.210	Ea	4.55	5.17	9.72
2"	P2@.250	Ea	6.64	6.16	12.80
3"	P2@.350	Ea	18.60	8.62	27.22
4"	P2@.475	Ea	29.90	11.70	41.60
ABS DWV combination with solvent-weld joints					
1½"	P2@.150	Ea	4.81	3.70	8.51
2"	P2@.190	Ea	6.26	4.68	10.94
3"	P2@.260	Ea	10.30	6.41	16.71
4"	P2@.350	Ea	20.20	8.62	28.82

Description	Craft@Hrs	Unit	Material $	Labor $	Total $
ABS DWV reducing combination with solvent-weld joints					
2 x 1½"	P2@.160	Ea	5.07	3.94	9.01
3 x 1½"	P2@.210	Ea	12.00	5.17	17.17
3 x 2"	P2@.235	Ea	7.50	5.79	13.29
4 x 2"	P2@.300	Ea	16.70	7.39	24.09
4 x 3"	P2@.325	Ea	18.10	8.01	26.11
ABS DWV bushing with solvent-weld joints					
2 x 1½"	P2@.062	Ea	.71	1.53	2.24
3 x 1½"	P2@.088	Ea	3.80	2.17	5.97
3 x 2"	P2@.088	Ea	2.42	2.17	4.59
4 x 2"	P2@.125	Ea	8.77	3.08	11.85
4 x 3"	P2@.125	Ea	5.08	3.08	8.16
6 x 4"	P2@.163	Ea	25.20	4.02	29.22
ABS DWV FTP adapter					
1½"	P2@.080	Ea	1.19	1.97	3.16
2"	P2@.085	Ea	1.92	2.09	4.01
3"	P2@.155	Ea	5.22	3.82	9.04
4"	P2@.220	Ea	6.57	5.42	11.99
6"	P2@.285	Ea	22.40	7.02	29.42
ABS DWV test cap					
1½"	P2@.030	Ea	.65	.74	1.39
2"	P2@.040	Ea	.68	.99	1.67
3"	P2@.050	Ea	.88	1.23	2.11
4"	P2@.090	Ea	1.04	2.22	3.26
ABS DWV cleanout plug					
1½"	P2@.090	Ea	.78	2.22	3.00
2"	P2@.120	Ea	.99	2.96	3.95
3"	P2@.240	Ea	1.70	5.91	7.61
4"	P2@.320	Ea	3.44	7.88	11.32
6"	P2@.450	Ea	10.50	11.10	21.60
ABS DWV cleanout tee with solvent-weld joints					
1½"	P2@.120	Ea	5.66	2.96	8.62
2"	P2@.125	Ea	7.10	3.08	10.18
3"	P2@.175	Ea	12.20	4.31	16.51
4"	P2@.250	Ea	22.80	6.16	28.96

ABS, DWV with Solvent-Weld Joints

Description	Craft@Hrs	Unit	Material $	Labor $	Total $
ABS DWV wye with solvent-weld joints					
1½"	P2@.150	Ea	3.01	3.70	6.71
2"	P2@.190	Ea	3.09	4.68	7.77
3"	P2@.260	Ea	6.85	6.41	13.26
4"	P2@.350	Ea	12.80	8.62	21.42
6"	P2@.450	Ea	64.00	11.10	75.10
ABS DWV reducing wye with solvent-weld joints					
2 x 1½"	P2@.160	Ea	5.32	3.94	9.26
3 x 1½"	P2@.210	Ea	7.04	5.17	12.21
3 x 2"	P2@.235	Ea	5.32	5.79	11.11
4 x 2"	P2@.300	Ea	10.30	7.39	17.69
4 x 3"	P2@.325	Ea	12.00	8.01	20.01
6 x 4"	P2@.400	Ea	52.90	9.86	62.76
ABS DWV reducer with solvent-weld joints					
2 x 1½"	P2@.125	Ea	1.70	3.08	4.78
3 x 1½"	P2@.150	Ea	5.38	3.70	9.08
3 x 2"	P2@.150	Ea	4.51	3.70	8.21
4 x 2"	P2@.185	Ea	8.89	4.56	13.45
4 x 3"	P2@.210	Ea	9.31	5.17	14.48
ABS DWV coupling with solvent-weld joints					
1½"	P2@.120	Ea	.67	2.96	3.63
2"	P2@.125	Ea	.71	3.08	3.79
3"	P2@.175	Ea	1.92	4.31	6.23
4"	P2@.250	Ea	3.16	6.16	9.32
6"	P2@.325	Ea	19.20	8.01	27.21
Hanger with swivel assembly					
1½"	P2@.450	Ea	1.02	11.10	12.12
2"	P2@.450	Ea	1.05	11.10	12.15
3"	P2@.450	Ea	3.45	11.10	14.55
4"	P2@.450	Ea	4.22	11.10	15.32
6"	P2@.450	Ea	7.07	11.10	18.17
Riser clamp					
1½"	P2@.110	Ea	3.90	2.71	6.61
2"	P2@.115	Ea	4.09	2.83	6.92
3"	P2@.120	Ea	4.70	2.96	7.66
4"	P2@.125	Ea	5.93	3.08	9.01
6"	P2@.130	Ea	9.86	3.20	13.06

Asbestos-Cement, Class 2400 or 3000 with Mechanical Joints

Description	Craft@Hrs	Unit	Material $	Labor $	Total $
Asbestos-cement pipe with mechanical joints					
6"	P2@.100	LF	2.94	2.46	5.40
8"	P2@.140	LF	4.66	3.45	8.11
10"	P2@.200	LF	7.23	4.93	12.16
12"	P2@.260	LF	9.53	6.41	15.94
14"	P2@.300	LF	13.00	7.39	20.39
16"	P2@.350	LF	16.00	8.62	24.62
Asbestos-cement 1/16 bend with mechanical joints					
4"	P2@1.00	Ea	69.60	24.60	94.20
6"	P2@1.25	Ea	93.70	30.80	124.50
8"	P2@1.70	Ea	129.00	41.90	170.90
10"	P2@2.20	Ea	196.00	54.20	250.20
12"	P2@2.70	Ea	251.00	66.50	317.50
14"	P2@3.40	Ea	449.00	83.80	532.80
16"	P2@3.90	Ea	647.00	96.10	743.10
Asbestos-cement 1/8 bend with mechanical joints					
4"	P2@1.00	Ea	67.90	24.60	92.50
6"	P2@1.25	Ea	92.00	30.80	122.80
8"	P2@1.70	Ea	129.00	41.90	170.90
10"	P2@2.20	Ea	188.00	54.20	242.20
12"	P2@2.70	Ea	250.00	66.50	316.50
14"	P2@3.40	Ea	442.00	83.80	525.80
16"	P2@3.90	Ea	634.00	96.10	730.10
Asbestos-cement 1/4 bend with mechanical joints					
4"	P2@1.00	Ea	75.90	24.60	100.50
6"	P2@1.25	Ea	102.00	30.80	132.80
8"	P2@1.70	Ea	147.00	41.90	188.90
10"	P2@2.20	Ea	225.00	54.20	279.20
12"	P2@2.70	Ea	297.00	66.50	363.50
14"	P2@3.40	Ea	559.00	83.80	642.80
16"	P2@3.90	Ea	821.00	96.10	917.10
Asbestos-cement tee or wye with mechanical joints					
4"	P2@1.40	Ea	114.00	34.50	148.50
6"	P2@1.75	Ea	152.00	43.10	195.10
8"	P2@2.40	Ea	217.00	59.10	276.10
10"	P2@3.10	Ea	374.00	76.40	450.40
12"	P2@3.80	Ea	476.00	93.60	569.60
14"	P2@4.75	Ea	858.00	117.00	975.00
16"	P2@5.45	Ea	1,240.00	134.00	1,374.00

Asbestos-Cement, Class 2400 or 3000 with Mechanical Joints

Description	Craft@Hrs	Unit	Material $	Labor $	Total $
Asbestos-cement reducer with mechanical joints					
6" x 4"	P2@1.20	Ea	80.50	29.60	110.10
8" x 6"	P2@1.50	Ea	118.00	37.00	155.00
10" x 8"	P2@1.95	Ea	172.00	48.00	220.00
12" x 10"	P2@2.45	Ea	239.00	60.40	299.40

Carbon Steel, Schedule 40 with 150# M.I. Fittings & Threaded Joints

Schedule 40 carbon steel (ASTM A-53 and A-120) pipe with 150 pound malleable iron threaded fittings is commonly used for heating hot water and chilled water systems where operating pressures do not exceed 125 PSIG and temperatures are below 250 degrees F., for pipe sizes 2 inches and smaller. For pipe sizes 2½ inches and larger, wrought steel fittings with either welded or grooved joints should be used.

This section has been arranged to save the estimator's time by including all normally-used system components such as pipe, fittings, valves, hanger assemblies, riser clamps and miscellaneous items under one heading. Other accessories can be found under "Plumbing and Piping Specialties." The cost estimates in this section are based on the conditions, limitations and wage rates described in the section "How to Use This Book" beginning on page 5.

Description	Craft@Hrs	Unit	Material $	Labor $	Total $

Schedule 40 carbon steel pipe with threaded joints

Description	Craft@Hrs	Unit	Material $	Labor $	Total $
1/2"	P2@.060	LF	.68	1.48	2.16
3/4"	P2@.070	LF	.77	1.72	2.49
1"	P2@.090	LF	1.16	2.22	3.38
1¼"	P2@.100	LF	1.51	2.46	3.97
1½"	P2@.110	LF	1.80	2.71	4.51
2"	P2@.120	LF	2.43	2.96	5.39
2½"	P2@.130	LF	3.75	3.20	6.95
3"	P2@.150	LF	4.96	3.70	8.66
4"	P2@.260	LF	7.06	6.41	13.47

150# malleable iron 45 degree ell with threaded joints

Description	Craft@Hrs	Unit	Material $	Labor $	Total $
1/2"	P2@.120	Ea	1.49	2.96	4.45
3/4"	P2@.130	Ea	1.83	3.20	5.03
1"	P2@.180	Ea	2.33	4.44	6.77
1¼"	P2@.240	Ea	4.10	5.91	10.01
1½"	P2@.300	Ea	5.07	7.39	12.46
2"	P2@.380	Ea	7.59	9.36	16.95
2½"	P2@.510	Ea	22.00	12.60	34.60
3"	P2@.640	Ea	28.50	15.80	44.30
4"	P2@.850	Ea	55.40	20.90	76.30

150# malleable iron 90 degree ell with threaded joints

Description	Craft@Hrs	Unit	Material $	Labor $	Total $
1/2"	P2@.120	Ea	.92	2.96	3.88
3/4"	P2@.130	Ea	1.12	3.20	4.32
1"	P2@.180	Ea	1.95	4.44	6.39
1¼"	P2@.240	Ea	3.18	5.91	9.09
1½"	P2@.300	Ea	4.18	7.39	11.57
2"	P2@.380	Ea	7.15	9.36	16.51
2½"	P2@.510	Ea	15.60	12.60	28.20
3"	P2@.640	Ea	23.30	15.80	39.10
4"	P2@.850	Ea	54.30	20.90	75.20

Carbon Steel, Schedule 40 with 150# M.I. Fittings & Threaded Joints

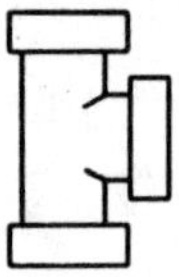

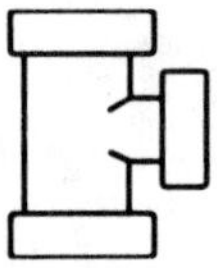

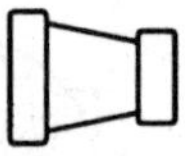

Description	Craft@Hrs	Unit	Material $	Labor $	Total $
150# malleable iron tee, threaded joints					
1/2"	P2@.180	Ea	1.24	4.44	5.68
3/4"	P2@.190	Ea	1.78	4.68	6.46
1"	P2@.230	Ea	3.02	5.67	8.69
1¼"	P2@.310	Ea	4.90	7.64	12.54
1½"	P2@.390	Ea	6.09	9.61	15.70
2"	P2@.490	Ea	10.30	12.10	22.40
2½"	P2@.660	Ea	21.70	16.30	38.00
3"	P2@.830	Ea	31.90	20.50	52.40
4"	P2@1.10	Ea	76.50	27.10	103.60
150# malleable iron reducing tee, threaded joints					
3/4 x 1/2"	P2@.180	Ea	2.63	4.44	7.07
1 x 1/2"	P2@.220	Ea	3.34	5.42	8.76
1 x 3/4"	P2@.220	Ea	3.34	5.42	8.76
1¼ x 1/2"	P2@.290	Ea	6.09	7.15	13.24
1¼ x 3/4"	P2@.290	Ea	5.35	7.15	12.50
1¼ x 1"	P2@.290	Ea	5.79	7.15	12.94
1½ x 1/2"	P2@.370	Ea	7.14	9.12	16.26
1½ x 3/4"	P2@.370	Ea	6.75	9.12	15.87
1½ x 1"	P2@.370	Ea	6.75	9.12	15.87
1½ x 1¼"	P2@.370	Ea	8.34	9.12	17.46
2 x 1"	P2@.460	Ea	10.60	11.30	21.90
2 x 1¼"	P2@.460	Ea	11.40	11.30	22.70
2 x 1½"	P2@.460	Ea	10.80	11.30	22.10
4 x 1½"	P2@1.05	Ea	79.20	25.90	105.10
4 x 2"	P2@1.05	Ea	76.30	25.90	102.20
4 x 3"	P2@1.05	Ea	79.20	25.90	105.10
150# malleable iron reducer, threaded joints					
3/4 x 1/2"	P2@.130	Ea	1.67	3.20	4.87
1 x 1/2"	P2@.160	Ea	2.59	3.94	6.53
1 x 3/4"	P2@.160	Ea	2.59	3.94	6.53
1¼ x 1/2"	P2@.210	Ea	3.63	5.17	8.80
1¼ x 1"	P2@.210	Ea	3.63	5.17	8.80
1½ x 1"	P2@.270	Ea	4.62	6.65	11.27
1½ x 1¼"	P2@.270	Ea	3.98	6.65	10.63
2 x 1¼"	P2@.340	Ea	6.81	8.38	15.19
2 x 1½"	P2@.340	Ea	5.95	8.38	14.33
3 x 2"	P2@.530	Ea	17.70	13.10	30.80
3 x 2½"	P2@.530	Ea	20.60	13.10	33.70

Carbon Steel, Schedule 40 with 150# M.I. Fittings & Threaded Joints

Description	Craft@Hrs	Unit	Material $	Labor $	Total $

150# malleable iron cross, threaded joints

Description	Craft@Hrs	Unit	Material $	Labor $	Total $
1/2"	P2@.280	Ea	4.48	6.90	11.38
3/4"	P2@.320	Ea	5.47	7.88	13.35
1"	P2@.360	Ea	6.72	8.87	15.59
1¼"	P2@.480	Ea	10.90	11.80	22.70
1½"	P2@.600	Ea	13.50	14.80	28.30
2"	P2@.700	Ea	22.20	17.20	39.40
2½"	P2@1.02	Ea	41.00	25.10	66.10
3"	P2@1.28	Ea	56.90	31.50	88.40

150# malleable iron cap, threaded joint

Description	Craft@Hrs	Unit	Material $	Labor $	Total $
1/2"	P2@.090	Ea	.94	2.22	3.16
3/4"	P2@.100	Ea	1.25	2.46	3.71
1"	P2@.140	Ea	1.54	3.45	4.99
1¼"	P2@.180	Ea	1.98	4.44	6.42
1½"	P2@.230	Ea	2.70	5.67	8.37
2"	P2@.290	Ea	3.99	7.15	11.14
2½"	P2@.380	Ea	8.90	9.36	18.26
3"	P2@.480	Ea	13.20	11.80	25.00
4"	P2@.640	Ea	22.40	15.80	38.20

150# malleable iron plug, threaded joint

Description	Craft@Hrs	Unit	Material $	Labor $	Total $
1/2"	P2@.090	Ea	.75	2.22	2.97
3/4"	P2@.100	Ea	.80	2.46	3.26
1"	P2@.140	Ea	.80	3.45	4.25
1¼"	P2@.180	Ea	1.28	4.44	5.72
1½"	P2@.230	Ea	1.81	5.67	7.48
2"	P2@.290	Ea	2.11	7.15	9.26
2½"	P2@.380	Ea	5.12	9.36	14.48
3"	P2@.480	Ea	5.95	11.80	17.75
4"	P2@.640	Ea	10.90	15.80	26.70

150# malleable iron union, threaded joints

Description	Craft@Hrs	Unit	Material $	Labor $	Total $
1/2"	P2@.140	Ea	4.06	3.45	7.51
3/4"	P2@.150	Ea	6.73	3.70	10.43
1"	P2@.210	Ea	6.10	5.17	11.27
1¼"	P2@.280	Ea	8.73	6.90	15.63
1½"	P2@.360	Ea	10.90	8.87	19.77
2"	P2@.450	Ea	12.50	11.10	23.60
2½"	P2@.610	Ea	37.40	15.00	52.40
3"	P2@.760	Ea	45.00	18.70	63.70

Carbon Steel, Schedule 40 with 150# M.I. Fittings & Threaded Joints

Description	Craft@Hrs	Unit	Material $	Labor $	Total $
150# malleable iron coupling, threaded joints					
1/2"	P2@.120	Ea	1.25	2.96	4.21
3/4"	P2@.130	Ea	1.48	3.20	4.68
1"	P2@.180	Ea	2.24	4.44	6.68
1¼"	P2@.240	Ea	2.86	5.91	8.77
1½"	P2@.300	Ea	3.83	7.39	11.22
2"	P2@.380	Ea	5.58	9.36	14.94
Class 125 bronze body gate valve, threaded					
1/2"	P2@.210	Ea	13.90	5.17	19.07
3/4"	P2@.250	Ea	17.10	6.16	23.26
1"	P2@.300	Ea	23.20	7.39	30.59
1¼"	P2@.400	Ea	33.00	9.86	42.86
1½"	P2@.450	Ea	40.10	11.10	51.20
2"	P2@.500	Ea	50.00	12.30	62.30
2½"	P2@.750	Ea	133.00	18.50	151.50
3"	P2@.950	Ea	184.00	23.40	207.40
Class 125 iron body gate valve, flanged					
2"	P2@.500	Ea	165.00	12.30	177.30
2½"	P2@.600	Ea	210.00	14.80	224.80
3"	P2@.750	Ea	228.00	18.50	246.50
4"	P2@1.35	Ea	359.00	33.30	392.30
Class 125 bronze body globe valve, threaded					
1/2"	P2@.210	Ea	24.10	5.17	29.27
3/4"	P2@.250	Ea	30.60	6.16	36.76
1"	P2@.300	Ea	40.80	7.39	48.19
1¼"	P2@.400	Ea	54.20	9.86	64.06
1½"	P2@.450	Ea	69.90	11.10	81.00
2"	P2@.500	Ea	106.00	12.30	118.30
Class 125 iron body globe valve, flanged					
2"	P2@.500	Ea	427.00	12.30	439.30
2½"	P2@.600	Ea	452.00	14.80	466.80
3"	P2@.750	Ea	522.00	18.50	540.50
4"	P2@1.35	Ea	746.00	33.30	779.30
200 PSI iron body butterfly valve, lug type					
2"	P2@.200	Ea	77.60	4.93	82.53
2½"	P2@.300	Ea	81.00	7.39	88.39
3"	P2@.400	Ea	89.80	9.86	99.66
4"	P2@.500	Ea	112.00	12.30	124.30

Carbon Steel, Schedule 40 with 150# M.I. Fittings & Threaded Joints

Description	Craft@Hrs	Unit	Material $	Labor $	Total $
200 PSI iron body butterfly valve, wafer type					
2"	P2@.200	Ea	66.80	4.93	71.73
2½"	P2@.300	Ea	69.50	7.39	76.89
3"	P2@.400	Ea	76.30	9.86	86.16
4"	P2@.500	Ea	95.20	12.30	107.50
Class 125 bronze body 2-piece ball valve, threaded					
1/2"	P2@.210	Ea	5.17	5.17	10.34
3/4"	P2@.250	Ea	8.53	6.16	14.69
1"	P2@.300	Ea	10.70	7.39	18.09
1¼"	P2@.400	Ea	18.00	9.86	27.86
1½"	P2@.450	Ea	23.00	11.10	34.10
2"	P2@.500	Ea	28.80	12.30	41.10
Class 125 bronze body swing check valve, threaded					
1/2"	P2@.210	Ea	17.50	5.17	22.67
3/4"	P2@.250	Ea	21.20	6.16	27.36
1"	P2@.300	Ea	29.20	7.39	36.59
1¼"	P2@.400	Ea	40.50	9.86	50.36
1½"	P2@.450	Ea	45.70	11.10	56.80
2"	P2@.500	Ea	69.70	12.30	82.00
Class 125 iron body swing check valve, flanged					
2"	P2@.500	Ea	174.00	12.30	186.30
2½"	P2@.600	Ea	221.00	14.80	235.80
3"	P2@.750	Ea	240.00	18.50	258.50
4"	P2@1.35	Ea	378.00	33.30	411.30
Class 125 iron body silent check valve, wafer type					
2"	P2@.500	Ea	97.20	12.30	109.50
2½"	P2@.600	Ea	112.00	14.80	126.80
3"	P2@.750	Ea	124.00	18.50	142.50
4"	P2@1.35	Ea	167.00	33.30	200.30
Class 150 bronze body wye strainer, threaded					
1/2"	P2@.210	Ea	21.30	5.17	26.47
3/4"	P2@.250	Ea	23.30	6.16	29.46
1"	P2@.300	Ea	34.90	7.39	42.29
1¼"	P2@.400	Ea	56.10	9.86	65.96
1½"	P2@.450	Ea	75.20	11.10	86.30
2"	P2@.500	Ea	100.00	12.30	112.30

Carbon Steel, Schedule 40 with 150# M.I. Fittings & Threaded Joints

Description	Craft@Hrs	Unit	Material $	Labor $	Total $
Class 125 iron body wye strainer, flanged					
2"	P2@.500	Ea	180.00	12.30	192.30
2½"	P2@.600	Ea	183.00	14.80	197.80
3"	P2@.750	Ea	212.00	18.50	230.50
4"	P2@1.35	Ea	382.00	33.30	415.30
Installation of 2-way control valve					
1/2"	P2@.210	Ea	--	5.17	5.17
3/4"	P2@.275	Ea	--	6.78	6.78
1"	P2@.350	Ea	--	8.62	8.62
1¼"	P2@.430	Ea	--	10.60	10.60
1½"	P2@.505	Ea	--	12.40	12.40
2"	P2@.675	Ea	--	16.60	16.60
2½"	P2@.830	Ea	--	20.50	20.50
3"	P2@.990	Ea	--	24.40	24.40
4"	P2@1.30	Ea	--	32.00	32.00
Installation of 3-way control valve					
1/2"	P2@.260	Ea	--	6.41	6.41
3/4"	P2@.365	Ea	--	8.99	8.99
1"	P2@.475	Ea	--	11.70	11.70
1¼"	P2@.575	Ea	--	14.20	14.20
1½"	P2@.680	Ea	--	16.80	16.80
2"	P2@.910	Ea	--	22.40	22.40
2½"	P2@1.12	Ea	--	27.60	27.60
3"	P2@1.33	Ea	--	32.80	32.80
4"	P2@2.00	Ea	--	49.30	49.30
Companion flange					
2"	P2@.290	Ea	14.40	7.15	21.55
2½"	P2@.380	Ea	16.90	9.36	26.26
3"	P2@.460	Ea	21.70	11.30	33.00
4"	P2@.600	Ea	29.30	14.80	44.10
Bolt and gasket set					
2"	P2@.500	Ea	3.85	12.30	16.15
2½"	P2@.650	Ea	4.08	16.00	20.08
3"	P2@.750	Ea	4.28	18.50	22.78
4"	P2@1.00	Ea	7.75	24.60	32.35

Carbon Steel, Schedule 40 with 150# M.I. Fittings & Threaded Joints

Description	Craft@Hrs	Unit	Material $	Labor $	Total $
Thermometer with well					
7"	P2@.250	Ea	14.30	6.16	20.46
9"	P2@.250	Ea	14.80	6.16	20.96
Pressure gauge					
2½"	P2@.200	Ea	15.60	4.93	20.53
3½"	P2@.200	Ea	27.10	4.93	32.03
Hanger with swivel assembly					
1/2"	P2@.450	Ea	.56	11.10	11.66
3/4"	P2@.450	Ea	.56	11.10	11.66
1"	P2@.450	Ea	.56	11.10	11.66
1¼"	P2@.450	Ea	.56	11.10	11.66
1½"	P2@.450	Ea	.62	11.10	11.72
2"	P2@.450	Ea	.65	11.10	11.75
2½"	P2@.450	Ea	3.95	11.10	15.05
3"	P2@.450	Ea	4.32	11.10	15.42
4"	P2@.450	Ea	5.87	11.10	16.97
Riser clamp					
1/2"	P2@.100	Ea	3.27	2.46	5.73
3/4"	P2@.100	Ea	3.27	2.46	5.73
1"	P2@.100	Ea	3.30	2.46	5.76
1¼"	P2@.105	Ea	4.01	2.59	6.60
1½"	P2@.110	Ea	4.23	2.71	6.94
2"	P2@.115	Ea	4.45	2.83	7.28
2½"	P2@.120	Ea	4.69	2.96	7.65
3"	P2@.120	Ea	5.13	2.96	8.09
4"	P2@.125	Ea	7.55	3.08	10.63

Carbon Steel, Schedule 80 with 300# M.I. Fittings & Butt-Welded Joints

Schedule 80 carbon steel (ASTM A-53 and A-120) pipe with 300 pound malleable iron threaded fittings is commonly used for heating hot water, chilled water, steam, and steam condensate systems where operating pressures do not exceed 250 PSIG for pipe sizes 2 inches and smaller. For pipe sizes 2½ inches and larger, wrought steel fittings with welded joints should be used.

This section has been arranged to save the estimator's time by including all normally-used system components such as pipe, fittings, valves, hanger assemblies, riser clamps and miscellaneous items under one heading. Additional items can be found under "Plumbing and Piping Specialties." The cost estimates in this section are based on the conditions, limitations and wage rates described in the section "How to Use This Book" beginning on page 5.

Description	Craft@Hrs	Unit	Material $	Labor $	Total $
Schedule 80 carbon steel pipe, threaded joints					
1/2"	P2@.070	LF	.63	1.72	2.35
3/4"	P2@.080	LF	.83	1.97	2.80
1"	P2@.100	LF	1.10	2.46	3.56
1¼"	P2@.110	LF	1.48	2.71	4.19
1½"	P2@.120	LF	1.95	2.96	4.91
2"	P2@.130	LF	2.47	3.20	5.67
2½"	P2@.190	LF	3.87	4.68	8.55
3"	P2@.220	LF	5.18	5.42	10.60
4"	P2@.290	LF	7.23	7.15	14.38
300# malleable iron 45 degree ell, threaded joints					
1/2"	P2@.132	Ea	8.45	3.25	11.70
3/4"	P2@.143	Ea	9.45	3.52	12.97
1"	P2@.198	Ea	10.50	4.88	15.38
1¼"	P2@.264	Ea	16.70	6.50	23.20
1½"	P2@.330	Ea	21.80	8.13	29.93
2"	P2@.418	Ea	32.90	10.30	43.20
2½"	P2@.561	Ea	62.60	13.80	76.40
3"	P2@.704	Ea	82.50	17.30	99.80
4"	P2@.935	Ea	159.00	23.00	182.00
300# malleable iron 90 degree ell, threaded joints					
1/2"	P2@.132	Ea	5.90	3.25	9.15
3/4"	P2@.143	Ea	6.80	3.52	10.32
1"	P2@.198	Ea	8.70	4.88	13.58
1¼"	P2@.264	Ea	12.50	6.50	19.00
1½"	P2@.330	Ea	14.60	8.13	22.73
2"	P2@.418	Ea	21.50	10.30	31.80
2½"	P2@.561	Ea	47.70	13.80	61.50
3"	P2@.704	Ea	62.60	17.30	79.90
4"	P2@.935	Ea	125.00	23.00	148.00

Carbon Steel, Schedule 80 with 300# M.I. Fittings & Threaded Joints

Description	Craft@Hrs	Unit	Material $	Labor $	Total $

300# malleable iron tee, threaded joints

Description	Craft@Hrs	Unit	Material $	Labor $	Total $
1/2"	P2@.198	Ea	8.70	4.88	13.58
3/4"	P2@.209	Ea	9.38	5.15	14.53
1"	P2@.253	Ea	11.30	6.23	17.53
1¼"	P2@.341	Ea	16.20	8.40	24.60
1½"	P2@.429	Ea	18.90	10.60	29.50
2"	P2@.539	Ea	27.80	13.30	41.10
2½"	P2@.726	Ea	57.50	17.90	75.40
3"	P2@.913	Ea	96.80	22.50	119.30
4"	P2@1.21	Ea	200.00	29.80	229.80

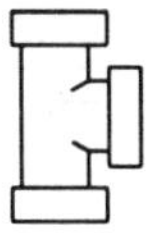

300# malleable iron reducing tee, threaded joints

Description	Craft@Hrs	Unit	Material $	Labor $	Total $
3/4 x 1/2"	P2@.205	Ea	13.50	5.05	18.55
1 x 1/2"	P2@.205	Ea	16.20	5.05	21.25
1 x 3/4"	P2@.205	Ea	16.20	5.05	21.25
1¼ x 1/2"	P2@.310	Ea	22.10	7.64	29.74
1¼ x 3/4"	P2@.310	Ea	21.60	7.64	29.24
1¼ x 1"	P2@.310	Ea	22.40	7.64	30.04
1½ x 1/2"	P2@.400	Ea	25.90	9.86	35.76
1½ x 3/4"	P2@.400	Ea	26.30	9.86	36.16
1½ x 1"	P2@.400	Ea	26.30	9.86	36.16
1½ x 1¼"	P2@.400	Ea	26.90	9.86	36.76
2 x 1"	P2@.500	Ea	38.70	12.30	51.00
2 x 1¼"	P2@.500	Ea	38.70	12.30	51.00
2 x 1½"	P2@.500	Ea	38.70	12.30	51.00
3 x 1¼"	P2@.830	Ea	140.00	20.50	160.50
3 x 1½"	P2@.830	Ea	140.00	20.50	160.50
3 x 2"	P2@.830	Ea	140.00	20.50	160.50

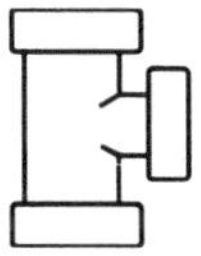

300# malleable iron reducer, threaded joints

Description	Craft@Hrs	Unit	Material $	Labor $	Total $
3/4 x 1/2"	P2@.135	Ea	8.93	3.33	12.26
1 x 1/2"	P2@.170	Ea	9.15	4.19	13.34
1 x 3/4"	P2@.170	Ea	9.00	4.19	13.19
1¼ x 3/4"	P2@.220	Ea	12.30	5.42	17.72
1¼ x 1"	P2@.220	Ea	11.90	5.42	17.32
1½ x 1"	P2@.285	Ea	16.20	7.02	23.22
1½ x 1¼"	P2@.285	Ea	16.80	7.02	23.82
2 x 1¼"	P2@.360	Ea	24.20	8.87	33.07
2 x 1½"	P2@.360	Ea	23.30	8.87	32.17
3 x 2"	P2@.550	Ea	56.40	13.60	70.00
3 x 2½"	P2@.550	Ea	58.60	13.60	72.20

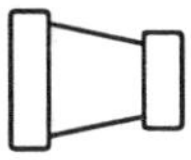

Carbon Steel, Schedule 80 with 300# M.I. Fittings & Threaded Joints

Description	Craft@Hrs	Unit	Material $	Labor $	Total $
300# malleable iron cross, threaded joints					
1/2"	P2@.310	Ea	16.40	7.64	24.04
3/4"	P2@.350	Ea	17.80	8.62	26.42
1"	P2@.400	Ea	19.20	9.86	29.06
1¼"	P2@.530	Ea	24.90	13.10	38.00
1½"	P2@.660	Ea	36.70	16.30	53.00
2"	P2@.770	Ea	46.80	19.00	65.80
300# malleable iron cap, threaded joint					
1/2"	P2@.100	Ea	4.71	2.46	7.17
3/4"	P2@.110	Ea	5.94	2.71	8.65
1"	P2@.155	Ea	7.80	3.82	11.62
1¼"	P2@.200	Ea	8.63	4.93	13.56
1½"	P2@.250	Ea	12.60	6.16	18.76
2"	P2@.320	Ea	17.50	7.88	25.38
2½"	P2@.420	Ea	28.70	10.30	39.00
3"	P2@.530	Ea	39.50	13.10	52.60
300# malleable iron plug, threaded joint					
1/2"	P2@.090	Ea	.89	2.22	3.11
3/4"	P2@.100	Ea	.89	2.46	3.35
1"	P2@.140	Ea	.89	3.45	4.34
1¼"	P2@.180	Ea	1.40	4.44	5.84
1½"	P2@.230	Ea	2.02	5.67	7.69
2"	P2@.290	Ea	2.36	7.15	9.51
2½"	P2@.380	Ea	5.52	9.36	14.88
3"	P2@.480	Ea	6.57	11.80	18.37
4"	P2@.640	Ea	12.10	15.80	27.90
300# malleable iron union, threaded joints					
1/2"	P2@.150	Ea	7.70	3.70	11.40
3/4"	P2@.165	Ea	8.51	4.07	12.58
1"	P2@.230	Ea	11.10	5.67	16.77
1¼"	P2@.310	Ea	18.20	7.64	25.84
1½"	P2@.400	Ea	19.00	9.86	28.86
2"	P2@.500	Ea	24.80	12.30	37.10
2½"	P2@.670	Ea	79.60	16.50	96.10
3"	P2@.840	Ea	104.00	20.70	124.70
4"	P2@.960	Ea	344.00	23.70	367.70

Carbon Steel, Schedule 80 with 300# M.I. Fittings & Threaded Joints

Description	Craft@Hrs	Unit	Material $	Labor $	Total $
300# malleable iron coupling, threaded joints					
1/2"	P2@.132	Ea	4.65	3.25	7.90
3/4"	P2@.143	Ea	5.34	3.52	8.86
1"	P2@.198	Ea	6.10	4.88	10.98
1¼"	P2@.264	Ea	7.26	6.50	13.76
1½"	P2@.330	Ea	10.60	8.13	18.73
2"	P2@.418	Ea	15.60	10.30	25.90
2½"	P2@.561	Ea	24.50	13.80	38.30
3"	P2@.704	Ea	35.30	17.30	52.60
Class 300 bronze body gate valve, threaded					
1/2"	P2@.210	Ea	40.00	5.17	45.17
3/4"	P2@.250	Ea	47.60	6.16	53.76
1"	P2@.300	Ea	67.80	7.39	75.19
1¼"	P2@.400	Ea	89.60	9.86	99.46
1½"	P2@.450	Ea	106.00	11.10	117.10
2"	P2@.500	Ea	131.00	12.30	143.30
Class 250 iron body gate valve, flanged					
2"	P2@.500	Ea	500.00	12.30	512.30
2½"	P2@.600	Ea	621.00	14.80	635.80
3"	P2@.750	Ea	688.00	18.50	706.50
4"	P2@1.35	Ea	1,020.00	33.30	1,053.30
Class 300 bronze body globe valve, threaded					
1/2"	P2@.210	Ea	54.60	5.17	59.77
3/4"	P2@.250	Ea	76.10	6.16	82.26
1"	P2@.300	Ea	92.10	7.39	99.49
1¼"	P2@.400	Ea	127.00	9.86	136.86
1½"	P2@.450	Ea	188.00	11.10	199.10
2"	P2@.500	Ea	292.00	12.30	304.30
Class 250 iron body globe valve, flanged					
2"	P2@.500	Ea	653.00	12.30	665.30
2½"	P2@.600	Ea	852.00	14.80	866.80
3"	P2@.750	Ea	908.00	18.50	926.50
4"	P2@1.35	Ea	1,290.00	33.30	1,323.30
200 PSI iron body butterfly valve, lug type					
2"	P2@.200	Ea	77.60	4.93	82.53
2½"	P2@.300	Ea	81.00	7.39	88.39
3"	P2@.400	Ea	89.80	9.86	99.66
4"	P2@.500	Ea	120.00	12.30	132.30

Carbon Steel, Schedule 80 with 300# M.I. Fittings & Threaded Joints

Description	Craft@Hrs	Unit	Material $	Labor $	Total $
200 PSI iron body butterfly valve, wafer type					
2"	P2@.200	Ea	55.40	4.93	60.33
2½"	P2@.300	Ea	58.30	7.39	65.69
3"	P2@.400	Ea	62.60	9.86	72.46
4"	P2@.500	Ea	78.50	12.30	90.80
Class 150 bronze body 2-piece ball valve, threaded					
1/2"	P2@.210	Ea	5.00	5.17	10.17
3/4"	P2@.250	Ea	7.51	6.16	13.67
1"	P2@.300	Ea	10.70	7.39	18.09
1¼"	P2@.400	Ea	18.50	9.86	28.36
1½"	P2@.450	Ea	23.00	11.10	34.10
2"	P2@.500	Ea	31.00	12.30	43.30
Class 300 bronze body swing check valve, threaded					
1/2"	P2@.210	Ea	31.20	5.17	36.37
3/4"	P2@.250	Ea	38.80	6.16	44.96
1"	P2@.300	Ea	52.30	7.39	59.69
1¼"	P2@.400	Ea	68.20	9.86	78.06
1½"	P2@.450	Ea	90.40	11.10	101.50
2"	P2@.500	Ea	134.00	12.30	146.30
Class 250 iron body swing check valve, flanged					
2"	P2@.500	Ea	531.00	12.30	543.30
2½"	P2@.600	Ea	623.00	14.80	637.80
3"	P2@.750	Ea	778.00	18.50	796.50
4"	P2@1.35	Ea	998.00	33.30	1,031.30
Class 250 iron body silent check valve, wafer type					
2"	P2@.500	Ea	104.00	12.30	116.30
2½"	P2@.600	Ea	120.00	14.80	134.80
3"	P2@.750	Ea	132.00	18.50	150.50
4"	P2@1.35	Ea	178.00	33.30	211.30
Class 250 bronze body wye strainer, threaded					
1/2"	P2@.210	Ea	20.80	5.17	25.97
3/4"	P2@.250	Ea	25.40	6.16	31.56
1"	P2@.300	Ea	30.30	7.39	37.69
1¼"	P2@.400	Ea	39.10	9.86	48.96
1½"	P2@.450	Ea	53.60	11.10	64.70
2"	P2@.500	Ea	81.80	12.30	94.10

Carbon Steel, Schedule 80 with 300# M.I. Fittings & Threaded Joints

Description	Craft@Hrs	Unit	Material $	Labor $	Total $
Class 250 iron body wye strainer, flanged					
2"	P2@.500	Ea	177.00	12.30	189.30
2½"	P2@.600	Ea	183.00	14.80	197.80
3"	P2@.750	Ea	220.00	18.50	238.50
4"	P2@1.35	Ea	400.00	33.30	433.30
Installation of 2-way control valve					
1/2"	P2@.210	Ea	--	5.17	5.17
3/4"	P2@.275	Ea	--	6.78	6.78
1"	P2@.350	Ea	--	8.62	8.62
1¼"	P2@.430	Ea	--	10.60	10.60
1½"	P2@.505	Ea	--	12.40	12.40
2"	P2@.675	Ea	--	16.60	16.60
2½"	P2@.830	Ea	--	20.50	20.50
3"	P2@.990	Ea	--	24.40	24.40
4"	P2@1.30	Ea	--	32.00	32.00
Installation of 3-way control valve					
1/2"	P2@.260	Ea	--	6.41	6.41
3/4"	P2@.365	Ea	--	8.99	8.99
1"	P2@.475	Ea	--	11.70	11.70
1¼"	P2@.575	Ea	--	14.20	14.20
1½"	P2@.680	Ea	--	16.80	16.80
2"	P2@.910	Ea	--	22.40	22.40
2½"	P2@1.12	Ea	--	27.60	27.60
3"	P2@1.33	Ea	--	32.80	32.80
4"	P2@2.00	Ea	--	49.30	49.30
Threaded companion flange, 300# malleable iron					
2"	P2@.290	Ea	47.00	7.15	54.15
2½"	P2@.380	Ea	64.80	9.36	74.16
3"	P2@.460	Ea	70.50	11.30	81.80
4"	P2@.600	Ea	99.80	14.80	114.60
Bolt and gasket set					
2"	P2@.500	Ea	3.85	12.30	16.15
2½"	P2@.650	Ea	4.08	16.00	20.08
3"	P2@.750	Ea	4.28	18.50	22.78
4"	P2@1.00	Ea	7.75	24.60	32.35

Carbon Steel, Schedule 80 with 300# M.I. Fittings & Threaded Joints

Description	Craft@Hrs	Unit	Material $	Labor $	Total $
Thermometer with well					
7"	P2@.250	Ea	14.30	6.16	20.46
9"	P2@.250	Ea	14.80	6.16	20.96
Pressure gauge					
2½"	P2@.200	Ea	15.60	4.93	20.53
3½"	P2@.200	Ea	27.10	4.93	32.03
Hanger with swivel assembly					
1/2"	P2@.450	Ea	.56	11.10	11.66
3/4"	P2@.450	Ea	.56	11.10	11.66
1"	P2@.450	Ea	.56	11.10	11.66
1¼"	P2@.450	Ea	.56	11.10	11.66
1½"	P2@.450	Ea	.62	11.10	11.72
2"	P2@.450	Ea	.65	11.10	11.75
2½"	P2@.450	Ea	3.95	11.10	15.05
3"	P2@.450	Ea	4.32	11.10	15.42
4"	P2@.450	Ea	5.87	11.10	16.97
Riser clamp					
1/2"	P2@.100	Ea	3.83	2.46	6.29
3/4"	P2@.100	Ea	3.83	2.46	6.29
1"	P2@.100	Ea	3.87	2.46	6.33
1¼"	P2@.105	Ea	4.70	2.59	7.29
1½"	P2@.110	Ea	4.95	2.71	7.66
2"	P2@.115	Ea	5.20	2.83	8.03
2½"	P2@.120	Ea	5.50	2.96	8.46
3"	P2@.120	Ea	6.00	2.96	8.96
4"	P2@.125	Ea	7.55	3.03	10.58

Carbon Steel, Schedule 40 with 150# Fittings & Butt-Welded Joints

Schedule 40 carbon steel (ASTM A-53 and A-120) pipe with 150 pound carbon steel welding fittings is commonly used for heating hot water, chilled water, steam, and steam condensate systems where operating pressures do not exceed 250 PSIG.

Schedule 40 (ASTM A-53B SML) steel pipe with carbon steel welding fittings can also be used for refrigerant piping systems.

This section has been arranged to save the estimator's time by including all normally-used system components such as pipe, fittings, valves, hanger assemblies, riser clamps and miscellaneous items under one heading. Additional items can be found under "Plumbing and Piping Specialties." The cost estimates in this section are based on the conditions, limitations and wage rates described in the section "How to Use This Book" beginning on page 5.

Description	Craft@Hrs	Unit	Material $	Labor $	Total $

Schedule 40 carbon steel pipe, plain end

Description	Craft@Hrs	Unit	Material $	Labor $	Total $
1/2"	P2@.070	LF	.51	1.72	2.23
3/4"	P2@.080	LF	.58	1.97	2.55
1"	P2@.100	LF	.87	2.46	3.33
1¼"	P2@.110	LF	1.13	2.71	3.84
1½"	P2@.120	LF	1.35	2.96	4.31
2"	P2@.140	LF	1.82	3.45	5.27
2½"	P2@.170	LF	2.81	4.19	7.00
3"	P2@.190	LF	3.72	4.68	8.40
4"	P2@.300	LF	5.30	7.39	12.69
6"	P2@.420	LF	9.11	10.30	19.41
8"	P2@.510	LF	13.60	12.60	26.20
10"	P2@.600	LF	19.30	14.80	34.10
12"	P2@.760	LF	23.40	18.70	42.10

Schedule 40 carbon steel 45 degree ell, butt-welded

Description	Craft@Hrs	Unit	Material $	Labor $	Total $
1/2"	P2@.220	Ea	12.30	5.42	17.72
3/4"	P2@.330	Ea	12.30	8.13	20.43
1"	P2@.440	Ea	6.32	10.80	17.12
1¼"	P2@.560	Ea	6.32	13.80	20.12
1½"	P2@.710	Ea	6.32	17.50	23.82
2"	P2@.890	Ea	6.32	21.90	28.22
2½"	P2@1.11	Ea	7.65	27.40	35.05
3"	P2@1.33	Ea	9.32	32.80	42.12
4"	P2@1.78	Ea	15.60	43.90	59.50
6"	P2@2.67	Ea	33.30	65.80	99.10
8"	P2@3.20	Ea	57.90	78.80	136.70
10"	P2@4.00	Ea	104.00	98.60	202.60
12"	P2@4.80	Ea	148.00	118.00	266.00

Description	Craft@Hrs	Unit	Material $	Labor $	Total $

Schedule 40 carbon steel 90 degree ell, butt-welded

Description	Craft@Hrs	Unit	Material $	Labor $	Total $
1/2"	P2@.220	Ea	13.00	5.42	18.42
3/4"	P2@.330	Ea	13.00	8.13	21.13
1"	P2@.440	Ea	6.65	10.80	17.45
1¼"	P2@.560	Ea	6.65	13.80	20.45
1½"	P2@.670	Ea	6.65	16.50	23.15
2"	P2@.890	Ea	6.65	21.90	28.55
2½"	P2@1.11	Ea	8.98	27.40	36.38
3"	P2@1.33	Ea	10.60	32.80	43.40
4"	P2@1.78	Ea	18.00	43.90	61.90
6"	P2@2.67	Ea	43.60	65.80	109.40
8"	P2@3.20	Ea	81.10	78.80	159.90
10"	P2@4.00	Ea	149.00	98.60	247.60
12"	P2@4.80	Ea	216.00	118.00	334.00

Schedule 40 carbon steel tee, butt-welded

Description	Craft@Hrs	Unit	Material $	Labor $	Total $
1/2"	P2@.330	Ea	30.60	8.13	38.73
3/4"	P2@.500	Ea	16.00	12.30	28.30
1"	P2@.670	Ea	16.00	16.50	32.50
1¼"	P2@.830	Ea	19.30	20.50	39.80
1½"	P2@1.00	Ea	19.30	24.60	43.90
2"	P2@1.33	Ea	17.30	32.80	50.10
2½"	P2@1.67	Ea	20.90	41.10	62.00
3"	P2@2.00	Ea	23.60	49.30	72.90
4"	P2@2.67	Ea	32.90	65.80	98.70
6"	P2@4.00	Ea	60.20	98.60	158.80
8"	P2@4.80	Ea	111.00	118.00	229.00
10"	P2@6.00	Ea	186.00	148.00	334.00
12"	P2@7.20	Ea	284.00	177.00	461.00

Carbon Steel, Schedule 40 with 150# Fittings & Butt-Welded Joints

Description	Craft@Hrs	Unit	Material $	Labor $	Total $
Schedule 40 carbon steel reducing tee, butt welded					
1½ x 1/2"	P2@.940	Ea	24.60	23.20	47.80
1½ x 3/4"	P2@.940	Ea	24.60	23.20	47.80
1½ x 1"	P2@.940	Ea	24.60	23.20	47.80
1½ x 1¼"	P2@.940	Ea	24.60	23.20	47.80
2 x 3/4"	P2@1.22	Ea	29.60	30.10	59.70
2 x 1"	P2@1.22	Ea	29.60	30.10	59.70
2 x 1¼"	P2@1.22	Ea	29.60	30.10	59.70
2 x 1½"	P2@1.22	Ea	29.60	30.10	59.70
2½ x 1"	P2@1.56	Ea	29.90	38.40	68.30
2½ x 1¼"	P2@1.56	Ea	29.90	38.40	68.30
2½ x 1½"	P2@1.56	Ea	29.90	38.40	68.30
2½ x 2"	P2@1.56	Ea	29.90	38.40	68.30
3 x 1¼"	P2@1.89	Ea	31.30	46.60	77.90
3 x 2"	P2@1.89	Ea	31.30	46.60	77.90
3 x 2½"	P2@1.89	Ea	31.30	46.60	77.90
4 x 1½"	P2@2.44	Ea	38.90	60.10	99.00
4 x 2"	P2@2.44	Ea	38.90	60.10	99.00
4 x 2½"	P2@2.44	Ea	38.90	60.10	99.00
4 x 3"	P2@2.44	Ea	38.90	60.10	99.00
6 x 2½"	P2@3.78	Ea	77.20	93.10	170.30
6 x 3"	P2@3.78	Ea	77.20	93.10	170.30
6 x 4"	P2@3.78	Ea	77.20	93.10	170.30
8 x 3"	P2@4.40	Ea	149.00	108.00	257.00
8 x 4"	P2@4.40	Ea	149.00	108.00	257.00
8 x 6"	P2@4.40	Ea	149.00	108.00	257.00
10 x 4"	P2@5.60	Ea	252.00	138.00	390.00
10 x 6"	P2@5.60	Ea	252.00	138.00	390.00
10 x 8"	P2@5.60	Ea	252.00	138.00	390.00
12 x 6"	P2@6.80	Ea	370.00	168.00	538.00
12 x 8"	P2@6.80	Ea	370.00	168.00	538.00
12 x 10"	P2@6.80	Ea	370.00	168.00	538.00

Carbon Steel, Schedule 40 with 150# Fittings & Butt-Welded Joints

Description	Craft@Hrs	Unit	Material $	Labor $	Total $

Schedule 40 carbon steel concentric reducer, butt welded

Description	Craft@Hrs	Unit	Material $	Labor $	Total $
3/4"	P2@.295	Ea	18.70	7.27	25.97
1"	P2@.395	Ea	23.10	9.73	32.83
1¼"	P2@.500	Ea	17.10	12.30	29.40
1½"	P2@.600	Ea	17.60	14.80	32.40
2"	P2@.800	Ea	18.20	19.70	37.90
2½"	P2@1.00	Ea	18.70	24.60	43.30
3"	P2@1.20	Ea	16.50	29.60	46.10
4"	P2@1.60	Ea	21.50	39.40	60.90
6"	P2@2.40	Ea	45.10	59.10	104.20
8"	P2@2.88	Ea	60.00	71.00	131.00
10"	P2@3.60	Ea	90.80	88.70	179.50
12"	P2@4.30	Ea	141.00	106.00	247.00

Schedule 40 carbon steel eccentric reducer, butt welded

Description	Craft@Hrs	Unit	Material $	Labor $	Total $
3/4"	P2@.295	Ea	25.20	7.27	32.47
1"	P2@.395	Ea	31.00	9.73	40.73
1¼"	P2@.500	Ea	28.00	12.30	40.30
1½"	P2@.600	Ea	23.10	14.80	37.90
2"	P2@.800	Ea	29.70	19.70	49.40
2½"	P2@1.00	Ea	25.30	24.60	49.90
3"	P2@1.20	Ea	23.70	29.60	53.30
4"	P2@1.60	Ea	29.20	39.40	68.60
6"	P2@2.40	Ea	58.30	59.10	117.40
8"	P2@2.88	Ea	87.50	71.00	158.50
10"	P2@3.60	Ea	103.00	88.70	191.70
12"	P2@4.30	Ea	162.00	106.00	268.00

Schedule 40 carbon steel cap, butt-welded

Description	Craft@Hrs	Unit	Material $	Labor $	Total $
1/2"	P2@.150	Ea	3.65	3.70	7.35
3/4"	P2@.230	Ea	3.65	5.67	9.32
1"	P2@.300	Ea	5.00	7.39	12.39
1¼"	P2@.390	Ea	5.32	9.61	14.93
1½"	P2@.470	Ea	5.32	11.60	16.92
2"	P2@.620	Ea	5.32	15.30	20.62
2½"	P2@.770	Ea	5.32	19.00	24.32
3"	P2@.930	Ea	5.32	22.90	28.22
4"	P2@1.25	Ea	7.00	30.80	37.80
6"	P2@1.87	Ea	13.00	46.10	59.10
8"	P2@2.24	Ea	19.60	55.20	74.80
10"	P2@2.80	Ea	35.00	69.00	104.00
12"	P2@3.36	Ea	43.20	82.80	126.00

Carbon Steel, Schedule 40 with 150# Fittings & Butt-Welded Joints

Description	Craft@Hrs	Unit	Material $	Labor $	Total $

Schedule 40 carbon steel weldolet

Description	Craft@Hrs	Unit	Material $	Labor $	Total $
1/2"	P2@.330	Ea	7.14	8.13	15.27
3/4"	P2@.500	Ea	7.31	12.30	19.61
1"	P2@.670	Ea	7.66	16.50	24.16
1¼"	P2@.830	Ea	9.31	20.50	29.81
1½"	P2@1.00	Ea	9.31	24.60	33.91
2"	P2@1.33	Ea	9.70	32.80	42.50
2½"	P2@1.67	Ea	22.00	41.10	63.10
3"	P2@2.00	Ea	25.60	49.30	74.90
4"	P2@2.67	Ea	32.70	65.80	98.50

Schedule 40 carbon steel threadolet

Description	Craft@Hrs	Unit	Material $	Labor $	Total $
1/2"	P2@.220	Ea	3.74	5.42	9.16
3/4"	P2@.330	Ea	4.38	8.13	12.51
1"	P2@.440	Ea	4.92	10.80	15.72
1¼"	P2@.560	Ea	6.97	13.80	20.77
1½"	P2@.670	Ea	7.58	16.50	24.08
2"	P2@.890	Ea	8.72	21.90	30.62
2½"	P2@1.11	Ea	27.50	27.40	54.90
3"	P2@1.33	Ea	31.80	32.80	64.60
4"	P2@1.78	Ea	62.30	43.90	106.20

150# forged steel companion flange

Description	Craft@Hrs	Unit	Material $	Labor $	Total $
2½"	P2@.610	Ea	21.30	15.00	36.30
3"	P2@.730	Ea	22.30	18.00	40.30
4"	P2@.980	Ea	26.60	24.10	50.70
6"	P2@1.47	Ea	40.90	36.20	77.10
8"	P2@1.77	Ea	71.50	43.60	115.10
10"	P2@2.20	Ea	117.00	54.20	171.20
12"	P2@2.64	Ea	171.00	65.00	236.00

Class 125 bronze body gate valve, threaded

Description	Craft@Hrs	Unit	Material $	Labor $	Total $
1/2"	P2@.210	Ea	15.60	5.17	20.77
3/4"	P2@.250	Ea	19.20	6.16	25.36
1"	P2@.300	Ea	26.00	7.39	33.39
1¼"	P2@.400	Ea	37.10	9.86	46.96
1½"	P2@.450	Ea	45.00	11.10	56.10
2"	P2@.500	Ea	56.20	12.30	68.50
2½"	P2@.750	Ea	150.00	18.50	168.50
3"	P2@.950	Ea	207.00	23.40	230.40

Carbon Steel, Schedule 40 with 150# Fittings & Butt-Welded Joints

Description	Craft@Hrs	Unit	Material $	Labor $	Total $
Class 125 iron body gate valve, flanged					
2½"	P2@.600	Ea	170.00	14.80	184.80
3"	P2@.750	Ea	192.00	18.50	210.50
4"	P2@1.35	Ea	274.00	33.30	307.30
6"	P2@2.50	Ea	467.00	61.60	528.60
8"	P2@3.00	Ea	801.00	73.90	874.90
10"	P2@4.00	Ea	1,410.00	98.60	1,508.60
12"	P2@4.50	Ea	1,940.00	111.00	2,051.00
Class 125 bronze body globe valve, threaded					
1/2"	P2@.210	Ea	19.40	5.17	24.57
3/4"	P2@.250	Ea	24.60	6.16	30.76
1"	P2@.300	Ea	32.90	7.39	40.29
1¼"	P2@.400	Ea	43.60	9.86	53.46
1½"	P2@.450	Ea	56.30	11.10	67.40
2"	P2@.500	Ea	85.70	12.30	98.00
Class 125 iron body globe valve, flanged					
2½"	P2@.600	Ea	359.00	14.80	373.80
3"	P2@.750	Ea	416.00	18.50	434.50
4"	P2@1.35	Ea	595.00	33.30	628.30
6"	P2@2.50	Ea	1,080.00	61.60	1,141.60
8"	P2@3.00	Ea	2,120.00	73.90	2,193.90
10"	P2@4.00	Ea	3,320.00	98.60	3,418.60
200 PSI iron body butterfly valve, lug type					
2"	P2@.200	Ea	73.40	4.93	78.33
2½"	P2@.300	Ea	74.20	7.39	81.59
3"	P2@.400	Ea	84.20	9.86	94.06
4"	P2@.500	Ea	100.00	12.30	112.30
6"	P2@.750	Ea	163.00	18.50	181.50
8"	P2@.950	Ea	213.00	23.40	236.40
10"	P2@1.40	Ea	328.00	34.50	362.50
12"	P2@1.80	Ea	477.00	44.40	521.40
200 PSI iron body butterfly valve, wafer type					
2"	P2@.200	Ea	73.40	4.93	78.33
2½"	P2@.300	Ea	73.40	7.39	80.79
3"	P2@.400	Ea	74.20	9.86	84.06
4"	P2@.500	Ea	89.30	12.30	101.60
6"	P2@.750	Ea	148.00	18.50	166.50
8"	P2@.950	Ea	210.00	23.40	233.40
10"	P2@1.40	Ea	295.00	34.50	329.50
12"	P2@1.80	Ea	414.00	44.40	458.40

Carbon Steel, Schedule 40 with 150# Fittings & Butt-Welded Joints

Description	Craft@Hrs	Unit	Material $	Labor $	Total $
Class 125 bronze body swing check valve, threaded					
1/2"	P2@.210	Ea	20.80	5.17	25.97
3/4"	P2@.250	Ea	25.10	6.16	31.26
1"	P2@.300	Ea	34.60	7.39	41.99
1¼"	P2@.400	Ea	47.90	9.86	57.76
1½"	P2@.450	Ea	57.10	11.10	68.20
2"	P2@.500	Ea	83.30	12.30	95.60
Class 125 iron body swing check valve, flanged					
2"	P2@.500	Ea	185.00	12.30	197.30
2½"	P2@.600	Ea	235.00	14.80	249.80
3"	P2@.750	Ea	256.00	18.50	274.50
4"	P2@1.35	Ea	403.00	33.30	436.30
6"	P2@2.50	Ea	691.00	61.60	752.60
8"	P2@3.00	Ea	1,310.00	73.90	1,383.90
10"	P2@4.00	Ea	2,220.00	98.60	2,318.60
12"	P2@4.50	Ea	3,470.00	111.00	3,581.00
Class 125 iron body silent check valve, flanged					
2"	P2@.500	Ea	104.00	12.30	116.30
2½"	P2@.600	Ea	120.00	14.80	134.80
3"	P2@.750	Ea	132.00	18.50	150.50
4"	P2@1.35	Ea	247.00	33.30	280.30
6"	P2@2.50	Ea	336.00	61.60	397.60
8"	P2@3.00	Ea	578.00	73.90	651.90
10"	P2@4.00	Ea	880.00	98.60	978.60
Class 150 bronze body strainer, threaded					
1/2"	P2@.210	Ea	15.10	5.17	20.27
3/4"	P2@.250	Ea	19.60	6.16	25.76
1"	P2@.300	Ea	24.80	7.39	32.19
1¼"	P2@.400	Ea	31.00	9.86	40.86
1½"	P2@.450	Ea	40.20	11.10	51.30
2"	P2@.500	Ea	65.20	12.30	77.50
Class 125 iron body strainer, flanged					
2"	P2@.500	Ea	121.00	12.30	133.30
2½"	P2@.600	Ea	122.00	14.80	136.80
3"	P2@.750	Ea	145.00	18.50	163.50
4"	P2@1.35	Ea	258.00	33.30	291.30
6"	P2@2.50	Ea	530.00	61.60	591.60
8"	P2@3.00	Ea	846.00	73.90	919.90

Description	Craft@Hrs	Unit	Material $	Labor $	Total $

Installation of bronze body 2-way threaded control valve

Description	Craft@Hrs	Unit	Material $	Labor $	Total $
1/2"	P2@.210	Ea	--	5.17	5.17
3/4"	P2@.250	Ea	--	6.16	6.16
1"	P2@.300	Ea	--	7.39	7.39
1¼"	P2@.400	Ea	--	9.86	9.86
1½"	P2@.450	Ea	--	11.10	11.10
2"	P2@.500	Ea	--	12.30	12.30

Installation of iron body 2-way flanged control valve

Description	Craft@Hrs	Unit	Material $	Labor $	Total $
2"	P2@.500	Ea	--	12.30	12.30
2½"	P2@.600	Ea	--	14.80	14.80
3"	P2@.750	Ea	--	18.50	18.50
4"	P2@1.35	Ea	--	33.30	33.30
6"	P2@2.50	Ea	--	61.60	61.60
8"	P2@3.00	Ea	--	73.90	73.90
10"	P2@4.00	Ea	--	98.60	98.60
12"	P2@4.50	Ea	--	111.00	111.00

Installation of bronze body 3-way threaded control valve

Description	Craft@Hrs	Unit	Material $	Labor $	Total $
1/2"	P2@.260	Ea	--	6.41	6.41
3/4"	P2@.365	Ea	--	8.99	8.99
1"	P2@.475	Ea	--	11.70	11.70
1¼"	P2@.575	Ea	--	14.20	14.20
1½"	P2@.680	Ea	--	16.80	16.80
2"	P2@.910	Ea	--	22.40	22.40

Installation of iron body 3-way flanged control valve

Description	Craft@Hrs	Unit	Material $	Labor $	Total $
2"	P2@.910	Ea	--	22.40	22.40
2½"	P2@1.12	Ea	--	27.60	27.60
3"	P2@1.33	Ea	--	32.80	32.80
4"	P2@2.00	Ea	--	49.30	49.30
6"	P2@3.70	Ea	--	91.20	91.20
8"	P2@4.40	Ea	--	108.00	108.00
10"	P2@5.90	Ea	--	145.00	145.00
12"	P2@6.50	Ea	--	160.00	160.00

Carbon Steel, Schedule 40 with 150# Fittings & Butt-Welded Joints

Description	Craft@Hrs	Unit	Material $	Labor $	Total $
Bolt and gasket set					
2"	P2@.500	Ea	3.85	12.30	16.15
2½"	P2@.650	Ea	4.08	16.00	20.08
3"	P2@.750	Ea	4.28	18.50	22.78
4"	P2@1.00	Ea	7.75	24.60	32.35
6"	P2@1.20	Ea	12.10	29.60	41.70
8"	P2@1.25	Ea	12.90	30.80	43.70
10"	P2@1.70	Ea	26.70	41.90	68.60
12"	P2@2.20	Ea	28.60	54.20	82.80
Thermometer with well					
7"	P2@.250	Ea	14.30	6.16	20.46
9"	P2@.250	Ea	14.10	6.16	20.26
Pressure gauge					
2½"	P2@.200	Ea	15.60	4.93	20.53
3½"	P2@.200	Ea	27.10	4.93	32.03
Hanger with swivel assembly					
1/2"	P2@.450	Ea	.56	11.10	11.66
3/4"	P2@.450	Ea	.56	11.10	11.66
1"	P2@.450	Ea	.56	11.10	11.66
1¼"	P2@.450	Ea	.56	11.10	11.66
1½"	P2@.450	Ea	.62	11.10	11.72
2"	P2@.450	Ea	.65	11.10	11.75
2½"	P2@.450	Ea	3.95	11.10	15.05
3"	P2@.450	Ea	4.32	11.10	15.42
4"	P2@.450	Ea	5.87	11.10	16.97
6"	P2@.500	Ea	11.70	12.30	24.00
8"	P2@.550	Ea	17.30	13.60	30.90
10"	P2@.650	Ea	21.00	16.00	37.00
12"	P2@.750	Ea	25.90	18.50	44.40
Riser clamp					
1/2"	P2@.100	Ea	3.27	2.46	5.73
3/4"	P2@.100	Ea	3.27	2.46	5.73
1"	P2@.100	Ea	3.30	2.46	5.76
1¼"	P2@.105	Ea	4.01	2.59	6.60
1½"	P2@.110	Ea	4.23	2.71	6.94
2"	P2@.115	Ea	4.45	2.83	7.28
2½"	P2@.120	Ea	4.69	2.96	7.65
3"	P2@.120	Ea	5.13	2.96	8.09
4"	P2@.125	Ea	6.45	3.08	9.53
6"	P2@.200	Ea	10.70	4.93	15.63
8"	P2@.200	Ea	17.40	4.93	22.33
10"	P2@.250	Ea	25.90	6.16	32.06
12"	P2@.250	Ea	30.70	6.16	36.86

Carbon Steel, Schedule 80 with 300# Fittings & Butt-Welded Joints

Schedule 80 carbon steel (ASTM A-53 and A-120) pipe with 300 pound carbon steel butt-welded fittings are commonly used for steam, steam condensate and re-circulating water systems where operating pressures do not exceed 700 PSIG.

This section has been arranged to save the estimator's time by including all normally-used system components such as pipe, fittings, valves, hanger assemblies, riser clamps and miscellaneous items under one heading. Additional items can be found under "Plumbing and Piping Specialties." The cost estimates in this section are based on the conditions, limitations and wage rates described in the section "How to Use This Book" beginning on page 5.

Description	Craft@Hrs	Unit	Material $	Labor $	Total $
Schedule 80 carbon steel plain end pipe					
1/2"	P2@.080	LF	.62	1.97	2.59
3/4"	P2@.090	LF	.83	2.22	3.05
1"	P2@.110	LF	1.10	2.71	3.81
1¼"	P2@.120	LF	1.48	2.96	4.44
1½"	P2@.130	LF	1.94	3.20	5.14
2"	P2@.170	LF	2.47	4.19	6.66
2½"	P2@.220	LF	3.87	5.42	9.29
3"	P2@.280	LF	5.18	6.90	12.08
4"	P2@.340	LF	7.23	8.38	15.61
6"	P2@.460	LF	18.70	11.30	30.00
8"	P2@.560	LF	28.50	13.80	42.30
10"	P2@.660	LF	50.20	16.30	66.50
12"	P2@.840	LF	67.70	20.70	88.40
Schedule 80 carbon steel 45 degree ell, butt-welded					
1/2"	P2@.300	Ea	16.30	7.39	23.69
3/4"	P2@.440	Ea	16.30	10.80	27.10
1"	P2@.590	Ea	8.65	14.50	23.15
1¼"	P2@.740	Ea	8.65	18.20	26.85
1½"	P2@.890	Ea	8.65	21.90	30.55
2"	P2@1.18	Ea	7.98	29.10	37.08
2½"	P2@1.49	Ea	20.30	36.70	57.00
3"	P2@1.77	Ea	12.30	43.60	55.90
4"	P2@2.37	Ea	18.60	58.40	77.00
6"	P2@3.55	Ea	49.50	87.50	137.00
8"	P2@4.26	Ea	80.50	105.00	185.50
10"	P2@5.32	Ea	130.00	131.00	261.00
12"	P2@6.38	Ea	195.00	157.00	352.00

Carbon Steel, Schedule 80 with 300# Fittings & Butt-Welded Joints

Description	Craft@Hrs	Unit	Material $	Labor $	Total $

Schedule 80 carbon steel 90 degree ell, butt-welded

Description	Craft@Hrs	Unit	Material $	Labor $	Total $
1/2"	P2@.300	Ea	16.30	7.39	23.69
3/4"	P2@.440	Ea	16.30	10.80	27.10
1"	P2@.590	Ea	8.65	14.50	23.15
1¼"	P2@.740	Ea	8.65	18.20	26.85
1½"	P2@.890	Ea	8.65	21.90	30.55
2"	P2@1.18	Ea	8.65	29.10	37.75
2½"	P2@1.49	Ea	11.60	36.70	48.30
3"	P2@1.77	Ea	16.00	43.60	59.60
4"	P2@2.37	Ea	26.60	58.40	85.00
6"	P2@3.55	Ea	64.80	87.50	152.30
8"	P2@4.26	Ea	121.00	105.00	226.00
10"	P2@5.32	Ea	216.00	131.00	347.00
12"	P2@6.38	Ea	305.00	157.00	462.00

Schedule 80 carbon steel tee, butt-welded

Description	Craft@Hrs	Unit	Material $	Labor $	Total $
1/2"	P2@.440	Ea	40.60	10.80	51.40
3/4"	P2@.670	Ea	20.60	16.50	37.10
1"	P2@.890	Ea	20.60	21.90	42.50
1¼"	P2@1.11	Ea	20.60	27.40	48.00
1½"	P2@1.33	Ea	23.30	32.80	56.10
2"	P2@1.77	Ea	19.60	43.60	63.20
2½"	P2@2.22	Ea	34.90	54.70	89.60
3"	P2@2.66	Ea	32.60	65.50	98.10
4"	P2@3.55	Ea	52.90	87.50	140.40
6"	P2@5.32	Ea	84.10	131.00	215.10
8"	P2@6.38	Ea	171.00	157.00	328.00
10"	P2@7.98	Ea	277.00	197.00	474.00
12"	P2@9.58	Ea	399.00	236.00	635.00

Carbon Steel, Schedule 80 with 300# Fittings & Butt-Welded Joints

Description	Craft@Hrs	Unit	Material $	Labor $	Total $
Schedule 80 carbon steel reducing tee, butt-welded					
1½ x 1/2"	P2@1.26	Ea	29.90	31.00	60.90
1½ x 3/4"	P2@1.26	Ea	29.90	31.00	60.90
1½ x 1"	P2@1.26	Ea	29.90	31.00	60.90
1½ x 1¼"	P2@1.26	Ea	29.90	31.00	60.90
2 x 3/4"	P2@1.63	Ea	29.30	40.20	69.50
2 x 1"	P2@1.63	Ea	29.30	40.20	69.50
2 x 1¼"	P2@1.63	Ea	29.30	40.20	69.50
2 x 1½"	P2@1.63	Ea	29.30	40.20	69.50
2½ x 1"	P2@2.07	Ea	44.20	51.00	95.20
2½ x 1¼"	P2@2.07	Ea	44.20	51.00	95.20
2½ x 1½"	P2@2.07	Ea	44.20	51.00	95.20
2½ x 2"	P2@2.07	Ea	44.20	51.00	95.20
3 x 1¼"	P2@2.51	Ea	33.30	61.80	95.10
3 x 2"	P2@2.51	Ea	33.30	61.80	95.10
3 x 2½"	P2@2.51	Ea	33.30	61.80	95.10
4 x 1½"	P2@3.25	Ea	53.20	80.10	133.30
4 x 2"	P2@3.25	Ea	53.20	80.10	133.30
4 x 2½"	P2@3.25	Ea	53.20	80.10	133.30
4 x 3"	P2@3.25	Ea	53.20	80.10	133.30
6 x 2½"	P2@5.03	Ea	186.00	124.00	310.00
6 x 3"	P2@5.03	Ea	186.00	124.00	310.00
6 x 4"	P2@5.03	Ea	186.00	124.00	310.00
8 x 3"	P2@5.85	Ea	176.00	144.00	320.00
8 x 4"	P2@5.85	Ea	176.00	144.00	320.00
8 x 6"	P2@5.85	Ea	176.00	144.00	320.00
10 x 4"	P2@7.45	Ea	282.00	184.00	466.00
10 x 6"	P2@7.45	Ea	282.00	184.00	466.00
10 x 8"	P2@7.45	Ea	282.00	184.00	466.00
12 x 6"	P2@9.04	Ea	507.00	223.00	730.00
12 x 8"	P2@9.04	Ea	507.00	223.00	730.00
12 x 10"	P2@9.04	Ea	507.00	223.00	730.00

Carbon Steel, Schedule 80 with 300# Fittings & Butt-Welded Joints

Description	Craft@Hrs	Unit	Material $	Labor $	Total $

Schedule 80 carbon steel concentric reducer, butt-welded

Description	Craft@Hrs	Unit	Material $	Labor $	Total $
3/4"	P2@.360	Ea	13.30	8.87	22.17
1"	P2@.500	Ea	13.60	12.30	25.90
1¼"	P2@.665	Ea	13.00	16.40	29.40
1½"	P2@.800	Ea	11.60	19.70	31.30
2"	P2@1.00	Ea	13.30	24.60	37.90
2½"	P2@1.33	Ea	12.30	32.80	45.10
3"	P2@1.63	Ea	12.30	40.20	52.50
4"	P2@2.00	Ea	15.60	49.30	64.90
6"	P2@2.96	Ea	33.60	72.90	106.50
8"	P2@3.90	Ea	47.90	96.10	144.00
10"	P2@4.79	Ea	66.20	118.00	184.20
12"	P2@5.85	Ea	91.10	144.00	235.10

Schedule 80 carbon steel eccentric reducer, butt-welded

Description	Craft@Hrs	Unit	Material $	Labor $	Total $
3/4"	P2@.360	Ea	14.00	8.87	22.87
1"	P2@.500	Ea	14.20	12.30	26.50
1¼"	P2@.665	Ea	15.30	16.40	31.70
1½"	P2@.800	Ea	16.30	19.70	36.00
2"	P2@1.00	Ea	23.30	24.60	47.90
2½"	P2@1.33	Ea	15.60	32.80	48.40
3"	P2@1.63	Ea	15.30	40.20	55.50
4"	P2@2.00	Ea	21.90	49.30	71.20
6"	P2@2.96	Ea	50.20	72.90	123.10
8"	P2@3.90	Ea	72.80	96.10	168.90
10"	P2@4.79	Ea	92.40	118.00	210.40
12"	P2@5.85	Ea	132.00	144.00	276.00

Schedule 80 carbon steel cap, butt-welded

Description	Craft@Hrs	Unit	Material $	Labor $	Total $
1/2"	P2@.180	Ea	4.00	4.44	8.44
3/4"	P2@.270	Ea	4.00	6.65	10.65
1"	P2@.360	Ea	5.32	8.87	14.19
1¼"	P2@.460	Ea	7.32	11.30	18.62
1½"	P2@.560	Ea	5.65	13.80	19.45
2"	P2@.740	Ea	5.32	18.20	23.52
2½"	P2@.920	Ea	5.65	22.70	28.35
3"	P2@1.11	Ea	6.32	27.40	33.72
4"	P2@1.50	Ea	9.31	37.00	46.31
6"	P2@2.24	Ea	23.90	55.20	79.10
8"	P2@2.68	Ea	32.30	66.00	98.30
10"	P2@3.36	Ea	50.20	82.80	133.00
12"	P2@4.00	Ea	72.80	98.60	171.40

Carbon Steel, Schedule 80 with 300# Fittings & Butt-Welded Joints

Description	Craft@Hrs	Unit	Material $	Labor $	Total $
Carbon steel weldolet					
1/2"	P2@.440	Ea	7.80	10.80	18.60
3/4"	P2@.670	Ea	8.02	16.50	24.52
1"	P2@.890	Ea	8.72	21.90	30.62
1¼"	P2@1.11	Ea	11.10	27.40	38.50
1½"	P2@1.33	Ea	11.10	32.80	43.90
2"	P2@1.77	Ea	11.80	43.60	55.40
2½"	P2@2.22	Ea	26.60	54.70	81.30
3"	P2@2.66	Ea	27.30	65.50	92.80
4"	P2@3.55	Ea	34.10	87.50	121.60
Carbon steel threadolet					
1/2"	P2@.300	Ea	3.74	7.39	11.13
3/4"	P2@.440	Ea	4.39	10.80	15.19
1"	P2@.590	Ea	4.92	14.50	19.42
1¼"	P2@.740	Ea	6.97	18.20	25.17
1½"	P2@.890	Ea	7.58	21.90	29.48
2"	P2@1.18	Ea	8.72	29.10	37.82
2½"	P2@1.48	Ea	27.50	36.50	64.00
3"	P2@1.77	Ea	31.80	43.60	75.40
4"	P2@2.37	Ea	62.30	58.40	120.70
300# forged steel companion flange					
2½"	P2@.810	Ea	28.30	20.00	48.30
3"	P2@.980	Ea	28.30	24.10	52.40
4"	P2@1.30	Ea	46.60	32.00	78.60
6"	P2@1.95	Ea	67.80	48.00	115.80
8"	P2@2.35	Ea	117.00	57.90	174.90
10"	P2@2.90	Ea	231.00	71.50	302.50
12"	P2@3.50	Ea	296.00	86.20	382.20
Class 300 bronze body gate valve, threaded					
1/2"	P2@.210	Ea	44.90	5.17	50.07
3/4"	P2@.250	Ea	53.40	6.16	59.56
1"	P2@.300	Ea	76.10	7.39	83.49
1¼"	P2@.400	Ea	101.00	9.86	110.86
1½"	P2@.450	Ea	119.00	11.10	130.10
2"	P2@.500	Ea	172.00	12.30	184.30
2½"	P2@.750	Ea	541.00	18.50	559.50
3"	P2@.950	Ea	691.00	23.40	714.40

Carbon Steel, Schedule 80 with 300# Fittings & Butt-Welded Joints

Description	Craft@Hrs	Unit	Material $	Labor $	Total $
Class 250 iron body gate valve, flanged					
2½"	P2@.600	Ea	476.00	14.80	490.80
3"	P2@.750	Ea	530.00	18.50	548.50
4"	P2@1.35	Ea	782.00	33.30	815.30
6"	P2@2.50	Ea	1,300.00	61.60	1,361.60
8"	P2@3.00	Ea	2,280.00	73.90	2,353.90
10"	P2@4.00	Ea	3,440.00	98.60	3,538.60
12"	P2@4.50	Ea	5,360.00	111.00	5,471.00
Class 300 bronze body globe valve, threaded					
1/2"	P2@.210	Ea	41.90	5.17	47.07
3/4"	P2@.250	Ea	58.40	6.16	64.56
1"	P2@.300	Ea	70.70	7.39	78.09
1¼"	P2@.400	Ea	97.90	9.86	107.76
1½"	P2@.450	Ea	138.00	11.10	149.10
2"	P2@.500	Ea	215.00	12.30	227.30
Class 250 iron body globe valve, flanged					
2½"	P2@.600	Ea	654.00	14.80	668.80
3"	P2@.750	Ea	678.00	18.50	696.50
4"	P2@1.35	Ea	990.00	33.30	1,023.30
6"	P2@2.50	Ea	1,780.00	61.60	1,841.60
8"	P2@3.00	Ea	3,020.00	73.90	3,093.90
200 PSI iron body butterfly valve, lug type					
2"	P2@.200	Ea	73.40	4.93	78.33
2½"	P2@.300	Ea	74.20	7.39	81.59
3"	P2@.400	Ea	84.20	9.86	94.06
4"	P2@.500	Ea	100.00	12.30	112.30
6"	P2@.750	Ea	163.00	18.50	181.50
8"	P2@.950	Ea	213.00	23.40	236.40
10"	P2@1.40	Ea	328.00	34.50	362.50
12"	P2@1.80	Ea	477.00	44.40	521.40
200 PSI iron body butterfly valve, wafer type					
2"	P2@.200	Ea	73.40	4.93	78.33
2½"	P2@.300	Ea	73.40	7.39	80.79
3"	P2@.400	Ea	74.20	9.86	84.06
4"	P2@.500	Ea	89.30	12.30	101.60
6"	P2@.750	Ea	148.00	18.50	166.50
8"	P2@.950	Ea	210.00	23.40	233.40
10"	P2@1.40	Ea	295.00	34.50	329.50
12"	P2@1.80	Ea	414.00	44.40	458.40

Carbon Steel, Schedule 80 with 300# Fittings & Butt-Welded Joints

Description	Craft@Hrs	Unit	Material $	Labor $	Total $
Class 300 bronze body swing check valve, threaded					
1/2"	P2@.210	Ea	32.80	5.17	37.97
3/4"	P2@.250	Ea	40.90	6.16	47.06
1"	P2@.300	Ea	55.10	7.39	62.49
1¼"	P2@.400	Ea	71.80	9.86	81.66
1½"	P2@.450	Ea	95.20	11.10	106.30
2"	P2@.500	Ea	141.00	12.30	153.30
Class 250 iron body swing check valve, flanged					
2"	P2@.500	Ea	559.00	12.30	571.30
2½"	P2@.600	Ea	656.00	14.80	670.80
3"	P2@.750	Ea	819.00	18.50	837.50
4"	P2@1.35	Ea	1,050.00	33.30	1,083.30
6"	P2@2.50	Ea	1,900.00	61.60	1,961.60
8"	P2@3.00	Ea	3,190.00	73.90	3,263.90
Class 250 iron body silent check valve, flanged					
2"	P2@.500	Ea	104.00	12.30	116.30
2½"	P2@.600	Ea	120.00	14.80	134.80
3"	P2@.750	Ea	132.00	18.50	150.50
4"	P2@1.35	Ea	178.00	33.30	211.30
6"	P2@2.50	Ea	321.00	61.60	382.60
8"	P2@3.00	Ea	706.00	73.90	779.90
10"	P2@4.00	Ea	1,070.00	98.60	1,168.60
Class 250 bronze body strainer, threaded					
1/2"	P2@.210	Ea	20.80	5.17	25.97
3/4"	P2@.250	Ea	25.40	6.16	31.56
1"	P2@.300	Ea	30.30	7.39	37.69
1¼"	P2@.400	Ea	39.10	9.86	48.96
1½"	P2@.450	Ea	53.60	11.10	64.70
2"	P2@.500	Ea	81.80	12.30	94.10
Class 250 iron body strainer, flanged					
2"	P2@.500	Ea	166.00	12.30	178.30
2½"	P2@.600	Ea	172.00	14.80	186.80
3"	P2@.750	Ea	207.00	18.50	225.50
4"	P2@1.35	Ea	376.00	33.30	409.30
6"	P2@2.50	Ea	663.00	61.60	724.60
8"	P2@3.00	Ea	1,090.00	73.90	1,163.90
Installation of bronze body 2-way threaded control valve					
1/2"	P2@.210	Ea	--	5.17	5.17
3/4"	P2@.250	Ea	--	6.16	6.16
1"	P2@.300	Ea	--	7.39	7.39
1¼"	P2@.400	Ea	--	9.86	9.86
1½"	P2@.450	Ea	--	11.10	11.10
2"	P2@.500	Ea	--	12.30	12.30

Carbon Steel, Schedule 80 with 300# Fittings & Butt-Welded Joints

Description	Craft@Hrs	Unit	Material $	Labor $	Total $
Installation of iron body 2-way flanged control valve					
2"	P2@.500	Ea	--	12.30	12.30
2½"	P2@.600	Ea	--	14.80	14.80
3"	P2@.750	Ea	--	18.50	18.50
4"	P2@1.35	Ea	--	33.30	33.30
6"	P2@2.50	Ea	--	61.60	61.60
8"	P2@3.00	Ea	--	73.90	73.90
10"	P2@4.00	Ea	--	98.60	98.60
12"	P2@4.50	Ea	--	111.00	111.00
Installation of bronze body 3-way threaded control valve					
1/2"	P2@.260	Ea	--	6.41	6.41
3/4"	P2@.365	Ea	--	8.99	8.99
1"	P2@.475	Ea	--	11.70	11.70
1¼"	P2@.575	Ea	--	14.20	14.20
1½"	P2@.680	Ea	--	16.80	16.80
2"	P2@.910	Ea	--	22.40	22.40
Installation of iron body 3-way flanged control valve					
2"	P2@.910	Ea	--	22.40	22.40
2½"	P2@1.12	Ea	--	27.60	27.60
3"	P2@1.33	Ea	--	32.80	32.80
4"	P2@2.00	Ea	--	49.30	49.30
6"	P2@3.70	Ea	--	91.20	91.20
8"	P2@4.40	Ea	--	108.00	108.00
10"	P2@5.90	Ea	--	145.00	145.00
12"	P2@6.50	Ea	--	160.00	160.00
Bolt and gasket set					
2"	P2@.500	Ea	3.85	12.30	16.15
2½"	P2@.650	Ea	4.08	16.00	20.08
3"	P2@.750	Ea	4.28	18.50	22.78
4"	P2@1.00	Ea	7.75	24.60	32.35
6"	P2@1.20	Ea	12.10	29.60	41.70
8"	P2@1.25	Ea	12.90	30.80	43.70
10"	P2@1.70	Ea	26.70	41.90	68.60
12"	P2@2.20	Ea	28.60	54.20	82.80
Thermometer with well					
7"	P2@.250	Ea	14.30	6.16	20.46
9"	P2@.250	Ea	14.80	6.16	20.96
Pressure gauge					
2½"	P2@.200	Ea	15.60	4.93	20.53
3½"	P2@.200	Ea	27.10	4.93	32.03

Carbon Steel, Schedule 80 with 300# Fittings & Butt-Welded Joints

Description	Craft@Hrs	Unit	Material $	Labor $	Total $
Hanger with swivel assembly					
1/2"	P2@.450	Ea	.56	11.10	11.66
3/4"	P2@.450	Ea	.56	11.10	11.66
1"	P2@.450	Ea	.56	11.10	11.66
1¼"	P2@.450	Ea	.56	11.10	11.66
1½"	P2@.450	Ea	.62	11.10	11.72
2"	P2@.450	Ea	.65	11.10	11.75
2½"	P2@.450	Ea	3.95	11.10	15.05
3"	P2@.450	Ea	4.32	11.10	15.42
4"	P2@.450	Ea	5.87	11.10	16.97
6"	P2@.500	Ea	11.70	12.30	24.00
8"	P2@.550	Ea	17.30	13.60	30.90
10"	P2@.650	Ea	21.00	16.00	37.00
12"	P2@.750	Ea	25.90	18.50	44.40

Description	Craft@Hrs	Unit	Material $	Labor $	Total $
Riser clamp					
1/2"	P2@.100	Ea	3.27	2.46	5.73
3/4"	P2@.100	Ea	3.27	2.46	5.73
1"	P2@.100	Ea	3.30	2.46	5.76
1¼"	P2@.105	Ea	4.01	2.59	6.60
1½"	P2@.110	Ea	4.23	2.71	6.94
2"	P2@.115	Ea	4.45	2.83	7.28
2½"	P2@.120	Ea	4.69	2.96	7.65
3"	P2@.120	Ea	5.13	2.96	8.09
4"	P2@.125	Ea	6.45	3.08	9.53
6"	P2@.200	Ea	10.70	4.93	15.63
8"	P2@.200	Ea	17.40	4.93	22.33
10"	P2@.250	Ea	25.90	6.16	32.06
12"	P2@.250	Ea	30.70	6.16	36.86

Schedule 160 carbon steel (ASTM A-106) pipe with steel fitting are commonly used for steam, steam condensate and re-circulating water systems.

This section has been arranged to save the estimator's time by including all normally-used system components such as pipe, fittings, valves, hanger assemblies, riser clamps and miscellaneous items under one heading. Additional items can be found under "Plumbing and Piping Specialties." The cost estimates in this section are based on the conditions, limitations and wage rates described in the section "How to Use This Book" beginning on page 5.

Description	Craft@Hrs	Unit	Material $	Labor $	Total $

Schedule 160 carbon steel plain end pipe

Description	Craft@Hrs	Unit	Material $	Labor $	Total $
1/2"	P2@.090	LF	4.07	2.22	6.29
3/4"	P2@.100	LF	4.96	2.46	7.42
1"	P2@.120	LF	10.30	2.96	13.26
1¼"	P2@.130	LF	10.70	3.20	13.90
1½"	P2@.140	LF	12.30	3.45	15.75
2"	P2@.190	LF	6.87	4.68	11.55
2½"	P2@.230	LF	10.40	5.67	16.07
3"	P2@.310	LF	14.10	7.64	21.74
4"	P2@.380	LF	20.90	9.36	30.26

3,000# carbon steel 45 degree ell, threaded

Description	Craft@Hrs	Unit	Material $	Labor $	Total $
1/2"	P2@.300	Ea	4.80	7.39	12.19
3/4"	P2@.440	Ea	5.47	10.80	16.27
1"	P2@.590	Ea	7.60	14.50	22.10
1¼"	P2@.740	Ea	10.30	18.20	28.50
1½"	P2@.890	Ea	15.80	21.90	37.70
2"	P2@1.18	Ea	21.80	29.10	50.90
2½"	P2@1.49	Ea	53.80	36.70	90.50
3"	P2@1.77	Ea	87.60	43.60	131.20
4"	P2@2.37	Ea	157.00	58.40	215.40

3,000# carbon steel 45 degree ell, socket-welded

Description	Craft@Hrs	Unit	Material $	Labor $	Total $
1/2"	P2@.070	Ea	3.94	1.72	5.66
3/4"	P2@.100	Ea	4.74	2.46	7.20
1"	P2@.130	Ea	6.21	3.20	9.41
1¼"	P2@.170	Ea	8.23	4.19	12.42
1½"	P2@.200	Ea	10.20	4.93	15.13
2"	P2@.270	Ea	17.70	6.65	24.35
2½"	P2@.340	Ea	45.90	8.38	54.28
3"	P2@.400	Ea	77.10	9.86	86.96
4"	P2@.530	Ea	152.00	13.10	165.10

Carbon Steel, Schedule 160 with 3,000-6,000# Fittings

Description	Craft@Hrs	Unit	Material $	Labor $	Total $
6,000# carbon steel 45 degree ell, threaded					
1/2"	P2@.360	Ea	7.98	8.87	16.85
3/4"	P2@.530	Ea	11.10	13.10	24.20
1"	P2@.710	Ea	11.90	17.50	29.40
1¼"	P2@.880	Ea	33.30	21.70	55.00
1½"	P2@1.10	Ea	33.30	27.10	60.40
2"	P2@1.40	Ea	46.50	34.50	81.00
2½"	P2@1.80	Ea	176.00	44.40	220.40
3"	P2@2.10	Ea	169.00	51.70	220.70
6,000# carbon steel 45 degree ell, socket-welded					
1/2"	P2@.080	Ea	6.13	1.97	8.10
3/4"	P2@.120	Ea	7.06	2.96	10.02
1"	P2@.160	Ea	9.49	3.94	13.43
1¼"	P2@.200	Ea	35.60	4.93	40.53
1½"	P2@.310	Ea	24.00	7.64	31.64
2"	P2@.400	Ea	29.80	9.86	39.66
2½"	P2@.470	Ea	128.00	11.60	139.60
3"	P2@.470	Ea	141.00	11.60	152.60
4"	P2@.630	Ea	165.00	15.50	180.50
3,000# carbon steel 90 degree ell, threaded					
1/2"	P2@.300	Ea	3.72	7.39	11.11
3/4"	P2@.440	Ea	4.52	10.80	15.32
1"	P2@.590	Ea	6.56	14.50	21.06
1¼"	P2@.740	Ea	11.10	18.20	29.30
1½"	P2@.890	Ea	16.50	21.90	38.40
2"	P2@1.18	Ea	20.10	29.10	49.20
2½"	P2@1.49	Ea	49.50	36.70	86.20
3"	P2@1.77	Ea	73.20	43.60	116.80
4"	P2@2.37	Ea	154.00	58.40	212.40
3,000# carbon steel 90 degree ell, socket-welded					
1/2"	P2@.070	Ea	3.73	1.72	5.45
3/4"	P2@.100	Ea	3.87	2.46	6.33
1"	P2@.130	Ea	4.99	3.20	8.19
1¼"	P2@.170	Ea	8.29	4.19	12.48
1½"	P2@.200	Ea	10.40	4.93	15.33
2"	P2@.270	Ea	15.90	6.65	22.55
2½"	P2@.340	Ea	37.80	8.38	46.18
3"	P2@.400	Ea	63.50	9.86	73.36
4"	P2@.530	Ea	151.00	13.10	164.10

Description	Craft@Hrs	Unit	Material $	Labor $	Total $

6,000# carbon steel 90 degree ell, threaded

Description	Craft@Hrs	Unit	Material $	Labor $	Total $
1/2"	P2@.080	Ea	6.15	1.97	8.12
3/4"	P2@.120	Ea	7.84	2.96	10.80
1"	P2@.160	Ea	10.00	3.94	13.94
1¼"	P2@.200	Ea	16.90	4.93	21.83
1½"	P2@.250	Ea	26.30	6.16	32.46
2"	P2@.310	Ea	46.10	7.64	53.74
2½"	P2@.400	Ea	68.10	9.86	77.96
3"	P2@.470	Ea	146.00	11.60	157.60

6,000# carbon steel 90 degree ell, socket-welded

Description	Craft@Hrs	Unit	Material $	Labor $	Total $
1/2"	P2@.360	Ea	5.21	8.87	14.08
3/4"	P2@.530	Ea	6.10	13.10	19.20
1"	P2@.710	Ea	8.36	17.50	25.86
1¼"	P2@.880	Ea	12.10	21.70	33.80
1½"	P2@1.10	Ea	18.20	27.10	45.30
2"	P2@1.40	Ea	27.30	34.50	61.80
2½"	P2@1.80	Ea	56.90	44.40	101.30
3"	P2@2.10	Ea	98.40	51.70	150.10
4"	P2@2.80	Ea	163.00	69.00	232.00

3,000# carbon steel tee, threaded

Description	Craft@Hrs	Unit	Material $	Labor $	Total $
1/2"	P2@.440	Ea	4.88	10.80	15.68
3/4"	P2@.670	Ea	7.13	16.50	23.63
1"	P2@.890	Ea	9.03	21.90	30.93
1¼"	P2@1.11	Ea	15.00	27.40	42.40
1½"	P2@1.33	Ea	20.60	32.80	53.40
2"	P2@1.77	Ea	27.30	43.60	70.90
2½"	P2@2.22	Ea	61.80	54.70	116.50
3"	P2@2.66	Ea	123.00	65.50	188.50
4"	P2@3.55	Ea	201.00	87.50	288.50

3,000# carbon steel tee, socket-welded

Description	Craft@Hrs	Unit	Material $	Labor $	Total $
1/2"	P2@.100	Ea	4.31	2.46	6.77
3/4"	P2@.150	Ea	5.43	3.70	9.13
1"	P2@.200	Ea	7.43	4.93	12.36
1¼"	P2@.250	Ea	10.10	6.16	16.26
1½"	P2@.300	Ea	15.70	7.39	23.09
2"	P2@.400	Ea	22.10	9.86	31.96
2½"	P2@.500	Ea	53.20	12.30	65.50
3"	P2@.600	Ea	126.00	14.80	140.80
4"	P2@.800	Ea	200.00	19.70	219.70

Carbon Steel, Schedule 160 with 3,000-6,000# Fittings

Description	Craft@Hrs	Unit	Material $	Labor $	Total $
6,000# carbon steel tee, threaded					
1/2"	P2@.530	Ea	7.78	13.10	20.88
3/4"	P2@.800	Ea	10.50	19.70	30.20
1"	P2@1.10	Ea	17.40	27.10	44.50
1¼"	P2@1.30	Ea	22.20	32.00	54.20
1½"	P2@1.60	Ea	30.20	39.40	69.60
2"	P2@2.10	Ea	57.70	51.70	109.40
2½"	P2@2.60	Ea	208.00	64.10	272.10
3"	P2@3.20	Ea	213.00	78.80	291.80
6,000# carbon steel tee, socket-welded					
1/2"	P2@.120	Ea	6.82	2.96	9.78
3/4"	P2@.180	Ea	9.01	4.44	13.45
1"	P2@.250	Ea	11.30	6.16	17.46
1¼"	P2@.290	Ea	32.50	7.15	39.65
1½"	P2@.360	Ea	29.10	8.87	37.97
2"	P2@.470	Ea	36.80	11.60	48.40
2½"	P2@.560	Ea	78.80	13.80	92.60
3"	P2@.720	Ea	129.00	17.70	146.70
4"	P2@.960	Ea	240.00	23.70	263.70
3,000# carbon steel threaded union					
1/2"	P2@.440	Ea	5.48	10.80	16.28
3/4"	P2@.670	Ea	5.48	16.50	21.98
1"	P2@.890	Ea	6.60	21.90	28.50
1¼"	P2@1.11	Ea	8.61	27.40	36.01
1½"	P2@1.33	Ea	14.50	32.80	47.30
2"	P2@1.77	Ea	15.80	43.60	59.40
2½"	P2@2.22	Ea	19.30	54.70	74.00
3"	P2@2.66	Ea	36.50	65.50	102.00
3,000# carbon steel socket-welded union					
1/2"	P2@.150	Ea	6.60	3.70	10.30
3/4"	P2@.230	Ea	7.50	5.67	13.17
1"	P2@.300	Ea	9.67	7.39	17.06
1¼"	P2@.380	Ea	15.60	9.36	24.96
1½"	P2@.450	Ea	16.00	11.10	27.10
2"	P2@.600	Ea	22.90	14.80	37.70
2½"	P2@.750	Ea	53.10	18.50	71.60
3"	P2@.900	Ea	67.90	22.20	90.10
6,000# carbon steel threaded union					
1/2"	P2@.530	Ea	12.90	13.10	26.00
3/4"	P2@.800	Ea	14.60	19.70	34.30
1"	P2@1.10	Ea	19.30	27.10	46.40
1¼"	P2@1.30	Ea	31.40	32.00	63.40
1½"	P2@1.60	Ea	36.80	39.40	76.20
2"	P2@2.10	Ea	50.80	51.70	102.50

Carbon Steel, Schedule 160 with 3,000-6,000# Fittings

Description	Craft@Hrs	Unit	Material $	Labor $	Total $
6,000# carbon steel socket-welded union					
1/2"	P2@.180	Ea	13.30	4.44	17.74
3/4"	P2@.280	Ea	15.20	6.90	22.10
1"	P2@.370	Ea	22.40	9.12	31.52
1¼"	P2@.440	Ea	33.90	10.80	44.70
1½"	P2@.540	Ea	35.60	13.30	48.90
2"	P2@.980	Ea	58.20	24.10	82.30
3,000# carbon steel reducer, socket-welded					
3/4"	P2@.360	Ea	3.56	8.87	12.43
1"	P2@.500	Ea	4.34	12.30	16.64
1¼"	P2@.665	Ea	5.04	16.40	21.44
1½"	P2@.800	Ea	6.36	19.70	26.06
2"	P2@1.00	Ea	8.62	24.60	33.22
2½"	P2@1.33	Ea	59.80	32.80	92.60
3"	P2@1.63	Ea	74.70	40.20	114.90
6,000# carbon steel reducer, socket-welded					
3/4"	P2@.430	Ea	13.70	10.60	24.30
1"	P2@.600	Ea	14.30	14.80	29.10
1¼"	P2@.800	Ea	30.50	19.70	50.20
1½"	P2@.960	Ea	30.50	23.70	54.20
2"	P2@1.20	Ea	40.50	29.60	70.10
3,000# carbon steel cap, socket-welded					
1/2"	P2@.180	Ea	2.14	4.44	6.58
3/4"	P2@.270	Ea	2.49	6.65	9.14
1"	P2@.360	Ea	3.82	8.87	12.69
1¼"	P2@.460	Ea	4.45	11.30	15.75
1½"	P2@.560	Ea	6.07	13.80	19.87
2"	P2@.740	Ea	9.46	18.20	27.66
2½"	P2@.920	Ea	51.00	22.70	73.70
3"	P2@1.11	Ea	52.70	27.40	80.10
4"	P2@1.50	Ea	79.20	37.00	116.20
6,000# carbon steel cap, socket-welded					
1/2"	P2@.220	Ea	4.10	5.42	9.52
3/4"	P2@.320	Ea	5.98	7.88	13.86
1"	P2@.430	Ea	6.35	10.60	16.95
1¼"	P2@.550	Ea	14.70	13.60	28.30
1½"	P2@.670	Ea	19.30	16.50	35.80
2"	P2@.890	Ea	28.60	21.90	50.50

Carbon Steel, Schedule 160 with 3,000-6,000# Fittings

Description	Craft@Hrs	Unit	Material $	Labor $	Total $

3,000# carbon steel weldolet

Description	Craft@Hrs	Unit	Material $	Labor $	Total $
1/2"	P2@.580	Ea	7.80	14.30	22.10
3/4"	P2@.890	Ea	8.02	21.90	29.92
1"	P2@1.18	Ea	8.72	29.10	37.82
1¼"	P2@1.48	Ea	11.10	36.50	47.60
1½"	P2@1.75	Ea	11.10	43.10	54.20
2"	P2@2.35	Ea	11.80	57.90	69.70
2½"	P2@2.90	Ea	26.60	71.50	98.10
3"	P2@3.50	Ea	27.30	86.20	113.50
4"	P2@4.75	Ea	34.10	117.00	151.10

6,000# carbon steel weldolet

Description	Craft@Hrs	Unit	Material $	Labor $	Total $
1/2"	P2@.690	Ea	34.10	17.00	51.10
3/4"	P2@1.05	Ea	35.50	25.90	61.40
1"	P2@1.40	Ea	37.20	34.50	71.70
1¼"	P2@1.75	Ea	44.70	43.10	87.80
1½"	P2@2.10	Ea	50.40	51.70	102.10
2"	P2@2.80	Ea	58.10	69.00	127.10
2½"	P2@3.50	Ea	85.20	86.20	171.40
3"	P2@4.20	Ea	95.10	103.00	198.10
4"	P2@5.70	Ea	161.00	140.00	301.00

3,000# carbon steel threadolet

Description	Craft@Hrs	Unit	Material $	Labor $	Total $
1/2"	P2@.400	Ea	3.74	9.86	13.60
3/4"	P2@.580	Ea	4.39	14.30	18.69
1"	P2@.790	Ea	4.92	19.50	24.42
1¼"	P2@.980	Ea	6.97	24.10	31.07
1½"	P2@1.20	Ea	7.58	29.60	37.18
2"	P2@1.60	Ea	8.72	39.40	48.12
2½"	P2@2.00	Ea	27.50	49.30	76.80
3"	P2@2.35	Ea	31.80	57.90	89.70
4"	P2@3.15	Ea	62.30	77.60	139.90

6,000# carbon steel threadolet

Description	Craft@Hrs	Unit	Material $	Labor $	Total $
1/2"	P2@.480	Ea	6.59	11.80	18.39
3/4"	P2@.690	Ea	7.29	17.00	24.29
1"	P2@.950	Ea	8.96	23.40	32.36
1¼"	P2@1.20	Ea	40.40	29.60	70.00
1½"	P2@1.45	Ea	40.40	35.70	76.10
2"	P2@1.90	Ea	46.90	46.80	93.70

Description	Craft@Hrs	Unit	Material $	Labor $	Total $
Class 300 forged steel companion flange					
1/2"	P2@.240	Ea	22.90	5.91	28.81
3/4"	P2@.320	Ea	22.90	7.88	30.78
1"	P2@.340	Ea	22.90	8.38	31.28
1¼"	P2@.425	Ea	22.90	10.50	33.40
1½"	P2@.560	Ea	22.90	13.80	36.70
2"	P2@.740	Ea	24.90	18.20	43.10
2½"	P2@.810	Ea	28.30	20.00	48.30
3"	P2@.980	Ea	28.30	24.10	52.40
4"	P2@1.30	Ea	46.60	32.00	78.60
Class 600 forged steel companion flange					
1/2"	P2@.290	Ea	27.60	7.15	34.75
3/4"	P2@.380	Ea	27.60	9.36	36.96
1"	P2@.410	Ea	27.60	10.10	37.70
1½"	P2@.670	Ea	27.60	16.50	44.10
2"	P2@.890	Ea	27.60	21.90	49.50
2½"	P2@.970	Ea	37.60	23.90	61.50
3"	P2@1.17	Ea	39.60	28.80	68.40
4"	P2@1.55	Ea	70.80	38.20	109.00
Class 300 cast steel gate valve, flanged					
2"	P2@.500	Ea	593.00	12.30	605.30
2½"	P2@.600	Ea	813.00	14.80	827.80
3"	P2@.700	Ea	813.00	17.20	830.20
4"	P2@.790	Ea	1,140.00	19.50	1,159.50
Class 600 cast steel gate valve, flanged					
2"	P2@.540	Ea	1,220.00	13.30	1,233.30
2½"	P2@.630	Ea	1,790.00	15.50	1,805.50
3"	P2@.730	Ea	1,790.00	18.00	1,808.00
4"	P2@.820	Ea	2,730.00	20.20	2,750.20
Class 300 cast steel body globe valve, flanged					
2"	P2@.500	Ea	931.00	12.30	943.30
2½"	P2@.600	Ea	1,320.00	14.80	1,334.80
3"	P2@.700	Ea	1,320.00	17.20	1,337.20
4"	P2@.790	Ea	1,800.00	19.50	1,819.50
Class 600 cast steel body globe valve, flanged					
2"	P2@.540	Ea	1,330.00	13.30	1,343.30
2½"	P2@.630	Ea	1,970.00	15.50	1,985.50
3"	P2@.730	Ea	1,970.00	18.00	1,988.00
4"	P2@.820	Ea	2,620.00	20.20	2,640.20

Description	Craft@Hrs	Unit	Material $	Labor $	Total $
Class 300 cast steel body swing check valve, flanged					
2"	P2@.500	Ea	656.00	12.30	668.30
2½"	P2@.600	Ea	925.00	14.80	939.80
3"	P2@.700	Ea	925.00	17.20	942.20
4"	P2@.790	Ea	1,200.00	19.50	1,219.50
Class 600 cast steel body swing check valve, flanged					
2"	P2@.540	Ea	961.00	13.30	974.30
2½"	P2@.630	Ea	1,320.00	15.50	1,335.50
3"	P2@.730	Ea	1,320.00	18.00	1,338.00
4"	P2@.820	Ea	1,930.00	20.20	1,950.20
Installation of cast steel body 2-way threaded control valve					
1/2"	P2@.510	Ea	--	12.60	12.60
3/4"	P2@.510	Ea	--	12.60	12.60
1"	P2@.540	Ea	--	13.30	13.30
1¼"	P2@.580	Ea	--	14.30	14.30
1½"	P2@.580	Ea	--	14.30	14.30
2"	P2@.660	Ea	--	16.30	16.30
Installation cast steel body 2-way flanged control valve					
2"	P2@.660	Ea	--	16.30	16.30
2½"	P2@.780	Ea	--	19.20	19.20
3"	P2@.890	Ea	--	21.90	21.90
4"	P2@1.60	Ea	--	39.40	39.40
Installation of cast steel body 3-way threaded control valve					
1/2"	P2@.630	Ea	--	15.50	15.50
3/4"	P2@.885	Ea	--	21.80	21.80
1"	P2@1.15	Ea	--	28.30	28.30
1¼"	P2@1.40	Ea	--	34.50	34.50
1½"	P2@1.65	Ea	--	40.70	40.70
2"	P2@2.20	Ea	--	54.20	54.20
Installation of cast steel body 3-way flanged control valve					
2"	P2@2.20	Ea	--	54.20	54.20
2½"	P2@2.70	Ea	--	66.50	66.50
3"	P2@3.22	Ea	--	79.30	79.30
4"	P2@4.85	Ea	--	120.00	120.00

Carbon Steel, Schedule 160 with 3,000-6,000# Fittings

Description	Craft@Hrs	Unit	Material $	Labor $	Total $
300# bolt and gasket set, ring face					
2"	P2@.570	Ea	6.15	14.00	20.15
2½"	P2@.740	Ea	9.79	18.20	27.99
3"	P2@.850	Ea	10.30	20.90	31.20
4"	P2@1.15	Ea	10.50	28.30	38.80
300# bolt and gasket set, flat face					
2"	P2@.640	Ea	6.45	15.80	22.25
2½"	P2@.850	Ea	10.30	20.90	31.20
3"	P2@.970	Ea	10.80	23.90	34.70
4"	P2@1.30	Ea	11.00	32.00	43.00
Thermometer with well					
7"	P2@.250	Ea	14.30	6.16	20.46
9"	P2@.250	Ea	14.80	6.16	20.96
Pressure gauge					
2½"	P2@.200	Ea	15.60	4.93	20.53
3½"	P2@.200	Ea	27.10	4.93	32.03
Hanger with swivel assembly					
1/2"	P2@.450	Ea	.56	11.10	11.66
3/4"	P2@.450	Ea	.56	11.10	11.66
1"	P2@.450	Ea	.56	11.10	11.66
1¼"	P2@.450	Ea	.56	11.10	11.66
1½"	P2@.450	Ea	.62	11.10	11.72
2"	P2@.450	Ea	.65	11.10	11.75
2½"	P2@.450	Ea	3.95	11.10	15.05
3"	P2@.450	Ea	4.32	11.10	15.42
4"	P2@.450	Ea	5.87	11.10	16.97
Riser clamp					
1/2"	P2@.100	Ea	3.27	2.46	5.73
3/4"	P2@.100	Ea	3.27	2.46	5.73
1"	P2@.100	Ea	3.30	2.46	5.76
1¼"	P2@.105	Ea	4.01	2.59	6.60
1½"	P2@.110	Ea	4.23	2.71	6.94
2"	P2@.115	Ea	4.45	2.83	7.28
2½"	P2@.120	Ea	4.69	2.96	7.65
3"	P2@.120	Ea	5.13	2.96	8.09
4"	P2@.125	Ea	6.45	3.08	9.53

Carbon Steel, Schedule 10 with Welded Joints

Schedule 10 carbon steel (A-106) pipe with welded steel fittings is widely used for fire protection systems inside of buildings. Because of its relatively low cost, it is also gaining acceptance for use in closed heating hot water and chilled water systems.

Consult the manufacturers for maximum operating temperature/pressure ratings for the various combinations of pipe and fittings.

This section has been arranged to save the estimator's time by including all normally-used system components such as pipe, fittings, hanger assemblies and riser clamps under one heading. Additional items can be found under "Plumbing and Piping Specialties." The cost estimates in this section are based on the conditions, limitations and wage rates described in the section "How to Use This Book" beginning on page 5.

Description	Craft@Hrs	Unit	Material $	Labor $	Total $

Schedule 10 carbon steel pipe

Description	Craft@Hrs	Unit	Material $	Labor $	Total $
2"	P2@.080	LF	1.85	1.97	3.82
2½"	P2@.100	LF	2.50	2.46	4.96
3"	P2@.115	LF	3.07	2.83	5.90
4"	P2@.160	LF	4.07	3.94	8.01
5"	P2@.200	LF	6.38	4.93	11.31
6"	P2@.260	LF	6.79	6.41	13.20
8"	P2@.300	LF	9.34	7.39	16.73
10"	P2@.380	LF	12.80	9.36	22.16
12"	P2@.450	LF	13.90	11.10	25.00

Schedule 10 carbon steel 45 degree ell, butt-welded

Description	Craft@Hrs	Unit	Material $	Labor $	Total $
2"	P2@.170	Ea	19.50	4.19	23.69
2½"	P2@.185	Ea	19.50	4.56	24.06
3"	P2@.200	Ea	34.60	4.93	39.53
4"	P2@.230	Ea	37.80	5.67	43.47
5"	P2@.250	Ea	91.20	6.16	97.36
6"	P2@.350	Ea	107.00	8.62	115.62
8"	P2@.500	Ea	224.00	12.30	236.30
10"	P2@.750	Ea	355.00	18.50	373.50
12"	P2@1.00	Ea	618.00	24.60	642.60

Schedule 10 carbon steel 90 degree ell, butt-welded

Description	Craft@Hrs	Unit	Material $	Labor $	Total $
2"	P2@.170	Ea	19.50	4.19	23.69
2½"	P2@.185	Ea	19.50	4.56	24.06
3"	P2@.200	Ea	34.60	4.93	39.53
4"	P2@.230	Ea	37.80	5.67	43.47
5"	P2@.250	Ea	91.20	6.16	97.36
6"	P2@.350	Ea	107.00	8.62	115.62
8"	P2@.500	Ea	224.00	12.30	236.30
10"	P2@.750	Ea	409.00	18.50	427.50
12"	P2@1.00	Ea	655.00	24.60	679.60

Description	Craft@Hrs	Unit	Material $	Labor $	Total $

Schedule 10 carbon steel tee, butt-welded

Description	Craft@Hrs	Unit	Material $	Labor $	Total $
2"	P2@.200	Ea	30.00	4.93	34.93
2½"	P2@.210	Ea	30.00	5.17	35.17
3"	P2@.230	Ea	42.10	5.67	47.77
4"	P2@.250	Ea	64.20	6.16	70.36
5"	P2@.300	Ea	151.00	7.39	158.39
6"	P2@.500	Ea	174.00	12.30	186.30
8"	P2@.750	Ea	383.00	18.50	401.50
10"	P2@1.00	Ea	796.00	24.60	820.60
12"	P2@1.50	Ea	1,110.00	37.00	1,147.00

Schedule 10 carbon steel reducing tee, butt-welded

Description	Craft@Hrs	Unit	Material $	Labor $	Total $
3 x 2"	P2@.230	Ea	57.20	5.67	62.87
4 x 2"	P2@.240	Ea	77.40	5.91	83.31
4 x 2½"	P2@.245	Ea	77.40	6.04	83.44
4 x 3"	P2@.250	Ea	77.40	6.16	83.56
5 x 2"	P2@.260	Ea	189.00	6.41	195.41
5 x 3"	P2@.280	Ea	154.00	6.90	160.90
5 x 4"	P2@.300	Ea	167.00	7.39	174.39
6 x 3"	P2@.400	Ea	183.00	9.86	192.86
6 x 4"	P2@.450	Ea	183.00	11.10	194.10
6 x 5"	P2@.500	Ea	183.00	12.30	195.30
8 x 3"	P2@.600	Ea	265.00	14.80	279.80
8 x 4"	P2@.650	Ea	383.00	16.00	399.00
8 x 5"	P2@.700	Ea	383.00	17.20	400.20
8 x 6"	P2@.750	Ea	383.00	18.50	401.50
10 x 4"	P2@.800	Ea	482.00	19.70	501.70
10 x 5"	P2@.900	Ea	493.00	22.20	515.20
10 x 6"	P2@1.00	Ea	493.00	24.60	517.60
10 x 8"	P2@1.10	Ea	502.00	27.10	529.10
12 x 6"	P2@1.20	Ea	712.00	29.60	741.60
12 x 8"	P2@1.35	Ea	733.00	33.30	766.30
12 x 10"	P2@1.50	Ea	766.00	37.00	803.00

Schedule 10 carbon steel reducer, butt-welded

Description	Craft@Hrs	Unit	Material $	Labor $	Total $
2"	P2@.160	Ea	22.70	3.94	26.64
3"	P2@.170	Ea	19.90	4.19	24.09
4"	P2@.210	Ea	26.40	5.17	31.57
5"	P2@.260	Ea	38.80	6.41	45.21
6"	P2@.310	Ea	53.70	7.64	61.34
8"	P2@.450	Ea	110.00	11.10	121.10
10"	P2@.690	Ea	262.00	17.00	279.00
12"	P2@.920	Ea	342.00	22.70	364.70

Carbon Steel, Schedule 10 with Welded Joints

Description	Craft@Hrs	Unit	Material $	Labor $	Total $

Schedule 10 carbon steel cap, butt-welded

Description	Craft@Hrs	Unit	Material $	Labor $	Total $
2"	P2@.110	Ea	16.00	2.71	18.71
2½"	P2@.120	Ea	16.00	2.96	18.96
3"	P2@.130	Ea	16.00	3.20	19.20
4"	P2@.140	Ea	17.40	3.45	20.85
5"	P2@.150	Ea	40.90	3.70	44.60
6"	P2@.170	Ea	41.90	4.19	46.09
8"	P2@.200	Ea	80.40	4.93	85.33
10"	P2@.225	Ea	171.00	5.54	176.54
12"	P2@.250	Ea	279.00	6.16	285.16

150# forged steel companion flange, weld neck

Description	Craft@Hrs	Unit	Material $	Labor $	Total $
2"	P2@.280	Ea	70.30	6.90	77.20
2½"	P2@.340	Ea	87.20	8.38	95.58
3"	P2@.390	Ea	94.40	9.61	104.01
4"	P2@.560	Ea	126.00	13.80	139.80
5"	P2@.670	Ea	147.00	16.50	163.50
6"	P2@.780	Ea	162.00	19.20	181.20
8"	P2@1.10	Ea	186.00	27.10	213.10
10"	P2@1.40	Ea	289.00	34.50	323.50
12"	P2@1.70	Ea	375.00	41.90	416.90

Class 125 iron body gate valve, flanged

Description	Craft@Hrs	Unit	Material $	Labor $	Total $
2"	P2@.500	Ea	164.00	12.30	176.30
2½"	P2@.600	Ea	168.00	14.80	182.80
3"	P2@.750	Ea	188.00	18.50	206.50
4"	P2@1.35	Ea	270.00	33.30	303.30
5"	P2@2.00	Ea	460.00	49.30	509.30
6"	P2@2.50	Ea	460.00	61.60	521.60
8"	P2@3.00	Ea	789.00	73.90	862.90
10"	P2@4.00	Ea	1,390.00	98.60	1,488.60
12"	P2@4.50	Ea	1,910.00	111.00	2,021.00

Class 125 iron body globe valve, flanged

Description	Craft@Hrs	Unit	Material $	Labor $	Total $
2"	P2@.500	Ea	319.00	12.30	331.30
2½"	P2@.600	Ea	337.00	14.80	351.80
3"	P2@.750	Ea	390.00	18.50	408.50
4"	P2@1.35	Ea	558.00	33.30	591.30
5"	P2@2.00	Ea	1,020.00	49.30	1,069.30
6"	P2@2.50	Ea	1,020.00	61.60	1,081.60
8"	P2@3.00	Ea	1,990.00	73.90	2,063.90
10"	P2@4.00	Ea	3,110.00	98.60	3,208.60

Description	Craft@Hrs	Unit	Material $	Labor $	Total $

200 PSI iron body butterfly valve, lug type

Description	Craft@Hrs	Unit	Material $	Labor $	Total $
2"	P2@.200	Ea	68.30	4.93	73.23
2½"	P2@.300	Ea	72.20	7.39	79.59
3"	P2@.400	Ea	76.10	9.86	85.96
4"	P2@.500	Ea	96.20	12.30	108.50
5"	P2@.650	Ea	118.00	16.00	134.00
6"	P2@.750	Ea	155.00	18.50	173.50
8"	P2@.950	Ea	231.00	23.40	254.40
10"	P2@1.40	Ea	391.00	34.50	425.50
12"	P2@1.80	Ea	501.00	44.40	545.40

200 PSI iron body butterfly valve, wafer type

Description	Craft@Hrs	Unit	Material $	Labor $	Total $
2"	P2@.200	Ea	73.50	4.93	78.43
2½"	P2@.300	Ea	81.30	7.39	88.69
3"	P2@.400	Ea	84.50	9.86	94.36
4"	P2@.500	Ea	108.00	12.30	120.30
5"	P2@.650	Ea	137.00	16.00	153.00
6"	P2@.750	Ea	183.00	18.50	201.50
8"	P2@.950	Ea	272.00	23.40	295.40
10"	P2@1.40	Ea	448.00	34.50	482.50
12"	P2@1.80	Ea	569.00	44.40	613.40

Class 125 iron body swing check valve, flanged

Description	Craft@Hrs	Unit	Material $	Labor $	Total $
2"	P2@.500	Ea	149.00	12.30	161.30
2½"	P2@.600	Ea	192.00	14.80	206.80
3"	P2@.750	Ea	205.00	18.50	223.50
4"	P2@1.35	Ea	322.00	33.30	355.30
5"	P2@2.00	Ea	549.00	49.30	598.30
6"	P2@2.50	Ea	549.00	61.60	610.60
8"	P2@3.00	Ea	1,030.00	73.90	1,103.90
10"	P2@4.00	Ea	1,750.00	98.60	1,848.60
12"	P2@4.50	Ea	2,740.00	111.00	2,851.00

Class 125 iron body silent check valve, flanged

Description	Craft@Hrs	Unit	Material $	Labor $	Total $
2"	P2@.500	Ea	124.00	12.30	136.30
2½"	P2@.600	Ea	145.00	14.80	159.80
3"	P2@.750	Ea	162.00	18.50	180.50
4"	P2@1.35	Ea	217.00	33.30	250.30
5"	P2@2.00	Ea	297.00	49.30	346.30
6"	P2@2.50	Ea	379.00	61.60	440.60
8"	P2@3.00	Ea	634.00	73.90	707.90
10"	P2@4.00	Ea	963.00	98.60	1,061.60

Description	Craft@Hrs	Unit	Material $	Labor $	Total $
Class 125 iron body strainer, flanged					
2"	P2@.500	Ea	83.00	12.30	95.30
2½"	P2@.600	Ea	91.80	14.80	106.60
3"	P2@.750	Ea	109.00	18.50	127.50
4"	P2@1.35	Ea	186.00	33.30	219.30
5"	P2@2.00	Ea	309.00	49.30	358.30
6"	P2@2.50	Ea	378.00	61.60	439.60
8"	P2@3.00	Ea	634.00	73.90	707.90
Installation of flanged 2-way control valve					
2"	P2@.500	Ea	--	12.30	12.30
2½"	P2@.600	Ea	--	14.80	14.80
3"	P2@.750	Ea	--	18.50	18.50
4"	P2@1.35	Ea	--	33.30	33.30
6"	P2@2.50	Ea	--	61.60	61.60
8"	P2@3.00	Ea	--	73.90	73.90
10"	P2@4.00	Ea	--	98.60	98.60
12"	P2@4.50	Ea	--	111.00	111.00
Installation of flanged 3-way control valve					
2"	P2@.910	Ea	--	22.40	22.40
2½"	P2@1.12	Ea	--	27.60	27.60
3"	P2@1.33	Ea	--	32.80	32.80
4"	P2@2.00	Ea	--	49.30	49.30
6"	P2@3.70	Ea	--	91.20	91.20
8"	P2@4.40	Ea	--	108.00	108.00
10"	P2@5.90	Ea	--	145.00	145.00
12"	P2@6.50	Ea	--	160.00	160.00
Bolt and nut set for flanges					
2"	P2@.500	Ea	3.85	12.30	16.15
2½"	P2@.650	Ea	4.08	16.00	20.08
3"	P2@.750	Ea	4.28	18.50	22.78
4"	P2@1.00	Ea	7.75	24.60	32.35
5"	P2@1.10	Ea	9.93	27.10	37.03
6"	P2@1.20	Ea	12.10	29.60	41.70
8"	P2@1.25	Ea	12.90	30.80	43.70
10"	P2@1.70	Ea	26.70	41.90	68.60
12"	P2@2.20	Ea	28.60	54.20	82.80

Description	Craft@Hrs	Unit	Material $	Labor $	Total $
Thermometer with well					
7"	P2@.250	Ea	14.30	6.16	20.46
9"	P2@.250	Ea	14.80	6.16	20.96
Pressure gauge					
2½"	P2@.200	Ea	15.60	4.93	20.53
3½"	P2@.200	Ea	27.10	4.93	32.03
Hanger with swivel assembly					
2"	P2@.450	Ea	.65	11.10	11.75
2½"	P2@.450	Ea	3.95	11.10	15.05
3"	P2@.450	Ea	4.32	11.10	15.42
4"	P2@.450	Ea	5.87	11.10	16.97
5"	P2@.500	Ea	8.78	12.30	21.08
6"	P2@.500	Ea	11.70	12.30	24.00
8"	P2@.550	Ea	17.30	13.60	30.90
10"	P2@.650	Ea	21.00	16.00	37.00
12"	P2@.750	Ea	25.90	18.50	44.40
Riser clamp					
2"	P2@.115	Ea	4.45	2.83	7.28
2½"	P2@.120	Ea	4.69	2.96	7.65
3"	P2@.120	Ea	5.13	2.96	8.09
4"	P2@.125	Ea	6.45	3.08	9.53
5"	P2@.180	Ea	8.58	4.44	13.02
6"	P2@.200	Ea	10.70	4.93	15.63
8"	P2@.200	Ea	17.40	4.93	22.33
10"	P2@.250	Ea	25.40	6.16	31.56
12"	P2@.250	Ea	30.70	6.16	36.86

Carbon Steel, Schedule 40 with Roll-Grooved Joints

Schedule 40 carbon steel (ASTM A-53 and A-120) roll-grooved pipe with factory-grooved steel fittings is commonly used for heating hot water, chilled water and condenser water systems.

Consult the manufacturers for maximum operating temperature/pressure ratings for the various combinations of pipe and fittings.

This section has been arranged to save the estimator's time by including all normally-used system components such as pipe, fittings, valves, hanger assemblies, riser clamps and miscellaneous items under one heading. Additional items can be found under "Plumbing and Piping Specialties." The cost estimates in this section are based on the conditions, limitations and wage rates described in the section "How to Use This Book" beginning on page 5.

Description	Craft@Hrs	Unit	Material $	Labor $	Total $

Schedule 40 carbon steel pipe with roll-grooved joints

Description	Craft@Hrs	Unit	Material $	Labor $	Total $
2"	P2@.100	LF	2.14	2.46	4.60
2½"	P2@.115	LF	3.33	2.83	6.16
3"	P2@.135	LF	4.33	3.33	7.66
4"	P2@.190	LF	6.22	4.68	10.90
5"	P2@.240	LF	14.00	5.91	19.91
6"	P2@.315	LF	14.90	7.76	22.66
8"	P2@.370	LF	22.30	9.12	31.42
10"	P2@.460	LF	31.90	11.30	43.20
12"	P2@.550	LF	41.80	13.60	55.40

Schedule 40 carbon steel 45 degree ell with roll-grooved joints

Description	Craft@Hrs	Unit	Material $	Labor $	Total $
2"	P2@.170	Ea	19.50	4.19	23.69
2½"	P2@.185	Ea	19.50	4.56	24.06
3"	P2@.200	Ea	34.60	4.93	39.53
4"	P2@.230	Ea	37.80	5.67	43.47
5"	P2@.250	Ea	91.20	6.16	97.36
6"	P2@.350	Ea	107.00	8.62	115.62
8"	P2@.500	Ea	224.00	12.30	236.30
10"	P2@.750	Ea	355.00	18.50	373.50
12"	P2@1.00	Ea	618.00	24.60	642.60

Schedule 40 carbon steel 90 degree ell with roll-grooved joints

Description	Craft@Hrs	Unit	Material $	Labor $	Total $
2"	P2@.170	Ea	19.50	4.19	23.69
2½"	P2@.185	Ea	19.50	4.56	24.06
3"	P2@.200	Ea	34.60	4.93	39.53
4"	P2@.230	Ea	37.80	5.67	43.47
5"	P2@.250	Ea	91.20	6.16	97.36
6"	P2@.350	Ea	107.00	8.62	115.62
8"	P2@.500	Ea	224.00	12.30	236.30
10"	P2@.750	Ea	409.00	18.50	427.50
12"	P2@1.00	Ea	655.00	24.60	679.60

Description	Craft@Hrs	Unit	Material $	Labor $	Total $

Schedule 40 carbon steel tee with roll-grooved joints

Description	Craft@Hrs	Unit	Material $	Labor $	Total $
2"	P2@.200	Ea	30.00	4.93	34.93
2½"	P2@.210	Ea	30.00	5.17	35.17
3"	P2@.230	Ea	42.10	5.67	47.77
4"	P2@.250	Ea	64.20	6.16	70.36
5"	P2@.300	Ea	151.00	7.39	158.39
6"	P2@.500	Ea	174.00	12.30	186.30
8"	P2@.750	Ea	383.00	18.50	401.50
10"	P2@1.00	Ea	796.00	24.60	820.60
12"	P2@1.50	Ea	1,110.00	37.00	1,147.00

Schedule 40 carbon steel reducing tee with roll-grooved joints

Description	Craft@Hrs	Unit	Material $	Labor $	Total $
3 x 2"	P2@.230	Ea	57.20	5.67	62.87
4 x 2"	P2@.240	Ea	77.40	5.91	83.31
4 x 2½"	P2@.245	Ea	77.40	6.04	83.44
4 x 3"	P2@.250	Ea	77.40	6.16	83.56
5 x 2"	P2@.260	Ea	189.00	6.41	195.41
5 x 3"	P2@.280	Ea	154.00	6.90	160.90
5 x 4"	P2@.300	Ea	167.00	7.39	174.39
6 x 3"	P2@.400	Ea	183.00	9.86	192.86
6 x 4"	P2@.450	Ea	183.00	11.10	194.10
6 x 5"	P2@.500	Ea	183.00	12.30	195.30
8 x 3"	P2@.600	Ea	265.00	14.80	279.80
8 x 4"	P2@.650	Ea	383.00	16.00	399.00
8 x 5"	P2@.700	Ea	383.00	17.20	400.20
8 x 6"	P2@.750	Ea	383.00	18.50	401.50
10 x 4"	P2@.800	Ea	482.00	19.70	501.70
10 x 5"	P2@.900	Ea	493.00	22.20	515.20
10 x 6"	P2@1.00	Ea	493.00	24.60	517.60
10 x 8"	P2@1.10	Ea	502.00	27.10	529.10
12 x 6"	P2@1.20	Ea	712.00	29.60	741.60
12 x 8"	P2@1.35	Ea	733.00	33.30	766.30
12 x 10"	P2@1.50	Ea	766.00	37.00	803.00

Schedule 40 carbon steel male adapter

Description	Craft@Hrs	Unit	Material $	Labor $	Total $
2"	P2@.170	Ea	10.70	4.19	14.89
2½"	P2@.185	Ea	12.50	4.56	17.06
3"	P2@.200	Ea	15.50	4.93	20.43
4"	P2@.230	Ea	25.90	5.67	31.57
5"	P2@.290	Ea	54.10	7.15	61.25
6"	P2@.400	Ea	70.00	9.86	79.86

Schedule 40 carbon steel female adapter

Description	Craft@Hrs	Unit	Material $	Labor $	Total $
2"	P2@.170	Ea	35.00	4.19	39.19
3"	P2@.200	Ea	40.90	4.93	45.83
4"	P2@.230	Ea	62.90	5.67	68.57

Carbon Steel, Schedule 40 with Roll-Grooved Joints

Description	Craft@Hrs	Unit	Material $	Labor $	Total $
Schedule 40 carbon steel reducer					
2"	P2@.160	Ea	22.70	3.94	26.64
3"	P2@.170	Ea	19.90	4.19	24.09
4"	P2@.210	Ea	26.40	5.17	31.57
5"	P2@.260	Ea	38.80	6.41	45.21
6"	P2@.310	Ea	53.70	7.64	61.34
8"	P2@.450	Ea	110.00	11.10	121.10
10"	P2@.690	Ea	262.00	17.00	279.00
12"	P2@.920	Ea	342.00	22.70	364.70
Schedule 40 carbon steel cap					
2"	P2@.110	Ea	16.00	2.71	18.71
2½"	P2@.120	Ea	16.00	2.96	18.96
3"	P2@.130	Ea	16.00	3.20	19.20
4"	P2@.140	Ea	17.40	3.45	20.85
5"	P2@.150	Ea	40.90	3.70	44.60
6"	P2@.170	Ea	41.90	4.19	46.09
8"	P2@.200	Ea	80.40	4.93	85.33
10"	P2@.225	Ea	171.00	5.54	176.54
12"	P2@.250	Ea	279.00	6.16	285.16
Flange with gasket					
2"	P2@.280	Ea	70.30	6.90	77.20
2½"	P2@.340	Ea	87.20	8.38	95.58
3"	P2@.390	Ea	94.40	9.61	104.01
4"	P2@.560	Ea	126.00	13.80	139.80
5"	P2@.670	Ea	147.00	16.50	163.50
6"	P2@.780	Ea	162.00	19.20	181.20
8"	P2@1.10	Ea	186.00	27.10	213.10
10"	P2@1.40	Ea	289.00	34.50	323.50
12"	P2@1.70	Ea	375.00	41.90	416.90
Roll grooved coupling with gasket					
2"	P2@.300	Ea	19.50	7.39	26.89
2½"	P2@.350	Ea	22.10	8.62	30.72
3"	P2@.400	Ea	24.70	9.86	34.56
4"	P2@.500	Ea	36.50	12.30	48.80
5"	P2@.600	Ea	51.40	14.80	66.20
6"	P2@.700	Ea	66.30	17.20	83.50
8"	P2@.900	Ea	107.00	22.20	129.20
10"	P2@1.10	Ea	393.00	27.10	420.10
12"	P2@1.35	Ea	287.00	33.30	320.30

Description	Craft@Hrs	Unit	Material $	Labor $	Total $

Class 125 iron body gate valve, flanged

Description	Craft@Hrs	Unit	Material $	Labor $	Total $
2"	P2@.500	Ea	164.00	12.30	176.30
2½"	P2@.600	Ea	168.00	14.80	182.80
3"	P2@.750	Ea	188.00	18.50	206.50
4"	P2@1.35	Ea	270.00	33.30	303.30
5"	P2@2.00	Ea	460.00	49.30	509.30
6"	P2@2.50	Ea	460.00	61.60	521.60
8"	P2@3.00	Ea	789.00	73.90	862.90
10"	P2@4.00	Ea	1,390.00	98.60	1,488.60
12"	P2@4.50	Ea	1,910.00	111.00	2,021.00

Class 125 iron body globe valve, flanged

Description	Craft@Hrs	Unit	Material $	Labor $	Total $
2"	P2@.500	Ea	319.00	12.30	331.30
2½"	P2@.600	Ea	337.00	14.80	351.80
3"	P2@.750	Ea	390.00	18.50	408.50
4"	P2@1.35	Ea	558.00	33.30	591.30
5"	P2@2.00	Ea	1,020.00	49.30	1,069.30
6"	P2@2.50	Ea	1,020.00	61.60	1,081.60
8"	P2@3.00	Ea	1,990.00	73.90	2,063.90
10"	P2@4.00	Ea	3,110.00	98.60	3,208.60

200 PSI iron body butterfly valve, lug type

Description	Craft@Hrs	Unit	Material $	Labor $	Total $
2"	P2@.200	Ea	68.30	4.93	73.23
2½"	P2@.300	Ea	72.20	7.39	79.59
3"	P2@.400	Ea	76.10	9.86	85.96
4"	P2@.500	Ea	96.20	12.30	108.50
5"	P2@.650	Ea	118.00	16.00	134.00
6"	P2@.750	Ea	155.00	18.50	173.50
8"	P2@.950	Ea	231.00	23.40	254.40
10"	P2@1.40	Ea	391.00	34.50	425.50
12"	P2@1.80	Ea	501.00	44.40	545.40

200 PSI iron body butterfly valve, wafer type

Description	Craft@Hrs	Unit	Material $	Labor $	Total $
2"	P2@.200	Ea	73.50	4.93	78.43
2½"	P2@.300	Ea	81.30	7.39	88.69
3"	P2@.400	Ea	84.50	9.86	94.36
4"	P2@.500	Ea	108.00	12.30	120.30
5"	P2@.650	Ea	137.00	16.00	153.00
6"	P2@.750	Ea	183.00	18.50	201.50
8"	P2@.950	Ea	272.00	23.40	295.40
10"	P2@1.40	Ea	448.00	34.50	482.50
12"	P2@1.80	Ea	569.00	44.40	613.40

Carbon Steel, Schedule 40 with Roll-Grooved Joints

Description	Craft@Hrs	Unit	Material $	Labor $	Total $
Class 125 iron body swing check valve, flanged					
2"	P2@.500	Ea	149.00	12.30	161.30
2½"	P2@.600	Ea	192.00	14.80	206.80
3"	P2@.750	Ea	205.00	18.50	223.50
4"	P2@1.35	Ea	322.00	33.30	355.30
5"	P2@2.00	Ea	549.00	49.30	598.30
6"	P2@2.50	Ea	549.00	61.60	610.60
8"	P2@3.00	Ea	1,030.00	73.90	1,103.90
10"	P2@4.00	Ea	1,750.00	98.60	1,848.60
12"	P2@4.50	Ea	2,740.00	111.00	2,851.00
Class 125 iron body silent check valve, flanged					
2"	P2@.500	Ea	124.00	12.30	136.30
2½"	P2@.600	Ea	145.00	14.80	159.80
3"	P2@.750	Ea	162.00	18.50	180.50
4"	P2@1.35	Ea	217.00	33.30	250.30
5"	P2@2.00	Ea	297.00	49.30	346.30
6"	P2@2.50	Ea	379.00	61.60	440.60
8"	P2@3.00	Ea	634.00	73.90	707.90
10"	P2@4.00	Ea	963.00	98.60	1,061.60
Class 125 iron body strainer, flanged					
2"	P2@.500	Ea	83.00	12.30	95.30
2½"	P2@.600	Ea	91.80	14.80	106.60
3"	P2@.750	Ea	109.00	18.50	127.50
4"	P2@1.35	Ea	186.00	33.30	219.30
5"	P2@2.00	Ea	309.00	49.30	358.30
6"	P2@2.50	Ea	378.00	61.60	439.60
8"	P2@3.00	Ea	634.00	73.90	707.90
Installation of flanged 2-way control valve					
2"	P2@.500	Ea	--	12.30	12.30
2½"	P2@.600	Ea	--	14.80	14.80
3"	P2@.750	Ea	--	18.50	18.50
4"	P2@1.35	Ea	--	33.30	33.30
6"	P2@2.50	Ea	--	61.60	61.60
8"	P2@3.00	Ea	--	73.90	73.90
10"	P2@4.00	Ea	--	98.60	98.60
12"	P2@4.50	Ea	--	111.00	111.00
Installation of flanged 3-way control valve					
2"	P2@.910	Ea	--	22.40	22.40
2½"	P2@1.12	Ea	--	27.60	27.60
3"	P2@1.33	Ea	--	32.80	32.80
4"	P2@2.00	Ea	--	49.30	49.30
6"	P2@3.70	Ea	--	91.20	91.20
8"	P2@4.40	Ea	--	108.00	108.00
10"	P2@5.90	Ea	--	145.00	145.00
12"	P2@6.50	Ea	--	160.00	160.00

Carbon Steel, Schedule 40 with Roll-Grooved Joints

Description	Craft@Hrs	Unit	Material $	Labor $	Total $
Bolt and nut set for flanges					
2"	P2@.500	Ea	3.85	12.30	16.15
2½"	P2@.650	Ea	4.08	16.00	20.08
3"	P2@.750	Ea	4.28	18.50	22.78
4"	P2@1.00	Ea	7.75	24.60	32.35
5"	P2@1.10	Ea	9.93	27.10	37.03
6"	P2@1.20	Ea	12.10	29.60	41.70
8"	P2@1.25	Ea	12.90	30.80	43.70
10"	P2@1.70	Ea	26.70	41.90	68.60
12"	P2@2.20	Ea	28.60	54.20	82.80
Thermometer with well					
7"	P2@.250	Ea	14.30	6.16	20.46
9"	P2@.250	Ea	14.80	6.16	20.96
Pressure gauge					
2½"	P2@.200	Ea	15.60	4.93	20.53
3½"	P2@.200	Ea	27.10	4.93	32.03
Hanger with swivel assembly					
2"	P2@.450	Ea	.65	11.10	11.75
2½"	P2@.450	Ea	3.95	11.10	15.05
3"	P2@.450	Ea	4.32	11.10	15.42
4"	P2@.450	Ea	5.87	11.10	16.97
5"	P2@.500	Ea	8.78	12.30	21.08
6"	P2@.500	Ea	11.70	12.30	24.00
8"	P2@.550	Ea	17.30	13.60	30.90
10"	P2@.650	Ea	21.00	16.00	37.00
12"	P2@.750	Ea	25.90	18.50	44.40
Riser clamp					
2"	P2@.115	Ea	4.45	2.83	7.28
2½"	P2@.120	Ea	4.69	2.96	7.65
3"	P2@.120	Ea	5.13	2.96	8.09
4"	P2@.125	Ea	6.45	3.08	9.53
5"	P2@.180	Ea	8.58	4.44	13.02
6"	P2@.200	Ea	10.70	4.93	15.63
8"	P2@.200	Ea	17.40	4.93	22.33
10"	P2@.250	Ea	25.40	6.16	31.56
12"	P2@.250	Ea	30.70	6.16	36.86

Carbon Steel, Schedule 40 with Cut-Grooved Joints

Schedule 40 carbon steel (ASTM A-53 and A-120) cut-grooved pipe with factory-grooved malleable iron, ductile iron or steel fittings is commonly used for heating hot water, chilled water and condenser water systems.

Consult the manufacturers for maximum operating temperature/pressure ratings for various combinations of pipe and fittings.

This section has been arranged to save the estimator's time by including all normally-used system components such as pipe, fittings, valves, hanger assemblies, riser clamps and miscellaneous items under one heading. Additional items can be found under "Plumbing and Piping Specialties." The cost estimates in this section are based on the conditions, limitations and wage rates described in the section "How to Use This Book" beginning on page 5.

Description	Craft@Hrs	Unit	Material $	Labor $	Total $

Schedule 40 carbon steel pipe with cut-grooved joints

Description	Craft@Hrs	Unit	Material $	Labor $	Total $
2"	P2@.110	LF	1.55	2.71	4.26
2½"	P2@.130	LF	2.47	3.20	5.67
3"	P2@.150	LF	3.33	3.70	7.03
4"	P2@.210	LF	4.60	5.17	9.77
5"	P2@.270	LF	9.98	6.65	16.63
6"	P2@.350	LF	10.60	8.62	19.22
8"	P2@.410	LF	15.80	10.10	25.90
10"	P2@.510	LF	22.80	12.60	35.40
12"	P2@.610	LF	30.00	15.00	45.00

Schedule 40 carbon steel 45 degree ell with cut-grooved joints

Description	Craft@Hrs	Unit	Material $	Labor $	Total $
2"	P2@.170	Ea	19.50	4.19	23.69
2½"	P2@.185	Ea	19.50	4.56	24.06
3"	P2@.200	Ea	34.60	4.93	39.53
4"	P2@.230	Ea	37.80	5.67	43.47
5"	P2@.250	Ea	91.20	6.16	97.36
6"	P2@.350	Ea	107.00	8.62	115.62
8"	P2@.500	Ea	224.00	12.30	236.30
10"	P2@.750	Ea	355.00	18.50	373.50
12"	P2@1.00	Ea	618.00	24.60	642.60

Schedule 40 carbon steel 90 degree ell with cut-grooved joints

Description	Craft@Hrs	Unit	Material $	Labor $	Total $
2"	P2@.170	Ea	19.50	4.19	23.69
2½"	P2@.185	Ea	19.50	4.56	24.06
3"	P2@.200	Ea	34.60	4.93	39.53
4"	P2@.230	Ea	37.80	5.67	43.47
5"	P2@.250	Ea	91.20	6.16	97.36
6"	P2@.350	Ea	107.00	8.62	115.62
8"	P2@.500	Ea	224.00	12.30	236.30
10"	P2@.750	Ea	409.00	18.50	427.50
12"	P2@1.00	Ea	655.00	24.60	679.60

Description	Craft@Hrs	Unit	Material $	Labor $	Total $

Schedule 40 carbon steel tee with cut-grooved joints

Description	Craft@Hrs	Unit	Material $	Labor $	Total $
2"	P2@.200	Ea	30.00	4.93	34.93
2½"	P2@.210	Ea	30.00	5.17	35.17
3"	P2@.230	Ea	42.10	5.67	47.77
4"	P2@.250	Ea	64.50	6.16	70.66
5"	P2@.300	Ea	151.00	7.39	158.39
6"	P2@.500	Ea	174.00	12.30	186.30
8"	P2@.750	Ea	383.00	18.50	401.50
10"	P2@1.00	Ea	796.00	24.60	820.60
12"	P2@1.50	Ea	1,110.00	37.00	1,147.00

Schedule 40 carbon steel reducing tee with cut-grooved joints

Description	Craft@Hrs	Unit	Material $	Labor $	Total $
3 x 2"	P2@.230	Ea	57.20	5.67	62.87
4 x 2"	P2@.240	Ea	77.40	5.91	83.31
4 x 2½"	P2@.245	Ea	77.40	6.04	83.44
4 x 3"	P2@.250	Ea	77.40	6.16	83.56
5 x 2"	P2@.260	Ea	189.00	6.41	195.41
5 x 3"	P2@.280	Ea	154.00	6.90	160.90
5 x 4"	P2@.300	Ea	167.00	7.39	174.39
6 x 3"	P2@.400	Ea	183.00	9.86	192.86
6 x 4"	P2@.450	Ea	183.00	11.10	194.10
6 x 5"	P2@.500	Ea	183.00	12.30	195.30
8 x 3"	P2@.600	Ea	265.00	14.80	279.80
8 x 4"	P2@.650	Ea	383.00	16.00	399.00
8 x 5"	P2@.700	Ea	383.00	17.20	400.20
8 x 6"	P2@.750	Ea	383.00	18.50	401.50
10 x 4"	P2@.800	Ea	482.00	19.70	501.70
10 x 5"	P2@.900	Ea	493.00	22.20	515.20
10 x 6"	P2@1.00	Ea	493.00	24.60	517.60
10 x 8"	P2@1.10	Ea	502.00	27.10	529.10
12 x 6"	P2@1.20	Ea	712.00	29.60	741.60
12 x 8"	P2@1.35	Ea	733.00	33.30	766.30
12 x 10"	P2@1.50	Ea	766.00	37.00	803.00

Schedule 40 carbon steel male adapter

Description	Craft@Hrs	Unit	Material $	Labor $	Total $
2"	P2@.170	Ea	10.70	4.19	14.89
2½"	P2@.185	Ea	12.50	4.56	17.06
3"	P2@.200	Ea	15.50	4.93	20.43
4"	P2@.230	Ea	25.90	5.67	31.57
5"	P2@.290	Ea	54.10	7.15	61.25
6"	P2@.400	Ea	70.00	9.86	79.86

Schedule 40 carbon steel female adapter

Description	Craft@Hrs	Unit	Material $	Labor $	Total $
2"	P2@.170	Ea	35.00	4.19	39.19
3"	P2@.200	Ea	40.90	4.93	45.83
4"	P2@.230	Ea	62.90	5.67	68.57

Carbon Steel, Schedule 40 with Cut-Grooved Joints

Description	Craft@Hrs	Unit	Material $	Labor $	Total $
Schedule 40 carbon steel reducer					
2"	P2@.160	Ea	22.70	3.94	26.64
3"	P2@.170	Ea	19.90	4.19	24.09
4"	P2@.210	Ea	26.40	5.17	31.57
5"	P2@.260	Ea	38.80	6.41	45.21
6"	P2@.310	Ea	53.70	7.64	61.34
8"	P2@.450	Ea	110.00	11.10	121.10
10"	P2@.690	Ea	262.00	17.00	279.00
12"	P2@.920	Ea	342.00	22.70	364.70
Schedule 40 carbon steel cap					
2"	P2@.110	Ea	16.00	2.71	18.71
2½"	P2@.120	Ea	16.00	2.96	18.96
3"	P2@.130	Ea	16.00	3.20	19.20
4"	P2@.140	Ea	17.40	3.45	20.85
5"	P2@.150	Ea	40.90	3.70	44.60
6"	P2@.170	Ea	41.90	4.19	46.09
8"	P2@.200	Ea	80.40	4.93	85.33
10"	P2@.225	Ea	171.00	5.54	176.54
12"	P2@.250	Ea	279.00	6.16	285.16
Flange with gasket					
2"	P2@.280	Ea	70.30	6.90	77.20
2½"	P2@.340	Ea	87.20	8.38	95.58
3"	P2@.390	Ea	94.40	9.61	104.01
4"	P2@.560	Ea	126.00	13.80	139.80
5"	P2@.670	Ea	147.00	16.50	163.50
6"	P2@.780	Ea	162.00	19.20	181.20
8"	P2@1.10	Ea	186.00	27.10	213.10
10"	P2@1.40	Ea	289.00	34.50	323.50
12"	P2@1.70	Ea	375.00	41.90	416.90
Cut grooved joint coupling with gasket					
2"	P2@.300	Ea	19.50	7.39	26.89
2½"	P2@.350	Ea	22.10	8.62	30.72
3"	P2@.400	Ea	24.70	9.86	34.56
4"	P2@.500	Ea	36.50	12.30	48.80
5"	P2@.600	Ea	51.40	14.80	66.20
6"	P2@.700	Ea	66.30	17.20	83.50
8"	P2@.900	Ea	107.00	22.20	129.20
10"	P2@1.10	Ea	393.00	27.10	420.10
12"	P2@1.35	Ea	287.00	33.30	320.30

Carbon Steel, Schedule 40 with Cut-Grooved Joints

Description	Craft@Hrs	Unit	Material $	Labor $	Total $
Class 125 iron body gate valve, flanged					
2"	P2@.500	Ea	164.00	12.30	176.30
2½"	P2@.600	Ea	168.00	14.80	182.80
3"	P2@.750	Ea	188.00	18.50	206.50
4"	P2@1.35	Ea	270.00	33.30	303.30
5"	P2@2.00	Ea	460.00	49.30	509.30
6"	P2@2.50	Ea	460.00	61.60	521.60
8"	P2@3.00	Ea	789.00	73.90	862.90
10"	P2@4.00	Ea	1,390.00	98.60	1,488.60
12"	P2@4.50	Ea	1,910.00	111.00	2,021.00
Class 125 iron body globe valve, flanged					
2"	P2@.500	Ea	319.00	12.30	331.30
2½"	P2@.600	Ea	337.00	14.80	351.80
3"	P2@.750	Ea	390.00	18.50	408.50
4"	P2@1.35	Ea	558.00	33.30	591.30
5"	P2@2.00	Ea	1,020.00	49.30	1,069.30
6"	P2@2.50	Ea	1,020.00	61.60	1,081.60
8"	P2@3.00	Ea	1,990.00	73.90	2,063.90
10"	P2@4.00	Ea	3,450.00	98.60	3,548.60
200 PSI iron body butterfly valve, lug type					
2"	P2@.200	Ea	68.30	4.93	73.23
2½"	P2@.300	Ea	72.20	7.39	79.59
3"	P2@.400	Ea	76.10	9.86	85.96
4"	P2@.500	Ea	96.20	12.30	108.50
5"	P2@.650	Ea	118.00	16.00	134.00
6"	P2@.750	Ea	155.00	18.50	173.50
8"	P2@.950	Ea	231.00	23.40	254.40
10"	P2@1.40	Ea	391.00	34.50	425.50
12"	P2@1.80	Ea	501.00	44.40	545.40
200 PSI iron body butterfly valve, wafer type					
2"	P2@.200	Ea	73.50	4.93	78.43
2½"	P2@.300	Ea	81.30	7.39	88.69
3"	P2@.400	Ea	84.50	9.86	94.36
4"	P2@.500	Ea	108.00	12.30	120.30
5"	P2@.650	Ea	137.00	16.00	153.00
6"	P2@.750	Ea	183.00	18.50	201.50
8"	P2@.950	Ea	272.00	23.40	295.40
10"	P2@1.40	Ea	448.00	34.50	482.50
12"	P2@1.80	Ea	569.00	44.40	613.40

Carbon Steel, Schedule 40 with Cut-Grooved Joints

Description	Craft@Hrs	Unit	Material $	Labor $	Total $
Class 125 iron body swing check valve, flanged					
2"	P2@.500	Ea	149.00	12.30	161.30
2½"	P2@.600	Ea	192.00	14.80	206.80
3"	P2@.750	Ea	205.00	18.50	223.50
4"	P2@1.35	Ea	322.00	33.30	355.30
5"	P2@2.00	Ea	549.00	49.30	598.30
6"	P2@2.50	Ea	549.00	61.60	610.60
8"	P2@3.00	Ea	1,030.00	73.90	1,103.90
10"	P2@4.00	Ea	1,750.00	98.60	1,848.60
12"	P2@4.50	Ea	2,740.00	111.00	2,851.00
Class 125 iron body silent check valve, flanged					
2"	P2@.500	Ea	124.00	12.30	136.30
2½"	P2@.600	Ea	145.00	14.80	159.80
3"	P2@.750	Ea	162.00	18.50	180.50
4"	P2@1.35	Ea	217.00	33.30	250.30
5"	P2@2.00	Ea	297.00	49.30	346.30
6"	P2@2.50	Ea	379.00	61.60	440.60
8"	P2@3.00	Ea	634.00	73.90	707.90
10"	P2@4.00	Ea	963.00	98.60	1,061.60
Class 125 iron body strainer, flanged					
2"	P2@.500	Ea	83.00	12.30	95.30
2½"	P2@.600	Ea	91.80	14.80	106.60
3"	P2@.750	Ea	109.00	18.50	127.50
4"	P2@1.35	Ea	186.00	33.30	219.30
5"	P2@2.00	Ea	309.00	49.30	358.30
6"	P2@2.50	Ea	378.00	61.60	439.60
8"	P2@3.00	Ea	634.00	73.90	707.90
Installation of flanged 2-way control valve					
2"	P2@.500	Ea	--	12.30	12.30
2½"	P2@.600	Ea	--	14.80	14.80
3"	P2@.750	Ea	--	18.50	18.50
4"	P2@1.35	Ea	--	33.30	33.30
6"	P2@2.50	Ea	--	61.60	61.60
8"	P2@3.00	Ea	--	73.90	73.90
10"	P2@4.00	Ea	--	98.60	98.60
12"	P2@4.50	Ea	--	111.00	111.00

Carbon Steel, Schedule 40 with Cut-Grooved Joints

Description	Craft@Hrs	Unit	Material $	Labor $	Total $
Installation of flanged 3-way control valve					
2"	P2@.910	Ea	--	22.40	22.40
2½"	P2@1.12	Ea	--	27.60	27.60
3"	P2@1.33	Ea	--	32.80	32.80
4"	P2@2.00	Ea	--	49.30	49.30
6"	P2@3.70	Ea	--	91.20	91.20
8"	P2@4.40	Ea	--	108.00	108.00
10"	P2@5.90	Ea	--	145.00	145.00
12"	P2@6.50	Ea	--	160.00	160.00
Bolt and nut set for flanges					
2"	P2@.500	Ea	3.85	12.30	16.15
2½"	P2@.650	Ea	4.08	16.00	20.08
3"	P2@.750	Ea	4.28	18.50	22.78
4"	P2@1.00	Ea	7.75	24.60	32.35
5"	P2@1.10	Ea	9.93	27.10	37.03
6"	P2@1.20	Ea	12.10	29.60	41.70
8"	P2@1.25	Ea	12.90	30.80	43.70
10"	P2@1.70	Ea	26.70	41.90	68.60
12"	P2@2.20	Ea	28.60	54.20	82.80
Thermometer with well					
7"	P2@.250	Ea	14.30	6.16	20.46
9"	P2@.250	Ea	14.80	6.16	20.96
Pressure gauge					
2½"	P2@.200	Ea	15.60	4.93	20.53
3½"	P2@.200	Ea	27.10	4.93	32.03
Hanger with swivel assembly					
2"	P2@.450	Ea	.65	11.10	11.75
2½"	P2@.450	Ea	3.95	11.10	15.05
3"	P2@.450	Ea	4.32	11.10	15.42
4"	P2@.450	Ea	5.87	11.10	16.97
5"	P2@.500	Ea	8.78	12.30	21.08
6"	P2@.500	Ea	11.70	12.30	24.00
8"	P2@.550	Ea	17.30	13.60	30.90
10"	P2@.650	Ea	21.00	16.00	37.00
12"	P2@.750	Ea	25.90	18.50	44.40

Carbon Steel, Schedule 40 with Cut-Grooved Joints

Description	Craft@Hrs	Unit	Material $	Labor $	Total $
Riser clamp					

2"	P2@.115	Ea	4.45	2.83	7.28
2½"	P2@.120	Ea	4.69	2.96	7.65
3"	P2@.120	Ea	5.13	2.96	8.09
4"	P2@.125	Ea	6.45	3.08	9.53
5"	P2@.180	Ea	8.58	4.44	13.02
6"	P2@.200	Ea	10.70	4.93	15.63
8"	P2@.200	Ea	17.40	4.93	22.33
10"	P2@.250	Ea	25.40	6.16	31.56
12"	P2@.250	Ea	30.70	6.16	36.86

Description	Craft@Hrs	Unit	Material $	Labor $	Total $
Class 150 cast iron pipe with mechanical joints					
1 ½"	P2@.080	LF	3.08	1.97	5.05
2"	P2@.100	LF	3.02	2.46	5.48
3"	P2@.120	LF	4.16	2.96	7.12
4"	P2@.140	LF	5.41	3.45	8.86
6"	P2@.180	LF	9.29	4.44	13.73
8"	P2@.200	LF	14.40	4.93	19.33
Class 150 cast iron 1/8 bend with mechanical joints					
1 ½"	P2@.750	Ea	3.37	18.50	21.87
2"	P2@.800	Ea	3.75	19.70	23.45
3"	P2@1.00	Ea	5.07	24.60	29.67
4"	P2@1.20	Ea	6.43	29.60	36.03
6"	P2@1.70	Ea	14.90	41.90	56.80
8"	P2@2.25	Ea	43.00	55.40	98.40
Class 150 cast iron 1/4 bend with mechanical joints					
2"	P2@.800	Ea	5.42	19.70	25.12
3"	P2@1.00	Ea	7.75	24.60	32.35
4"	P2@1.20	Ea	12.20	29.60	41.80
6"	P2@1.70	Ea	25.40	41.90	67.30
8"	P2@2.25	Ea	68.40	55.40	123.80
Class 150 cast iron sanitary tee with mechanical joints					
1 ½"	P2@.750	Ea	5.61	18.50	24.11
2"	P2@1.00	Ea	6.11	24.60	30.71
3"	P2@1.20	Ea	7.43	29.60	37.03
4"	P2@1.65	Ea	11.50	40.70	52.20
6"	P2@2.35	Ea	33.40	57.90	91.30
8"	P2@3.15	Ea	104.00	77.60	181.60
Class 150 cast iron wye with mechanical joints					
1 ½"	P2@.750	Ea	5.73	18.50	24.23
2"	P2@1.00	Ea	5.73	24.60	30.33
3"	P2@1.20	Ea	8.13	29.60	37.73
4"	P2@1.65	Ea	13.20	40.70	53.90
6"	P2@2.35	Ea	35.20	57.90	93.10
8"	P2@3.15	Ea	125.00	77.60	202.60
Class 150 cast iron reducer					
2 x 1 ½"	P2@.600	Ea	3.18	14.80	17.98
3 x 2"	P2@.800	Ea	3.06	19.70	22.76
4 x 2"	P2@1.00	Ea	4.76	24.60	29.36
6 x 4"	P2@1.45	Ea	12.70	35.70	48.40
8 x 6"	P2@2.00	Ea	21.80	49.30	71.10

Cast Iron, DWV, Service Weight, Hub & Spigot with Gasketed Joints

Service weight hub and spigot cast iron soil pipe with gasketed joints is widely used for non-pressurized underground and in-building drain, waste and vent (DWV) systems.

This section has been arranged to save the estimator's time by including all normally-used system components such as pipe, fittings, valves, hanger assemblies and riser clamps under one heading. Additional items can be found under "Plumbing and Piping Specialties." The cost estimates in this section are based on the conditions, limitations and wage rates described in the section "How to Use This Book" beginning on page 5.

Description	Craft@Hrs	Unit	Material $	Labor $	Total $

DWV cast iron pipe, hub and spigot

Description	Craft@Hrs	Unit	Material $	Labor $	Total $
2"	P2@.060	LF	3.11	1.48	4.59
3"	P2@.070	LF	4.29	1.72	6.01
4"	P2@.090	LF	5.58	2.22	7.80
5"	P2@.100	LF	7.87	2.46	10.33
6"	P2@.110	LF	9.59	2.71	12.30
8"	P2@.140	LF	15.40	3.45	18.85
10"	P2@.180	LF	25.60	4.44	30.04
12"	P2@.230	LF	37.20	5.67	42.87
15"	P2@.280	LF	54.40	6.90	61.30

DWV cast iron 1/8 bend, hub and spigot

Description	Craft@Hrs	Unit	Material $	Labor $	Total $
2"	P2@.300	Ea	3.85	7.39	11.24
3"	P2@.400	Ea	7.00	9.86	16.86
4"	P2@.550	Ea	10.20	13.60	23.80
5"	P2@.650	Ea	14.10	16.00	30.10
6"	P2@.700	Ea	17.30	17.20	34.50
8"	P2@1.15	Ea	51.60	28.30	79.90
10"	P2@1.25	Ea	74.20	30.80	105.00
12"	P2@2.25	Ea	140.00	55.40	195.40
15"	P2@2.50	Ea	280.00	61.60	341.60

DWV cast iron 1/4 bend, hub and spigot

Description	Craft@Hrs	Unit	Material $	Labor $	Total $
2"	P2@.300	Ea	4.81	7.39	12.20
3"	P2@.400	Ea	8.33	9.86	18.19
4"	P2@.550	Ea	13.10	13.60	26.70
5"	P2@.650	Ea	18.20	16.00	34.20
6"	P2@.900	Ea	22.70	22.20	44.90
8"	P2@1.15	Ea	68.40	28.30	96.70
10"	P2@1.25	Ea	100.00	30.80	130.80
12"	P2@2.25	Ea	135.00	55.40	190.40
15"	P2@2.50	Ea	434.00	61.60	495.60

DWV cast iron heel inlet 1/4 bend, hub and spigot

Description	Craft@Hrs	Unit	Material $	Labor $	Total $
3 x 2"	P2@.550	Ea	12.80	13.60	26.40
4 x 2"	P2@.700	Ea	15.30	17.20	32.50

Cast Iron, DWV, Service Weight, Hub & Spigot with Gasketed Joints

Description	Craft@Hrs	Unit	Material $	Labor $	Total $
DWV cast iron 1/16 bend, hub and spigot					
2"	P2@.300	Ea	3.85	7.39	11.24
3"	P2@.400	Ea	6.40	9.86	16.26
4"	P2@.550	Ea	8.94	13.60	22.54
5"	P2@.650	Ea	13.40	16.00	29.40
6"	P2@.700	Ea	15.00	17.20	32.20
8"	P2@1.15	Ea	52.80	28.30	81.10
DWV cast iron closet bend, hub and spigot					
4 x 4"	P2@.400	Ea	22.30	9.86	32.16
DWV cast iron closet flange, hub and spigot					
4 x 2"	P2@.075	Ea	7.21	1.85	9.06
4 x 3"	P2@.100	Ea	9.83	2.46	12.29
4 x 4"	P2@.125	Ea	12.90	3.08	15.98
DWV cast iron regular P-trap, hub and spigot					
2"	P2@.300	Ea	9.93	7.39	17.32
3"	P2@.400	Ea	14.70	9.86	24.56
4"	P2@.550	Ea	21.40	13.60	35.00
5"	P2@.650	Ea	45.20	16.00	61.20
6"	P2@.700	Ea	66.00	17.20	83.20
8"	P2@1.15	Ea	188.00	28.30	216.30
10"	P2@2.00	Ea	309.00	49.30	358.30
DWV cast iron sanitary tee, hub and spigot					
2"	P2@.400	Ea	7.67	9.86	17.33
3"	P2@.600	Ea	14.10	14.80	28.90
4"	P2@.750	Ea	17.30	18.50	35.80
5"	P2@.850	Ea	34.30	20.90	55.20
6"	P2@.900	Ea	49.80	22.20	72.00
8"	P2@1.50	Ea	103.00	37.00	140.00
10"	P2@2.00	Ea	172.00	49.30	221.30
12"	P2@2.50	Ea	287.00	61.60	348.60
15"	P2@3.00	Ea	645.00	73.90	718.90
DWV cast iron sanitary reducing tee, hub and spigot					
3 x 2"	P2@.400	Ea	12.20	9.86	22.06
4 x 2"	P2@.650	Ea	14.70	16.00	30.70
4 x 3"	P2@.700	Ea	15.80	17.20	33.00
5 x 4"	P2@.800	Ea	32.30	19.70	52.00
6 x 2"	P2@.750	Ea	29.00	18.50	47.50
6 x 3"	P2@.800	Ea	30.00	19.70	49.70
6 x 4"	P2@.850	Ea	44.10	20.90	65.00

Cast Iron, DWV, Service Weight, Hub & Spigot with Gasketed Joints

Description	Craft@Hrs	Unit	Material $	Labor $	Total $
DWV cast iron sanitary tapped tee, hub and spigot					
2"	P2@.300	Ea	10.60	7.39	17.99
3"	P2@.400	Ea	15.50	9.86	25.36
4"	P2@.550	Ea	15.30	13.60	28.90
DWV cast iron combination, hub and spigot					
2"	P2@.400	Ea	10.90	9.86	20.76
3"	P2@.600	Ea	16.60	14.80	31.40
4"	P2@.750	Ea	23.00	18.50	41.50
5"	P2@.850	Ea	43.50	20.90	64.40
6"	P2@.900	Ea	55.00	22.20	77.20
DWV cast iron reducing combination, hub and spigot					
3 x 2"	P2@.550	Ea	12.50	13.60	26.10
4 x 2"	P2@.650	Ea	16.60	16.00	32.60
4 x 3"	P2@.700	Ea	19.80	17.20	37.00
5 x 2"	P2@.700	Ea	35.00	17.20	52.20
5 x 4"	P2@.800	Ea	41.60	19.70	61.30
6 x 2"	P2@.750	Ea	35.50	18.50	54.00
6 x 3"	P2@.750	Ea	37.70	18.50	56.20
6 x 4"	P2@.800	Ea	39.60	19.70	59.30
DWV cast iron double combination, hub and spigot					
2"	P2@.500	Ea	21.10	12.30	33.40
3"	P2@.750	Ea	28.70	18.50	47.20
4"	P2@.950	Ea	40.90	23.40	64.30
6"	P2@1.10	Ea	141.00	27.10	168.10
DWV cast iron reducing double combination, hub and spigot					
3 x 2"	P2@.550	Ea	23.10	13.60	36.70
4 x 2"	P2@.650	Ea	29.00	16.00	45.00
4 x 3"	P2@.700	Ea	33.00	17.20	50.20
6 x 4"	P2@.800	Ea	92.40	19.70	112.10
DWV cast iron wye, hub and spigot					
2"	P2@.400	Ea	7.98	9.86	17.84
3"	P2@.600	Ea	14.70	14.80	29.50
4"	P2@.750	Ea	19.80	18.50	38.30
5"	P2@.850	Ea	33.20	20.90	54.10
6"	P2@.900	Ea	45.40	22.20	67.60
8"	P2@1.50	Ea	111.00	37.00	148.00
10"	P2@2.00	Ea	179.00	49.30	228.30
12"	P2@2.50	Ea	308.00	61.60	369.60
15"	P2@3.00	Ea	650.00	73.90	723.90

Cast Iron, DWV, Service Weight, Hub & Spigot with Gasketed Joints

Description	Craft@Hrs	Unit	Material $	Labor $	Total $
DWV cast iron reducing wye, hub and spigot					
4 x 2"	P2@.650	Ea	15.30	16.00	31.30
4 x 3"	P2@.700	Ea	16.90	17.20	34.10
5 x 2"	P2@.700	Ea	22.40	17.20	39.60
5 x 3"	P2@.750	Ea	24.10	18.50	42.60
5 x 4"	P2@.800	Ea	26.90	19.70	46.60
6 x 2"	P2@.750	Ea	28.70	18.50	47.20
6 x 3"	P2@.750	Ea	29.40	18.50	47.90
6 x 4"	P2@.800	Ea	30.70	19.70	50.40
8 x 4"	P2@1.35	Ea	55.50	33.30	88.80
8 x 5"	P2@1.40	Ea	65.70	34.50	100.20
8 x 6"	P2@1.45	Ea	69.10	35.70	104.80
10 x 4"	P2@1.85	Ea	100.00	45.60	145.60
10 x 5"	P2@1.85	Ea	110.00	45.60	155.60
10 x 6"	P2@1.90	Ea	109.00	46.80	155.80
10 x 8"	P2@1.95	Ea	150.00	48.00	198.00
12 x 5"	P2@2.35	Ea	180.00	57.90	237.90
12 x 6"	P2@2.35	Ea	174.00	57.90	231.90
12 x 8"	P2@2.40	Ea	209.00	59.10	268.10
12 x 10"	P2@2.45	Ea	267.00	60.40	327.40
15 x 6"	P2@2.85	Ea	538.00	70.20	608.20
15 x 8"	P2@2.85	Ea	548.00	70.20	618.20
15 x 10"	P2@2.90	Ea	559.00	71.50	630.50
15 x 12"	P2@2.95	Ea	573.00	72.70	645.70
DWV cast iron double wye, hub and spigot					
2"	P2@.500	Ea	13.20	12.30	25.50
3"	P2@.750	Ea	22.80	18.50	41.30
4"	P2@.950	Ea	31.60	23.40	55.00
6"	P2@1.10	Ea	73.90	27.10	101.00
8"	P2@1.85	Ea	181.00	45.60	226.60
10"	P2@2.50	Ea	254.00	61.60	315.60
12"	P2@2.75	Ea	367.00	67.80	434.80
15"	P2@3.50	Ea	1,110.00	86.20	1,196.20
DWV cast iron reducing double wye, hub and spigot					
3 x 2"	P2@.550	Ea	19.50	13.60	33.10
4 x 2"	P2@.650	Ea	23.50	16.00	39.50
4 x 3"	P2@.700	Ea	26.20	17.20	43.40
6 x 4"	P2@.800	Ea	55.80	19.70	75.50

Cast Iron, DWV, Service Weight, Hub & Spigot with Gasketed Joints

Description	Craft@Hrs	Unit	Material $	Labor $	Total $
DWV cast iron reducer, hub and spigot					
3"	P2@.150	Ea	6.07	3.70	9.77
4"	P2@.200	Ea	6.41	4.93	11.34
5"	P2@.270	Ea	10.20	6.65	16.85
6"	P2@.320	Ea	16.00	7.88	23.88
8"	P2@.350	Ea	26.80	8.62	35.42
10"	P2@.570	Ea	42.60	14.00	56.60
12"	P2@.620	Ea	70.30	15.30	85.60
15"	P2@1.12	Ea	137.00	27.60	164.60
Gaskets for hub and spigot joints					
2"	--	Ea	1.79	--	1.79
3"	--	Ea	2.37	--	2.37
4"	--	Ea	3.02	--	3.02
5"	--	Ea	4.65	--	4.65
6"	--	Ea	4.88	--	4.88
8"	--	Ea	10.70	--	10.70
10"	--	Ea	16.50	--	16.50
12"	--	Ea	21.00	--	21.00
15"	--	Ea	25.00	--	25.00
Hanger with swivel assembly					
2"	P2@.400	Ea	.65	9.86	10.51
3"	P2@.450	Ea	3.95	11.10	15.05
4"	P2@.450	Ea	4.32	11.10	15.42
5"	P2@.500	Ea	5.87	12.30	18.17
6"	P2@.500	Ea	8.78	12.30	21.08
8"	P2@.550	Ea	11.70	13.60	25.30
10"	P2@.650	Ea	17.30	16.00	33.30
12"	P2@.750	Ea	21.00	18.50	39.50
15"	P2@.850	Ea	25.90	20.90	46.80
Riser clamp					
2"	P2@.115	Ea	4.45	2.83	7.28
3"	P2@.120	Ea	4.69	2.96	7.65
4"	P2@.125	Ea	5.13	3.08	8.21
5"	P2@.180	Ea	6.45	4.44	10.89
6"	P2@.200	Ea	8.58	4.93	13.51
8"	P2@.200	Ea	10.70	4.93	15.63
10"	P2@.250	Ea	17.40	6.16	23.56
12"	P2@.250	Ea	25.40	6.16	31.56
15"	P2@.300	Ea	30.70	7.39	38.09

Cast Iron, DWV, Service Weight, No-Hub with Coupled Joints

Service weight cast iron soil pipe with hub-less coupled joints is widely used for non-pressurized underground and in-building drain, waste and vent (DWV) systems.

This section has been arranged to save the estimator's time by including all normally-used system components such as pipe, fittings, hanger assemblies and riser clamps under one heading. Additional items can be found under "Plumbing and Piping Specialties." The cost estimates in this section are based on the conditions, limitations and wage rates described in the section "How to Use This Book" beginning on page 5.

Description	Craft@Hrs	Unit	Material $	Labor $	Total $
DWV cast iron no-hub pipe					
2"	P2@.050	LF	2.63	1.23	3.86
3"	P2@.060	LF	3.63	1.48	5.11
4"	P2@.080	LF	4.72	1.97	6.69
5"	P2@.090	LF	6.66	2.22	8.88
6"	P2@.110	LF	8.11	2.71	10.82
8"	P2@.130	LF	13.00	3.20	16.20
10"	P2@.140	LF	21.70	3.45	25.15
DWV cast iron no-hub 1/8 bend					
2"	P2@.250	Ea	3.27	6.16	9.43
3"	P2@.350	Ea	4.43	8.62	13.05
4"	P2@.500	Ea	5.61	12.30	17.91
5"	P2@.580	Ea	12.00	14.30	26.30
6"	P2@.650	Ea	13.00	16.00	29.00
8"	P2@.710	Ea	37.50	17.50	55.00
10"	P2@.800	Ea	69.90	19.70	89.60
DWV cast iron no-hub 1/4 bend					
2"	P2@.250	Ea	3.85	6.16	10.01
3"	P2@.350	Ea	5.34	8.62	13.96
4"	P2@.500	Ea	7.62	12.30	19.92
5"	P2@.580	Ea	17.20	14.30	31.50
6"	P2@.650	Ea	19.40	16.00	35.40
8"	P2@.710	Ea	53.40	17.50	70.90
10"	P2@.800	Ea	78.20	19.70	97.90
DWV cast iron no-hub long sweep 1/4 bend					
2"	P2@.250	Ea	8.27	6.16	14.43
3"	P2@.350	Ea	9.85	8.62	18.47
4"	P2@.500	Ea	15.70	12.30	28.00
5"	P2@.580	Ea	28.90	14.30	43.20
6"	P2@.650	Ea	35.30	16.00	51.30
DWV cast iron no-hub low heel outlet 1/4 bend					
3"	P2@.470	Ea	8.40	11.60	20.00
4"	P2@.650	Ea	10.20	16.00	26.20

Cast Iron, DWV, Service Weight, No-Hub with Coupled Joints

Description	Craft@Hrs	Unit	Material $	Labor $	Total $
DWV cast iron no-hub closet bend					
3"	P2@.350	Ea	19.40	8.62	28.02
4"	P2@.500	Ea	19.80	12.30	32.10
DWV cast iron no-hub closet flange					
4"	P2@.125	Ea	11.20	3.08	14.28
DWV cast iron no-hub P-trap					
2"	P2@.250	Ea	5.61	6.16	11.77
3"	P2@.350	Ea	12.40	8.62	21.02
4"	P2@.500	Ea	21.50	12.30	33.80
6"	P2@.650	Ea	52.20	16.00	68.20
DWV cast iron no-hub wye					
2"	P2@.380	Ea	5.00	9.36	14.36
3"	P2@.520	Ea	7.10	12.80	19.90
4"	P2@.700	Ea	11.50	17.20	28.70
5"	P2@.800	Ea	26.70	19.70	46.40
6"	P2@.900	Ea	30.70	22.20	52.90
8"	P2@1.15	Ea	72.30	28.30	100.60
10"	P2@1.50	Ea	157.00	37.00	194.00
DWV cast iron no-hub reducing wye					
3 x 2"	P2@.470	Ea	5.34	11.60	16.94
4 x 2"	P2@.600	Ea	7.40	14.80	22.20
4 x 3"	P2@.650	Ea	10.00	16.00	26.00
5 x 2"	P2@.720	Ea	16.60	17.70	34.30
5 x 3"	P2@.770	Ea	23.30	19.00	42.30
5 x 4"	P2@.770	Ea	29.40	19.00	48.40
6 x 2"	P2@.770	Ea	18.50	19.00	37.50
6 x 3"	P2@.770	Ea	23.30	19.00	42.30
6 x 4"	P2@.800	Ea	24.00	19.70	43.70
6 x 5"	P2@1.00	Ea	29.20	24.60	53.80
8 x 2"	P2@1.00	Ea	28.60	24.60	53.20
8 x 3"	P2@1.00	Ea	37.70	24.60	62.30
8 x 4"	P2@1.00	Ea	41.30	24.60	65.90
8 x 5"	P2@1.05	Ea	50.00	25.90	75.90
8 x 6"	P2@1.10	Ea	50.80	27.10	77.90
DWV cast iron no-hub upright wye					
2"	P2@.380	Ea	7.40	9.36	16.76
3"	P2@.520	Ea	9.82	12.80	22.62
4"	P2@.700	Ea	30.90	17.20	48.10
5"	P2@.840	Ea	55.80	20.70	76.50

Cast Iron, DWV, Service Weight, No-Hub with Coupled Joints

Description	Craft@Hrs	Unit	Material $	Labor $	Total $
DWV cast iron no-hub reducing upright wye					
3 x 2"	P2@.470	Ea	8.30	11.60	19.90
4 x 2"	P2@.600	Ea	11.00	14.80	25.80
4 x 3"	P2@.650	Ea	15.60	16.00	31.60
5 x 2"	P2@.720	Ea	50.60	17.70	68.30
5 x 3"	P2@.770	Ea	51.90	19.00	70.90
DWV cast iron no-hub double wye					
2"	P2@.500	Ea	7.67	12.30	19.97
3"	P2@.700	Ea	14.10	17.20	31.30
4"	P2@.950	Ea	28.90	23.40	52.30
6"	P2@1.10	Ea	50.90	27.10	78.00
8"	P2@1.45	Ea	145.00	35.70	180.70
DWV cast iron no-hub double reducing wye					
3 x 2"	P2@.520	Ea	19.50	12.80	32.30
4 x 2"	P2@.740	Ea	23.50	18.20	41.70
4 x 3"	P2@1.00	Ea	19.80	24.60	44.40
6 x 4"	P2@1.30	Ea	39.60	32.00	71.60
8 x 4"	P2@1.75	Ea	65.80	43.10	108.90
DWV cast iron no-hub combination					
2"	P2@.380	Ea	5.60	9.36	14.96
3"	P2@.520	Ea	8.55	12.80	21.35
4"	P2@.700	Ea	15.60	17.20	32.80
5"	P2@.800	Ea	34.30	19.70	54.00
6"	P2@.900	Ea	41.40	22.20	63.60
8"	P2@1.15	Ea	108.00	28.30	136.30
DWV cast iron no-hub reducing combination					
3 x 2"	P2@.470	Ea	5.90	11.60	17.50
4 x 2"	P2@.600	Ea	8.85	14.80	23.65
4 x 3"	P2@.650	Ea	11.30	16.00	27.30
5 x 2"	P2@.720	Ea	21.40	17.70	39.10
5 x 3"	P2@.770	Ea	27.30	19.00	46.30
5 x 4"	P2@.770	Ea	31.60	19.00	50.60
6 x 2"	P2@.770	Ea	21.20	19.00	40.20
6 x 3"	P2@.770	Ea	26.70	19.00	45.70
6 x 4"	P2@.800	Ea	27.60	19.70	47.30
6 x 5"	P2@1.00	Ea	37.00	24.60	61.60
8 x 4"	P2@1.20	Ea	56.20	29.60	85.80
8 x 5"	P2@1.50	Ea	84.00	37.00	121.00
DWV cast iron no-hub double combination					
2"	P2@.500	Ea	13.20	12.30	25.50
3"	P2@.700	Ea	20.20	17.20	37.40
4"	P2@.950	Ea	39.90	23.40	63.30

Cast Iron, DWV, Service Weight, No-Hub with Coupled Joints

Description	Craft@Hrs	Unit	Material $	Labor $	Total $
DWV cast iron no-hub reducing double combination					
3 x 2"	P2@.600	Ea	15.30	14.80	30.10
4 x 2"	P2@1.25	Ea	25.20	30.80	56.00
4 x 3"	P2@1.35	Ea	26.40	33.30	59.70
DWV cast iron no-hub sanitary tee					
2"	P2@.380	Ea	5.34	9.36	14.70
3"	P2@.520	Ea	6.49	12.80	19.29
4"	P2@.700	Ea	10.00	17.20	27.20
5"	P2@.800	Ea	28.60	19.70	48.30
6"	P2@.900	Ea	29.20	22.20	51.40
8"	P2@1.15	Ea	91.00	28.30	119.30
DWV cast iron no-hub reducing sanitary tee					
3 x 2"	P2@.470	Ea	5.92	11.60	17.52
4 x 2"	P2@.600	Ea	7.67	14.80	22.47
4 x 3"	P2@.650	Ea	9.21	16.00	25.21
5 x 2"	P2@.720	Ea	20.50	17.70	38.20
5 x 3"	P2@.770	Ea	21.50	19.00	40.50
5 x 4"	P2@.770	Ea	22.10	19.00	41.10
6 x 2"	P2@.770	Ea	21.10	19.00	40.10
6 x 3"	P2@.770	Ea	21.50	19.00	40.50
6 x 4"	P2@.800	Ea	22.10	19.70	41.80
8 x 3"	P2@1.00	Ea	60.70	24.60	85.30
8 x 4"	P2@1.00	Ea	62.40	24.60	87.00
8 x 5"	P2@1.05	Ea	66.90	25.90	92.80
8 x 6"	P2@1.10	Ea	68.20	27.10	95.30
DWV cast iron no-hub tapped sanitary tee					
2 x 1½"	P2@.200	Ea	5.85	4.93	10.78
2 x 2"	P2@.250	Ea	6.49	6.16	12.65
3 x 2"	P2@.350	Ea	8.30	8.62	16.92
4 x 2"	P2@.425	Ea	8.30	10.50	18.80
5 x 2"	P2@.550	Ea	16.60	13.60	30.20
6 x 2"	P2@.700	Ea	17.60	17.20	34.80
DWV cast iron no-hub test tee					
2"	P2@.370	Ea	4.73	9.12	13.85
3"	P2@.520	Ea	7.40	12.80	20.20
4"	P2@.750	Ea	12.40	18.50	30.90
5"	P2@.900	Ea	23.90	22.20	46.10
6"	P2@1.12	Ea	29.80	27.60	57.40
8"	P2@1.25	Ea	64.70	30.80	95.50

Cast Iron, DWV, Service Weight, No-Hub with Coupled Joints

Description	Craft@Hrs	Unit	Material $	Labor $	Total $
DWV cast iron no-hub cross					
2"	P2@.500	Ea	8.91	12.30	21.21
3"	P2@.700	Ea	13.80	17.20	31.00
4"	P2@.950	Ea	26.10	23.40	49.50
6"	P2@1.30	Ea	55.20	32.00	87.20
DWV cast iron no-hub reducing cross					
3 x 2"	P2@.600	Ea	12.30	14.80	27.10
4 x 2"	P2@.900	Ea	17.50	22.20	39.70
4 x 3"	P2@.900	Ea	21.20	22.20	43.40
5 x 4"	P2@1.05	Ea	37.70	25.90	63.60
6 x 4"	P2@1.20	Ea	37.70	29.60	67.30
8 x 4"	P2@1.50	Ea	88.00	37.00	125.00
DWV cast iron no-hub reducer					
2"	P2@.120	Ea	2.77	2.96	5.73
3"	P2@.130	Ea	2.67	3.20	5.87
4"	P2@.150	Ea	4.15	3.70	7.85
5"	P2@.180	Ea	7.67	4.44	12.11
6"	P2@.200	Ea	11.10	4.93	16.03
8"	P2@.400	Ea	17.10	9.86	26.96
10"	P2@.650	Ea	35.70	16.00	51.70
DWV cast iron no-hub cap					
2"	P2@.125	Ea	1.22	3.08	4.30
3"	P2@.175	Ea	2.12	4.31	6.43
4"	P2@.250	Ea	2.80	6.16	8.96
5"	P2@.300	Ea	5.17	7.39	12.56
6"	P2@.375	Ea	5.17	9.24	14.41
8"	P2@.500	Ea	16.50	12.30	28.80
10"	P2@.750	Ea	20.70	18.50	39.20
DWV cast iron no-hub coupling					
2"	--	Ea	3.30	--	3.30
3"	--	Ea	3.93	--	3.93
4"	--	Ea	4.68	--	4.68
5"	--	Ea	11.40	--	11.40
6"	--	Ea	11.90	--	11.90
8"	--	Ea	18.40	--	18.40
10"	--	Ea	24.00	--	24.00

Cast Iron, DWV, Service Weight, No-Hub with Coupled Joints

Description	Craft@Hrs	Unit	Material $	Labor $	Total $
Hanger with swivel assembly					
2"	P2@.400	Ea	.65	9.86	10.51
3"	P2@.450	Ea	3.95	11.10	15.05
4"	P2@.450	Ea	4.32	11.10	15.42
5"	P2@.500	Ea	5.87	12.30	18.17
6"	P2@.500	Ea	8.78	12.30	21.08
8"	P2@.550	Ea	11.70	13.60	25.30
10"	P2@.650	Ea	17.30	16.00	33.30
Riser clamp					
2"	P2@.115	Ea	4.45	2.83	7.28
3"	P2@.120	Ea	4.69	2.96	7.65
4"	P2@.125	Ea	5.13	3.08	8.21
5"	P2@.180	Ea	6.45	4.44	10.89
6"	P2@.200	Ea	8.58	4.93	13.51
8"	P2@.200	Ea	10.70	4.93	15.63
10"	P2@.250	Ea	17.40	6.16	23.56

Copper drainage tubing is used for non-pressurized underground and in-building drain, waste and vent (DWV) systems.

Because of its thin walls, it must be handled carefully and protected from damage during building construction.

This section has been arranged to save the estimator's time by including all normally-used system components such as pipe, fittings, hanger assemblies and riser clamps under one heading. Additional items can be found under "Plumbing and Piping Specialties." The cost estimates in this section are based on the conditions, limitations and wage rates described in the section "How to Use This Book" beginning on page 5.

Description	Craft@Hrs	Unit	Material $	Labor $	Total $

DWV copper pipe, soft-soldered joints

Description	Craft@Hrs	Unit	Material $	Labor $	Total $
1¼"	P2@.040	LF	2.30	.99	3.29
1½"	P2@.045	LF	2.91	1.11	4.02
2"	P2@.050	LF	3.86	1.23	5.09
3"	P2@.060	LF	6.10	1.48	7.58
4"	P2@.070	LF	11.00	1.72	12.72

DWV copper 1/8 bend, soft-soldered joints

Description	Craft@Hrs	Unit	Material $	Labor $	Total $
1¼"	P2@.170	Ea	3.35	4.19	7.54
1½"	P2@.200	Ea	2.86	4.93	7.79
2"	P2@.270	Ea	6.64	6.65	13.29
3"	P2@.400	Ea	13.50	9.86	23.36
4"	P2@.540	Ea	54.00	13.30	67.30

DWV copper 1/4 bend, soft-soldered joints

Description	Craft@Hrs	Unit	Material $	Labor $	Total $
1¼"	P2@.170	Ea	6.24	4.19	10.43
1½"	P2@.200	Ea	5.16	4.93	10.09
2"	P2@.270	Ea	10.00	6.65	16.65
3"	P2@.400	Ea	21.00	9.86	30.86
4"	P2@.540	Ea	110.00	13.30	123.30

DWV copper closet flange, soft-soldered joints

Description	Craft@Hrs	Unit	Material $	Labor $	Total $
3"	P2@.680	Ea	35.80	16.80	52.60
4"	P2@.980	Ea	23.00	24.10	47.10

DWV copper P-trap, soft-soldered joints

Description	Craft@Hrs	Unit	Material $	Labor $	Total $
1½"	P2@.240	Ea	29.60	5.91	35.51
2"	P2@.320	Ea	45.20	7.88	53.08

DWV copper sanitary tee, soft-soldered joints

Description	Craft@Hrs	Unit	Material $	Labor $	Total $
1½"	P2@.240	Ea	9.16	5.91	15.07
2"	P2@.320	Ea	10.70	7.88	18.58
3"	P2@.490	Ea	36.60	12.10	48.70
4"	P2@.650	Ea	88.80	16.00	104.80

Copper, DWV, with Soft-Soldered Joints

Description	Craft@Hrs	Unit	Material $	Labor $	Total $
DWV copper sanitary cross, soft-soldered joints					
1½"	P2@.640	Ea	15.00	15.80	30.80
2"	P2@.900	Ea	36.40	22.20	58.60
3"	P2@1.36	Ea	54.00	33.50	87.50
DWV copper combination, soft-soldered joints					
1½"	P2@.240	Ea	15.60	5.91	21.51
2"	P2@.320	Ea	36.30	7.88	44.18
3"	P2@.490	Ea	60.80	12.10	72.90
4"	P2@.650	Ea	161.00	16.00	177.00
DWV copper wye, soft-soldered joints					
1½"	P2@.240	Ea	14.70	5.91	20.61
2"	P2@.320	Ea	20.20	7.88	28.08
3"	P2@.490	Ea	45.60	12.10	57.70
4"	P2@.650	Ea	88.80	16.00	104.80
DWV copper cleanout with plug, soft-soldered joints					
1¼"	P2@.100	Ea	4.52	2.46	6.98
1½"	P2@.160	Ea	4.32	3.94	8.26
2"	P2@.230	Ea	5.72	5.67	11.39
3"	P2@.340	Ea	13.70	8.38	22.08
DWV copper female adapter					
1½"	P2@.130	Ea	4.80	3.20	8.00
2"	P2@.170	Ea	9.12	4.19	13.31
3"	P2@.230	Ea	36.30	5.67	41.97
4"	P2@.300	Ea	56.40	7.39	63.79
DWV copper male adapter					
1½"	P2@.130	Ea	5.88	3.20	9.08
2"	P2@.170	Ea	7.44	4.19	11.63
3"	P2@.230	Ea	33.20	5.67	38.87
4"	P2@.300	Ea	53.20	7.39	60.59
DWV copper soil adapter					
2"	P2@.170	Ea	9.64	4.19	13.83
3"	P2@.250	Ea	23.80	6.16	29.96
4"	P2@.320	Ea	36.70	7.88	44.58
DWV copper test cap, soft-soldered joint					
1½"	P2@.100	Ea	.67	2.46	3.13
2"	P2@.130	Ea	1.38	3.20	4.58
3"	P2@.200	Ea	2.67	4.93	7.60
4"	P2@.250	Ea	5.44	6.16	11.60

Description	Craft@Hrs	Unit	Material $	Labor $	Total $

DWV copper test tee, soft-soldered joints

Description	Craft@Hrs	Unit	Material $	Labor $	Total $
1½"	P2@.200	Ea	20.40	4.93	25.33
2"	P2@.270	Ea	26.60	6.65	33.25
3"	P2@.400	Ea	62.40	9.86	72.26

DWV copper reducer, soft-soldered joints

Description	Craft@Hrs	Unit	Material $	Labor $	Total $
1½"	P2@.100	Ea	3.09	2.46	5.55
2"	P2@.140	Ea	3.60	3.45	7.05
3"	P2@.200	Ea	12.80	4.93	17.73
4"	P2@.270	Ea	20.80	6.65	27.45

DWV copper coupling, soft-soldered joints

Description	Craft@Hrs	Unit	Material $	Labor $	Total $
1¼"	P2@.170	Ea	1.63	4.19	5.82
1½"	P2@.200	Ea	2.18	4.93	7.11
2"	P2@.270	Ea	2.99	6.65	9.64
3"	P2@.400	Ea	5.00	9.86	14.86
4"	P2@.540	Ea	12.70	13.30	26.00

Hanger with swivel assembly

Description	Craft@Hrs	Unit	Material $	Labor $	Total $
1¼"	P2@.450	Ea	.56	11.10	11.66
1½"	P2@.450	Ea	.62	11.10	11.72
2"	P2@.450	Ea	.65	11.10	11.75
3"	P2@.450	Ea	3.95	11.10	15.05
4"	P2@.450	Ea	4.32	11.10	15.42

Riser clamp

Description	Craft@Hrs	Unit	Material $	Labor $	Total $
1¼"	P2@.105	Ea	4.01	2.59	6.60
1½"	P2@.110	Ea	4.23	2.71	6.94
2"	P2@.115	Ea	4.45	2.83	7.28
3"	P2@.120	Ea	5.13	2.96	8.09
4"	P2@.125	Ea	6.45	3.08	9.53

Copper, Type K with Brazed Joints

Type K hard-drawn copper pipe with wrought copper fittings and brazed joints is used in a wide variety of plumbing and HVAC systems such as heating hot water, chilled water, potable water, compressed air, refrigerant and A.C. condensate.

Brazed joints are those made with silver or other alloy filler metals having melting points at, or above, 1,000 degrees F. Maximum working pressure/temperature relationships for brazed joints are approximately as follows:

Maximum Working Pressure (PSIG)*			
Water Temperature (degrees F.)	Nominal Pipe Size (inches)		
	Up to 1	1¼ to 2	2½ to 4
Up to 350	270	190	150

*For copper pipe and solder-type fittings using brazing alloys melting at, or above 1,000 degrees F.

This section has been arranged to save the estimator's time by including all normally-used system components such as pipe, fittings, valves, hanger assemblies, riser clamps and miscellaneous items under one heading. Additional items can be found under "Plumbing and Piping Specialties." The cost estimates in this section are based on the conditions, limitations and wage rates described in the section "How to Use This Book" beginning on page 5.

Description	Craft@Hrs	Unit	Material $	Labor $	Total $

Type K copper pipe with brazed joints

Description	Craft@Hrs	Unit	Material $	Labor $	Total $
1/2"	P2@.032	LF	1.22	.79	2.01
3/4"	P2@.035	LF	2.24	.86	3.10
1"	P2@.038	LF	2.87	.94	3.81
1¼"	P2@.042	LF	3.58	1.03	4.61
1½"	P2@.046	LF	4.60	1.13	5.73
2"	P2@.053	LF	7.36	1.31	8.67
2½"	P2@.060	LF	10.70	1.48	12.18
3"	P2@.066	LF	14.70	1.63	16.33
4"	P2@.080	LF	23.00	1.97	24.97

Type K copper 45 degree ell C x C with brazed joints

Description	Craft@Hrs	Unit	Material $	Labor $	Total $
1/2"	P2@.129	Ea	.72	3.18	3.90
3/4"	P2@.181	Ea	1.22	4.46	5.68
1"	P2@.232	Ea	3.08	5.72	8.80
1¼"	P2@.284	Ea	4.20	7.00	11.20
1½"	P2@.335	Ea	5.04	8.25	13.29
2"	P2@.447	Ea	8.44	11.00	19.44
2½"	P2@.550	Ea	18.00	13.60	31.60
3"	P2@.654	Ea	26.70	16.10	42.80
4"	P2@.860	Ea	55.60	21.20	76.80

Description	Craft@Hrs	Unit	Material $	Labor $	Total $

Type K copper 90 degree ell C x C with brazed joints

1/2"	P2@.129	Ea	.43	3.18	3.61
3/4"	P2@.181	Ea	.91	4.46	5.37
1"	P2@.232	Ea	2.11	5.72	7.83
1¼"	P2@.284	Ea	3.21	7.00	10.21
1½"	P2@.335	Ea	5.00	8.25	13.25
2"	P2@.447	Ea	9.12	11.00	20.12
2½"	P2@.550	Ea	17.40	13.60	31.00
3"	P2@.654	Ea	24.30	16.10	40.40
4"	P2@.860	Ea	56.00	21.20	77.20

Type K copper 90 degree ell Ftg x C with brazed joint

1/2"	P2@.129	Ea	.60	3.18	3.78
3/4"	P2@.181	Ea	1.33	4.46	5.79
1"	P2@.232	Ea	3.20	5.72	8.92
1¼"	P2@.284	Ea	4.88	7.00	11.88
1½"	P2@.335	Ea	6.40	8.25	14.65
2"	P2@.447	Ea	13.90	11.00	24.90
2½"	P2@.550	Ea	29.10	13.60	42.70
3"	P2@.654	Ea	34.70	16.10	50.80
4"	P2@.860	Ea	80.00	21.20	101.20

Type K copper tee C x C x C with brazed joints

1/2"	P2@.156	Ea	.70	3.84	4.54
3/4"	P2@.204	Ea	1.70	5.03	6.73
1"	P2@.263	Ea	4.92	6.48	11.40
1¼"	P2@.322	Ea	7.44	7.93	15.37
1½"	P2@.380	Ea	10.30	9.36	19.66
2"	P2@.507	Ea	16.10	12.50	28.60
2½"	P2@.624	Ea	32.50	15.40	47.90
3"	P2@.741	Ea	49.60	18.30	67.90
4"	P2@.975	Ea	107.00	24.00	131.00

Copper, Type K with Brazed Joints

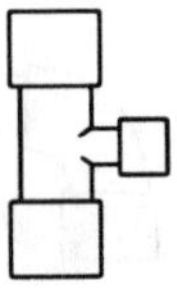

Description	Craft@Hrs	Unit	Material $	Labor $	Total $
Type K copper branch reducing tee C x C x C with brazed joints					
1/2 x 3/8"	P2@.129	Ea	3.30	3.18	6.48
3/4 x 1/2"	P2@.182	Ea	1.54	4.48	6.02
1 x 1/2"	P2@.209	Ea	4.92	5.15	10.07
1 x 3/4"	P2@.234	Ea	4.92	5.77	10.69
1¼ x 1/2"	P2@.241	Ea	7.24	5.94	13.18
1¼ x 3/4"	P2@.264	Ea	7.24	6.50	13.74
1¼ x 1"	P2@.287	Ea	7.24	7.07	14.31
1½ x 1/2"	P2@.291	Ea	7.68	7.17	14.85
1½ x 3/4"	P2@.307	Ea	7.68	7.56	15.24
1½ x 1"	P2@.323	Ea	7.68	7.96	15.64
1½ x 1¼"	P2@.339	Ea	7.68	8.35	16.03
2 x 1/2"	P2@.348	Ea	12.10	8.57	20.67
2 x 3/4"	P2@.373	Ea	12.10	9.19	21.29
2 x 1"	P2@.399	Ea	12.10	9.83	21.93
2 x 1¼"	P2@.426	Ea	12.10	10.50	22.60
2 x 1½"	P2@.452	Ea	12.10	11.10	23.20
2½ x 1/2"	P2@.468	Ea	37.00	11.50	48.50
2½ x 3/4"	P2@.485	Ea	37.00	12.00	49.00
2½ x 1"	P2@.502	Ea	37.00	12.40	49.40
2½ x 1½"	P2@.537	Ea	37.00	13.20	50.20
2½ x 2"	P2@.555	Ea	37.00	13.70	50.70
3 x 1¼"	P2@.562	Ea	40.80	13.80	54.60
3 x 2"	P2@.626	Ea	40.80	15.40	56.20
3 x 2½"	P2@.660	Ea	40.80	16.30	57.10
4 x 1¼"	P2@.669	Ea	73.20	16.50	89.70
4 x 1½"	P2@.718	Ea	73.20	17.70	90.90
4 x 2"	P2@.768	Ea	73.20	18.90	92.10
4 x 2½"	P2@.819	Ea	73.20	20.20	93.40
4 x 3"	P2@.868	Ea	73.20	21.40	94.60

Description	Craft@Hrs	Unit	Material $	Labor $	Total $

Type K copper reducer with brazed joints

Description	Craft@Hrs	Unit	Material $	Labor $	Total $
1/2 x 3/8"	P2@.148	Ea	.71	3.65	4.36
3/4 x 1/2"	P2@.155	Ea	1.06	3.82	4.88
1 x 3/4"	P2@.206	Ea	1.72	5.08	6.80
1¼ x 1/2"	P2@.206	Ea	2.96	5.08	8.04
1¼ x 3/4"	P2@.232	Ea	2.96	5.72	8.68
1¼ x 1"	P2@.258	Ea	2.96	6.36	9.32
1½ x 1/2"	P2@.232	Ea	3.90	5.72	9.62
1½ x 3/4"	P2@.257	Ea	3.90	6.33	10.23
1½ x 1"	P2@.283	Ea	3.90	6.97	10.87
1½ x 1¼"	P2@.308	Ea	3.90	7.59	11.49
2 x 1/2"	P2@.287	Ea	7.33	7.07	14.40
2 x 3/4"	P2@.277	Ea	7.33	6.83	14.16
2 x 1"	P2@.338	Ea	7.33	8.33	15.66
2 x 1¼"	P2@.365	Ea	7.33	8.99	16.32
2 x 1½"	P2@.390	Ea	7.33	9.61	16.94
2½ x 1"	P2@.390	Ea	14.60	9.61	24.21
2½ x 1¼"	P2@.416	Ea	14.60	10.30	24.90
2½ x 1½"	P2@.442	Ea	14.60	10.90	25.50
2½ x 2"	P2@.497	Ea	14.60	12.20	26.80
3 x 1¼"	P2@.468	Ea	18.30	11.50	29.80
3 x 1½"	P2@.493	Ea	18.30	12.10	30.40
3 x 2"	P2@.548	Ea	18.30	13.50	31.80
3 x 2½"	P2@.600	Ea	18.30	14.80	33.10
4 x 2"	P2@.652	Ea	36.60	16.10	52.70
4 x 2½"	P2@.704	Ea	36.60	17.30	53.90
4 x 3"	P2@.755	Ea	36.60	18.60	55.20

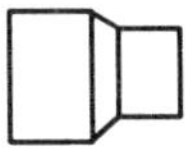

Type K copper adapter C x MPT

Description	Craft@Hrs	Unit	Material $	Labor $	Total $
1/2"	P2@.090	Ea	.87	2.22	3.09
3/4"	P2@.126	Ea	1.45	3.10	4.55
1"	P2@.162	Ea	3.54	3.99	7.53
1¼"	P2@.198	Ea	5.22	4.88	10.10
1½"	P2@.234	Ea	5.99	5.77	11.76
2"	P2@.312	Ea	10.20	7.69	17.89
2½"	P2@.384	Ea	31.20	9.46	40.66
3"	P2@.456	Ea	37.30	11.20	48.50
4"	P2@.600	Ea	63.60	14.80	78.40

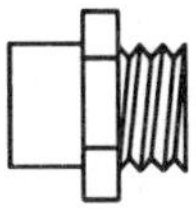

Type K copper adapter C x FPT

Description	Craft@Hrs	Unit	Material $	Labor $	Total $
1/2"	P2@.090	Ea	1.30	2.22	3.52
3/4"	P2@.126	Ea	1.79	3.10	4.89
1"	P2@.162	Ea	3.66	3.99	7.65
1¼"	P2@.198	Ea	5.99	4.88	10.87
1½"	P2@.234	Ea	9.40	5.77	15.17
2"	P2@.312	Ea	12.80	7.69	20.49
2½"	P2@.384	Ea	35.40	9.46	44.86
3"	P2@.456	Ea	51.20	11.20	62.40
4"	P2@.600	Ea	87.10	14.80	101.90

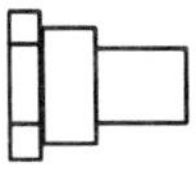

Description	Craft@Hrs	Unit	Material $	Labor $	Total $

Type K copper flush bushing with brazed joints

Description	Craft@Hrs	Unit	Material $	Labor $	Total $
1/2 x 3/8"	P2@.148	Ea	.82	3.65	4.47
3/4 x 1/2"	P2@.155	Ea	1.50	3.82	5.32
1 x 3/4"	P2@.206	Ea	2.32	5.08	7.40
1 x 1/2"	P2@.206	Ea	2.32	5.08	7.40
1¼ x 1"	P2@.258	Ea	2.81	6.36	9.17
1½ x 1¼"	P2@.308	Ea	3.55	7.59	11.14
2 x 1½"	P2@.390	Ea	10.20	9.61	19.81

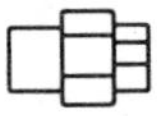

Type K copper union with brazed joint

Description	Craft@Hrs	Unit	Material $	Labor $	Total $
1/2"	P2@.146	Ea	3.82	3.60	7.42
3/4"	P2@.205	Ea	4.80	5.05	9.85
1"	P2@.263	Ea	8.52	6.48	15.00
1¼"	P2@.322	Ea	14.80	7.93	22.73
1½"	P2@.380	Ea	19.40	9.36	28.76
2"	P2@.507	Ea	33.10	12.50	45.60

Type K copper dielectric union with brazed joint

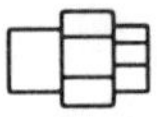

Description	Craft@Hrs	Unit	Material $	Labor $	Total $
1/2"	P2@.146	Ea	3.39	3.60	6.99
3/4"	P2@.205	Ea	3.39	5.05	8.44
1"	P2@.263	Ea	7.76	6.48	14.24
1¼"	P2@.322	Ea	12.70	7.93	20.63
1½"	P2@.380	Ea	18.60	9.36	27.96
2"	P2@.507	Ea	26.30	12.50	38.80

Type K copper cap with brazed joint

Description	Craft@Hrs	Unit	Material $	Labor $	Total $
1/2"	P2@.083	Ea	.29	2.05	2.34
3/4"	P2@.116	Ea	.50	2.86	3.36
1"	P2@.149	Ea	1.18	3.67	4.85
1¼"	P2@.182	Ea	1.68	4.48	6.16
1½"	P2@.215	Ea	2.46	5.30	7.76
2"	P2@.286	Ea	4.50	7.05	11.55
2½"	P2@.352	Ea	13.90	8.67	22.57
3"	P2@.418	Ea	19.00	10.30	29.30
4"	P2@.550	Ea	37.30	13.60	50.90

Type K copper coupling with brazed joints

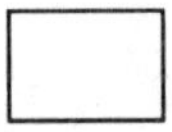

Description	Craft@Hrs	Unit	Material $	Labor $	Total $
1/2"	P2@.129	Ea	.32	3.18	3.50
3/4"	P2@.181	Ea	.61	4.46	5.07
1"	P2@.232	Ea	1.60	5.72	7.32
1¼"	P2@.284	Ea	2.11	7.00	9.11
1½"	P2@.335	Ea	2.79	8.25	11.04
2"	P2@.447	Ea	4.62	11.00	15.62
2½"	P2@.550	Ea	9.84	13.60	23.44
3"	P2@.654	Ea	15.10	16.10	31.20
4"	P2@.860	Ea	29.60	21.20	50.80

Description	Craft@Hrs	Unit	Material $	Labor $	Total $

Class 125 bronze body gate valve, soldered

Description	Craft@Hrs	Unit	Material $	Labor $	Total $
1/2"	P2@.240	Ea	12.20	5.91	18.11
3/4"	P2@.300	Ea	14.70	7.39	22.09
1"	P2@.360	Ea	18.30	8.87	27.17
1¼"	P2@.480	Ea	31.30	11.80	43.10
1½"	P2@.540	Ea	34.50	13.30	47.80
2"	P2@.600	Ea	48.40	14.80	63.20
2½"	P2@1.00	Ea	113.00	24.60	137.60
3"	P2@1.50	Ea	161.00	37.00	198.00

Class 125 iron body gate valve, flanged

Description	Craft@Hrs	Unit	Material $	Labor $	Total $
2"	P2@.500	Ea	164.00	12.30	176.30
2½"	P2@.600	Ea	168.00	14.80	182.80
3"	P2@.750	Ea	288.00	18.50	306.50
4"	P2@1.35	Ea	270.00	33.30	303.30
5"	P2@2.00	Ea	460.00	49.30	509.30
6"	P2@2.50	Ea	460.00	61.60	521.60

Class 125 bronze body globe valve, soldered

Description	Craft@Hrs	Unit	Material $	Labor $	Total $
1/2"	P2@.240	Ea	18.30	5.91	24.21
3/4"	P2@.300	Ea	23.30	7.39	30.69
1"	P2@.360	Ea	33.80	8.87	42.67
1¼"	P2@.480	Ea	41.30	11.80	53.10
1½"	P2@.540	Ea	53.30	13.30	66.60
2"	P2@.600	Ea	81.30	14.80	96.10

Class 125 iron body globe valve, flanged

Description	Craft@Hrs	Unit	Material $	Labor $	Total $
2½"	P2@.600	Ea	195.00	14.80	209.80
3"	P2@.750	Ea	289.00	18.50	307.50
4"	P2@1.35	Ea	328.00	33.30	361.30

200 PSI iron body butterfly valve, lug type

Description	Craft@Hrs	Unit	Material $	Labor $	Total $
2"	P2@.200	Ea	77.60	4.93	82.53
2½"	P2@.300	Ea	81.00	7.39	88.39
3"	P2@.400	Ea	89.80	9.86	99.66
4"	P2@.500	Ea	120.00	12.30	132.30

200 PSI iron body butterfly valve, wafer type

Description	Craft@Hrs	Unit	Material $	Labor $	Total $
2"	P2@.200	Ea	66.80	4.93	71.73
2½"	P2@.300	Ea	69.50	7.39	76.89
3"	P2@.400	Ea	76.30	9.86	86.16
4"	P2@.500	Ea	95.20	12.30	107.50

Description	Craft@Hrs	Unit	Material $	Labor $	Total $

Class 125 bronze body 2- piece ball valve, soldered

Description	Craft@Hrs	Unit	Material $	Labor $	Total $
1/2"	P2@.240	Ea	6.08	5.91	11.99
3/4"	P2@.300	Ea	9.70	7.39	17.09
1"	P2@.360	Ea	12.40	8.87	21.27
1¼"	P2@.480	Ea	21.70	11.80	33.50
1½"	P2@.540	Ea	27.10	13.30	40.40
2"	P2@.600	Ea	34.60	14.80	49.40

Class 125 bronze body swing check valve, soldered

Description	Craft@Hrs	Unit	Material $	Labor $	Total $
1/2"	P2@.240	Ea	20.50	5.91	26.41
3/4"	P2@.300	Ea	24.90	7.39	32.29
1"	P2@.360	Ea	34.30	8.87	43.17
1¼"	P2@.480	Ea	47.40	11.80	59.20
1½"	P2@.540	Ea	56.30	13.30	69.60
2"	P2@.600	Ea	82.20	14.80	97.00
2½"	P2@1.00	Ea	203.00	24.60	227.60
3"	P2@1.50	Ea	300.00	37.00	337.00

Class 125 iron body check valve, flanged

Description	Craft@Hrs	Unit	Material $	Labor $	Total $
2"	P2@.500	Ea	174.00	12.30	186.30
2½"	P2@.600	Ea	221.00	14.80	235.80
3"	P2@.750	Ea	240.00	18.50	258.50
4"	P2@1.35	Ea	378.00	33.30	411.30

Class 125 iron body check valve, wafer type

Description	Craft@Hrs	Unit	Material $	Labor $	Total $
2"	P2@.500	Ea	97.20	12.30	109.50
2½"	P2@.600	Ea	112.00	14.80	126.80
3"	P2@.750	Ea	124.00	18.50	142.50
4"	P2@1.35	Ea	167.00	33.30	200.30

Class 150 bronze body strainer, threaded

Description	Craft@Hrs	Unit	Material $	Labor $	Total $
1/2"	P2@.230	Ea	21.30	5.67	26.97
3/4"	P2@.260	Ea	23.20	6.41	29.61
1"	P2@.330	Ea	34.90	8.13	43.03
1¼"	P2@.440	Ea	56.10	10.80	66.90
1½"	P2@.495	Ea	75.20	12.20	87.40
2"	P2@.550	Ea	100.00	13.60	113.60

Class 125 iron body strainer, flanged

Description	Craft@Hrs	Unit	Material $	Labor $	Total $
2"	P2@.500	Ea	180.00	12.30	192.30
2½"	P2@.600	Ea	183.00	14.80	197.80
3"	P2@.750	Ea	212.00	18.50	230.50
4"	P2@1.35	Ea	382.00	33.30	415.30

Description	Craft@Hrs	Unit	Material $	Labor $	Total $
Installation of 2-way control valve					
1/2"	P2@.210	Ea	--	5.17	5.17
3/4"	P2@.275	Ea	--	6.78	6.78
1"	P2@.350	Ea	--	8.62	8.62
1¼"	P2@.430	Ea	--	10.60	10.60
1½"	P2@.505	Ea	--	12.40	12.40
2"	P2@.675	Ea	--	16.60	16.60
2½"	P2@.830	Ea	--	20.50	20.50
3"	P2@.990	Ea	--	24.40	24.40
4"	P2@1.30	Ea	--	32.00	32.00
Installation of 3-way control valve					
1/2"	P2@.260	Ea	--	6.41	6.41
3/4"	P2@.365	Ea	--	8.99	8.99
1"	P2@.475	Ea	--	11.70	11.70
1¼"	P2@.575	Ea	--	14.20	14.20
1½"	P2@.680	Ea	--	16.80	16.80
2"	P2@.910	Ea	--	22.40	22.40
2½"	P2@1.12	Ea	--	27.60	27.60
3"	P2@1.33	Ea	--	32.80	32.80
4"	P2@2.00	Ea	--	49.30	49.30
Companion flange					
2"	P2@.290	Ea	47.50	7.15	54.65
2½"	P2@.380	Ea	61.80	9.36	71.16
3"	P2@.460	Ea	61.40	11.30	72.70
4"	P2@.600	Ea	91.70	14.80	106.50
Bolt and gasket sets					
2"	P2@.500	Ea	3.85	12.30	16.15
2½"	P2@.650	Ea	4.08	16.00	20.08
3"	P2@.750	Ea	4.28	18.50	22.78
4"	P2@1.00	Ea	7.75	24.60	32.35
Thermometer with well					
7"	P2@.250	Ea	14.30	6.16	20.46
9"	P2@.250	Ea	14.80	6.16	20.96
Pressure gauge					
2½"	P2@.200	Ea	15.60	4.93	20.53
3½"	P2@.200	Ea	27.10	4.93	32.03

Copper, Type K with Brazed Joints

Description	Craft@Hrs	Unit	Material $	Labor $	Total $
Hanger with swivel assembly					
1/2"	P2@.450	Ea	.56	11.10	11.66
3/4"	P2@.450	Ea	.56	11.10	11.66
1"	P2@.450	Ea	.56	11.10	11.66
1¼"	P2@.450	Ea	.56	11.10	11.66
1½"	P2@.450	Ea	.62	11.10	11.72
2"	P2@.450	Ea	.65	11.10	11.75
2½"	P2@.450	Ea	3.95	11.10	15.05
3"	P2@.450	Ea	4.32	11.10	15.42
4"	P2@.450	Ea	5.87	11.10	16.97
Riser clamp					
1/2"	P2@.100	Ea	3.27	2.46	5.73
3/4"	P2@.100	Ea	3.27	2.46	5.73
1"	P2@.100	Ea	3.30	2.46	5.76
1¼"	P2@.105	Ea	4.01	2.59	6.60
1½"	P2@.110	Ea	4.23	2.71	6.94
2"	P2@.115	Ea	4.45	2.83	7.28
2½"	P2@.120	Ea	4.69	2.96	7.65
3"	P2@.120	Ea	5.13	2.96	8.09
4"	P2@.125	Ea	6.45	3.08	9.53

Type K hard-drawn copper pipe with wrought copper fittings and soft-soldered joints is used in a wide variety of plumbing and HVAC systems such as heating hot water, chilled water, potable water and A.C. condensate.

Soft-soldered joints are those made with solders having melting points in the 350 degree F. to 500 degree F. range. Maximum working pressure/temperature relationships for soft-soldered joints are approximately as follows:

Maximum Working Pressures (PSIG)*			
Soft-solder Type	Water Temperature (degrees F.)	Nominal Pipe Size (inches)		
		Up to 1	1¼ to 2	2½ to 4
50-50 tin-lead**	100	200	175	150
	150	150	125	100
	200	100	90	75
	250	85	75	50
95-5 tin-antimony	100	500	400	300
	150	400	350	275
	200	300	250	200
	250	200	175	150

*For copper pipe and solder-type fittings using soft-solders melting at approximately 350 degrees F. to 500 degrees F.
The use of any solder containing lead is **not allowed in potable water systems.

This section has been arranged to save the estimator's time by including all normally-used system components such as pipe, fittings, valves, hanger assemblies, riser clamps and miscellaneous items under one heading. Additional items can be found under "Plumbing and Piping Specialties." The cost estimates in this section are based on the conditions, limitations and wage rates described in the section "How to Use This Book" beginning on page 5.

Description	Craft@Hrs	Unit	Material $	Labor $	Total $

Type K copper pipe with soft soldered joints

1/2"	P2@.032	LF	1.22	.79	2.01
3/4"	P2@.035	LF	2.24	.86	3.10
1"	P2@.038	LF	2.87	.94	3.81
1¼"	P2@.042	LF	3.58	1.03	4.61
1½"	P2@.046	LF	4.60	1.13	5.73
2"	P2@.053	LF	7.36	1.31	8.67
2½"	P2@.060	LF	10.70	1.48	12.18
3"	P2@.066	LF	14.70	1.63	16.33
4"	P2@.080	LF	23.00	1.97	24.97

Description	Craft@Hrs	Unit	Material $	Labor $	Total $

Type K copper 45 degree ell with soft soldered joints C x C

Description	Craft@Hrs	Unit	Material $	Labor $	Total $
1/2"	P2@.107	Ea	.72	2.64	3.36
3/4"	P2@.150	Ea	1.22	3.70	4.92
1"	P2@.193	Ea	3.08	4.76	7.84
1¼"	P2@.236	Ea	4.20	5.82	10.02
1½"	P2@.278	Ea	5.04	6.85	11.89
2"	P2@.371	Ea	8.44	9.14	17.58
2½"	P2@.457	Ea	18.00	11.30	29.30
3"	P2@.543	Ea	26.70	13.40	40.10
4"	P2@.714	Ea	55.60	17.60	73.20

Type K copper 90 degree ell with soft soldered joints C x C

Description	Craft@Hrs	Unit	Material $	Labor $	Total $
1/2"	P2@.107	Ea	.43	2.64	3.07
3/4"	P2@.150	Ea	.91	3.70	4.61
1"	P2@.193	Ea	2.11	4.76	6.87
1¼"	P2@.236	Ea	3.21	5.82	9.03
1½"	P2@.278	Ea	5.00	6.85	11.85
2"	P2@.371	Ea	9.12	9.14	18.26
2½"	P2@.457	Ea	17.40	11.30	28.70
3"	P2@.543	Ea	24.30	13.40	37.70
4"	P2@.714	Ea	56.00	17.60	73.60

Type K copper 90 degree ell with soft soldered joint Ftg. x C

Description	Craft@Hrs	Unit	Material $	Labor $	Total $
1/2"	P2@.107	Ea	.60	2.64	3.24
3/4"	P2@.150	Ea	1.33	3.70	5.03
1"	P2@.193	Ea	3.20	4.76	7.96
1¼"	P2@.236	Ea	4.88	5.82	10.70
1½"	P2@.278	Ea	6.40	6.85	13.25
2"	P2@.371	Ea	13.90	9.14	23.04
2½"	P2@.457	Ea	29.10	11.30	40.40
3"	P2@.543	Ea	34.70	13.40	48.10
4"	P2@.714	Ea	80.00	17.60	97.60

Type K copper tee with soft soldered joints C x C x C

Description	Craft@Hrs	Unit	Material $	Labor $	Total $
1/2"	P2@.129	Ea	.70	3.18	3.88
3/4"	P2@.181	Ea	1.70	4.46	6.16
1"	P2@.233	Ea	4.92	5.74	10.66
1¼"	P2@.285	Ea	7.44	7.02	14.46
1½"	P2@.337	Ea	10.30	8.30	18.60
2"	P2@.449	Ea	16.10	11.10	27.20
2½"	P2@.552	Ea	32.50	13.60	46.10
3"	P2@.656	Ea	49.60	16.20	65.80
4"	P2@.863	Ea	107.00	21.30	128.30

Description	Craft@Hrs	Unit	Material $	Labor $	Total $

Type K copper branch reducing tee with soft soldered joints CxCxC

Description	Craft@Hrs	Unit	Material $	Labor $	Total $
1/2 x 3/8"	P2@.121	Ea	3.30	2.98	6.28
3/4 x 1/2"	P2@.170	Ea	1.54	4.19	5.73
1 x 1/2"	P2@.195	Ea	4.92	4.80	9.72
1 x 3/4"	P2@.219	Ea	4.92	5.40	10.32
1¼ x 1/2"	P2@.225	Ea	7.24	5.54	12.78
1¼ x 3/4"	P2@.247	Ea	7.24	6.09	13.33
1¼ x 1"	P2@.268	Ea	7.24	6.60	13.84
1½ x 1/2"	P2@.272	Ea	7.68	6.70	14.38
1½ x 3/4"	P2@.287	Ea	7.68	7.07	14.75
1½ x 1"	P2@.302	Ea	7.68	7.44	15.12
1½ x 1¼"	P2@.317	Ea	7.68	7.81	15.49
2 x 1/2"	P2@.325	Ea	12.10	8.01	20.11
2 x 3/4"	P2@.348	Ea	12.10	8.57	20.67
2 x 1"	P2@.373	Ea	12.10	9.19	21.29
2 x 1¼"	P2@.398	Ea	12.10	9.81	21.91
2 x 1½"	P2@.422	Ea	12.10	10.40	22.50
2½ x 1/2"	P2@.437	Ea	37.00	10.80	47.80
2½ x 3/4"	P2@.453	Ea	37.00	11.20	48.20
2½ x 1"	P2@.469	Ea	37.00	11.60	48.60
2½ x 1½"	P2@.502	Ea	37.00	12.40	49.40
2½ x 2"	P2@.519	Ea	37.00	12.80	49.80
3 x 1¼"	P2@.525	Ea	40.80	12.90	53.70
3 x 2"	P2@.585	Ea	40.80	14.40	55.20
3 x 2½"	P2@.617	Ea	40.80	15.20	56.00
4 x 1¼"	P2@.625	Ea	73.20	15.40	88.60
4 x 1½"	P2@.671	Ea	73.20	16.50	89.70
4 x 2"	P2@.718	Ea	73.20	17.70	90.90
4 x 2½"	P2@.765	Ea	73.20	18.80	92.00
4 x 3"	P2@.811	Ea	73.20	20.00	93.20

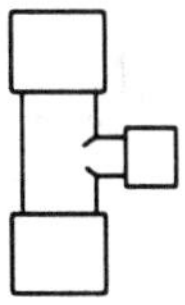

Copper, Type K with Soft-Soldered Joints

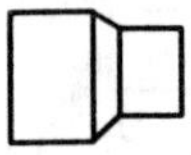

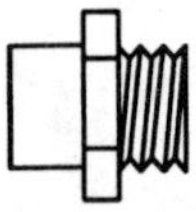

Description	Craft@Hrs	Unit	Material $	Labor $	Total $
Type K copper reducer with soft soldered joints					
1/2 x 3/8"	P2@.124	Ea	.71	3.06	3.77
3/4 x 1/2"	P2@.129	Ea	1.06	3.18	4.24
1 x 3/4"	P2@.172	Ea	1.72	4.24	5.96
1¼ x 1/2"	P2@.172	Ea	2.96	4.24	7.20
1¼ x 3/4"	P2@.193	Ea	2.96	4.76	7.72
1¼ x 1"	P2@.215	Ea	2.96	5.30	8.26
1½ x 1/2"	P2@.193	Ea	3.90	4.76	8.66
1½ x 3/4"	P2@.214	Ea	3.90	5.27	9.17
1½ x 1"	P2@.236	Ea	3.90	5.82	9.72
1½ x 1¼"	P2@.257	Ea	3.90	6.33	10.23
2 x 1/2"	P2@.239	Ea	7.33	5.89	13.22
2 x 3/4"	P2@.261	Ea	7.33	6.43	13.76
2 x 1"	P2@.282	Ea	7.33	6.95	14.28
2 x 1¼"	P2@.304	Ea	7.33	7.49	14.82
2 x 1½"	P2@.325	Ea	7.33	8.01	15.34
2½ x 1"	P2@.325	Ea	14.60	8.01	22.61
2½ x 1¼"	P2@.347	Ea	14.60	8.55	23.15
2½ x 1½"	P2@.368	Ea	14.60	9.07	23.67
2½ x 2"	P2@.414	Ea	14.60	10.20	24.80
3 x 2"	P2@.457	Ea	18.30	11.30	29.60
3 x 2½"	P2@.500	Ea	18.30	12.30	30.60
4 x 2"	P2@.543	Ea	36.60	13.40	50.00
4 x 2½"	P2@.586	Ea	36.60	14.40	51.00
4 x 3"	P2@.629	Ea	36.60	15.50	52.10
Type K copper adapter with soft soldered joint C x MPT					
1/2"	P2@.075	Ea	.87	1.85	2.72
3/4"	P2@.105	Ea	1.45	2.59	4.04
1"	P2@.134	Ea	3.54	3.30	6.84
1¼"	P2@.164	Ea	5.22	4.04	9.26
1½"	P2@.194	Ea	5.99	4.78	10.77
2"	P2@.259	Ea	10.20	6.38	16.58
2½"	P2@.319	Ea	31.20	7.86	39.06
3"	P2@.378	Ea	37.30	9.31	46.61
4"	P2@.498	Ea	63.60	12.30	75.90

Description	Craft@Hrs	Unit	Material $	Labor $	Total $

Type K copper adapter with soft soldered joint C x FPT

Description	Craft@Hrs	Unit	Material $	Labor $	Total $
1/2"	P2@.075	Ea	1.30	1.85	3.15
3/4"	P2@.105	Ea	1.79	2.59	4.38
1"	P2@.134	Ea	3.66	3.30	6.96
1¼"	P2@.164	Ea	5.99	4.04	10.03
1½"	P2@.194	Ea	9.40	4.78	14.18
2"	P2@.259	Ea	12.80	6.38	19.18
2½"	P2@.319	Ea	35.40	7.86	43.26
3"	P2@.378	Ea	51.20	9.31	60.51
4"	P2@.498	Ea	87.10	12.30	99.40

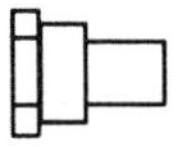

Type K copper flush bushing with soft soldered joints

Description	Craft@Hrs	Unit	Material $	Labor $	Total $
1/2 x 3/8"	P2@.124	Ea	.82	3.06	3.88
3/4 x 1/2"	P2@.129	Ea	1.50	3.18	4.68
1 x 3/4"	P2@.172	Ea	2.32	4.24	6.56
1 x 1/2"	P2@.172	Ea	2.32	4.24	6.56
1¼ x 1"	P2@.215	Ea	2.81	5.30	8.11
1½ x 1¼"	P2@.257	Ea	3.55	6.33	9.88
2 x 1½"	P2@.325	Ea	10.20	8.01	18.21

Type K copper union with soft soldered joint

Description	Craft@Hrs	Unit	Material $	Labor $	Total $
1/2"	P2@.121	Ea	3.82	2.98	6.80
3/4"	P2@.170	Ea	4.80	4.19	8.99
1"	P2@.218	Ea	8.52	5.37	13.89
1¼"	P2@.267	Ea	14.80	6.58	21.38
1½"	P2@.315	Ea	19.40	7.76	27.16
2"	P2@.421	Ea	33.10	10.40	43.50

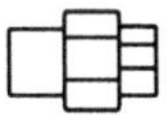

Type K copper dielectric union with soft soldered joint

Description	Craft@Hrs	Unit	Material $	Labor $	Total $
1/2"	P2@.121	Ea	3.39	2.98	6.37
3/4"	P2@.170	Ea	3.39	4.19	7.58
1"	P2@.218	Ea	7.76	5.37	13.13
1¼"	P2@.267	Ea	12.70	6.58	19.28
1½"	P2@.315	Ea	18.60	7.76	26.36
2"	P2@.421	Ea	26.30	10.40	36.70

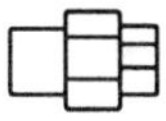

Type K copper cap with soft soldered joint

Description	Craft@Hrs	Unit	Material $	Labor $	Total $
1/2"	P2@.069	Ea	.29	1.70	1.99
3/4"	P2@.096	Ea	.50	2.37	2.87
1"	P2@.124	Ea	1.18	3.06	4.24
1¼"	P2@.151	Ea	1.68	3.72	5.40
1½"	P2@.178	Ea	2.46	4.39	6.85
2"	P2@.237	Ea	4.50	5.84	10.34
2½"	P2@.292	Ea	13.90	7.19	21.09
3"	P2@.347	Ea	19.00	8.55	27.55
4"	P2@.457	Ea	37.30	11.30	48.60

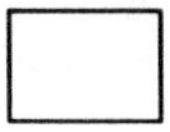

Description	Craft@Hrs	Unit	Material $	Labor $	Total $
Type K copper coupling with soft soldered joints					
1/2"	P2@.107	Ea	.32	2.64	2.96
3/4"	P2@.150	Ea	.61	3.70	4.31
1"	P2@.193	Ea	1.60	4.76	6.36
1¼"	P2@.236	Ea	2.11	5.82	7.93
1½"	P2@.278	Ea	2.79	6.85	9.64
2"	P2@.371	Ea	4.62	9.14	13.76
2½"	P2@.457	Ea	9.84	11.30	21.14
3"	P2@.543	Ea	15.10	13.40	28.50
4"	P2@.714	Ea	29.60	17.60	47.20
Class 125 bronze gate valve, solder ends					
1/2"	P2@.200	Ea	12.20	4.93	17.13
3/4"	P2@.249	Ea	14.70	6.14	20.84
1"	P2@.299	Ea	18.30	7.37	25.67
1¼"	P2@.398	Ea	31.30	9.81	41.11
1½"	P2@.448	Ea	34.50	11.00	45.50
2"	P2@.498	Ea	48.40	12.30	60.70
2½"	P2@.830	Ea	113.00	20.50	133.50
3"	P2@1.24	Ea	161.00	30.60	191.60
Class 125 iron body gate valve, flanged ends					
2"	P2@.500	Ea	164.00	12.30	176.30
2½"	P2@.600	Ea	168.00	14.80	182.80
3"	P2@.750	Ea	188.00	18.50	206.50
4"	P2@1.35	Ea	270.00	33.30	303.30
5"	P2@2.00	Ea	460.00	49.30	509.30
6"	P2@2.50	Ea	460.00	61.60	521.60
Class 125 bronze body globe valve, solder ends					
1/2"	P2@.200	Ea	18.30	4.93	23.23
3/4"	P2@.249	Ea	23.30	6.14	29.44
1"	P2@.299	Ea	33.80	7.37	41.17
1¼"	P2@.398	Ea	41.30	9.81	51.11
1½"	P2@.448	Ea	53.30	11.00	64.30
2"	P2@.498	Ea	81.30	12.30	93.60
Class 125 iron body globe valve, flanged ends					
2½"	P2@.600	Ea	195.00	14.80	209.80
3"	P2@.750	Ea	229.00	18.50	247.50
4"	P2@1.35	Ea	328.00	33.30	361.30

Description	Craft@Hrs	Unit	Material $	Labor $	Total $

200 PSI iron body butterfly valve, lug type

2"	P2@.200	Ea	77.60	4.93	82.53
2½"	P2@.300	Ea	81.00	7.39	88.39
3"	P2@.400	Ea	89.60	9.86	99.46
4"	P2@.500	Ea	120.00	12.30	132.30

200 PSI iron body butterfly valve, wafer type

2"	P2@.200	Ea	66.80	4.93	71.73
2½"	P2@.300	Ea	69.50	7.39	76.89
3"	P2@.400	Ea	76.30	9.86	86.16
4"	P2@.500	Ea	95.20	12.30	107.50

Class 125 bronze body ball valve, solder ends

1/2"	P2@.200	Ea	6.08	4.93	11.01
3/4"	P2@.249	Ea	9.70	6.14	15.84
1"	P2@.299	Ea	12.40	7.37	19.77
1¼"	P2@.398	Ea	21.70	9.81	31.51
1½"	P2@.448	Ea	27.10	11.00	38.10
2"	P2@.498	Ea	34.60	12.30	46.90

Class 125 bronze swing check valve, solder ends

1/2"	P2@.200	Ea	20.50	4.93	25.43
3/4"	P2@.249	Ea	24.90	6.14	31.04
1"	P2@.299	Ea	34.30	7.37	41.67
1¼"	P2@.398	Ea	47.40	9.81	57.21
1½"	P2@.448	Ea	56.30	11.00	67.30
2"	P2@.498	Ea	82.20	12.30	94.50
2½"	P2@.830	Ea	203.00	20.50	223.50
3"	P2@1.24	Ea	300.00	30.60	330.60

Class 125 iron body swing check valve, flanged ends

2"	P2@.500	Ea	174.00	12.30	186.30
2½"	P2@.600	Ea	221.00	14.80	235.80
3"	P2@.750	Ea	240.00	18.50	258.50
4"	P2@1.35	Ea	378.00	33.30	411.30

Class 125 iron body silent check valve, wafer type

2"	P2@.500	Ea	97.20	12.30	109.50
2½"	P2@.600	Ea	112.00	14.80	126.80
3"	P2@.750	Ea	124.00	18.50	142.50
4"	P2@1.35	Ea	167.00	33.30	200.30

Copper, Type K with Soft-Soldered Joints

Description	Craft@Hrs	Unit	Material $	Labor $	Total $

Class 150 bronze body strainer, threaded ends

Description	Craft@Hrs	Unit	Material $	Labor $	Total $
1/2"	P2@.230	Ea	21.30	5.67	26.97
3/4"	P2@.260	Ea	23.20	6.41	29.61
1"	P2@.330	Ea	34.90	8.13	43.03
1¼"	P2@.440	Ea	56.10	10.80	66.90
1½"	P2@.495	Ea	75.20	12.20	87.40
2"	P2@.550	Ea	100.00	13.60	113.60

Class 125 iron body strainer, flanged ends

Description	Craft@Hrs	Unit	Material $	Labor $	Total $
2"	P2@.500	Ea	180.00	12.30	192.30
2½"	P2@.600	Ea	183.00	14.80	197.80
3"	P2@.750	Ea	212.00	18.50	230.50
4"	P2@1.35	Ea	382.00	33.30	415.30

Installation of copper 2-way control valve

Description	Craft@Hrs	Unit	Material $	Labor $	Total $
1/2"	P2@.210	Ea	--	5.17	5.17
3/4"	P2@.275	Ea	--	6.78	6.78
1"	P2@.350	Ea	--	8.62	8.62
1¼"	P2@.430	Ea	--	10.60	10.60
1½"	P2@.505	Ea	--	12.40	12.40
2"	P2@.675	Ea	--	16.60	16.60
2½"	P2@.830	Ea	--	20.50	20.50
3"	P2@.990	Ea	--	24.40	24.40
4"	P2@1.30	Ea	--	32.00	32.00

Installation of copper 3-way control valve

Description	Craft@Hrs	Unit	Material $	Labor $	Total $
1/2"	P2@.260	Ea	--	6.41	6.41
3/4"	P2@.365	Ea	--	8.99	8.99
1"	P2@.475	Ea	--	11.70	11.70
1¼"	P2@.575	Ea	--	14.20	14.20
1½"	P2@.680	Ea	--	16.80	16.80
2"	P2@.910	Ea	--	22.40	22.40
2½"	P2@1.12	Ea	--	27.60	27.60
3"	P2@1.33	Ea	--	32.80	32.80
4"	P2@2.00	Ea	--	49.30	49.30

Companion flange

Description	Craft@Hrs	Unit	Material $	Labor $	Total $
2"	P2@.290	Ea	47.50	7.15	54.65
2½"	P2@.380	Ea	61.80	9.36	71.16
3"	P2@.460	Ea	61.40	11.30	72.70
4"	P2@.600	Ea	91.70	14.80	106.50

Description	Craft@Hrs	Unit	Material $	Labor $	Total $
Bolt and gasket sets					
2"	P2@.500	Ea	3.85	12.30	16.15
2½"	P2@.650	Ea	4.08	16.00	20.08
3"	P2@.750	Ea	4.28	18.50	22.78
4"	P2@1.00	Ea	7.75	24.60	32.35
Thermometer with well					
7"	P2@.250	Ea	14.30	6.16	20.46
9"	P2@.250	Ea	14.80	6.16	20.96
Pressure gauge					
2½"	P2@.200	Ea	15.60	4.93	20.53
3½"	P2@.200	Ea	27.10	4.93	32.03
Hanger with swivel assembly					
1/2"	P2@.450	Ea	.56	11.10	11.66
3/4"	P2@.450	Ea	.56	11.10	11.66
1"	P2@.450	Ea	.56	11.10	11.66
1¼"	P2@.450	Ea	.56	11.10	11.66
1½"	P2@.450	Ea	.62	11.10	11.72
2"	P2@.450	Ea	.65	11.10	11.75
2½"	P2@.450	Ea	3.95	11.10	15.05
3"	P2@.450	Ea	4.32	11.10	15.42
4"	P2@.450	Ea	5.87	11.10	16.97
Riser clamp					
1/2"	P2@.100	Ea	3.27	2.46	5.73
3/4"	P2@.100	Ea	3.27	2.46	5.73
1"	P2@.100	Ea	3.30	2.46	5.76
1¼"	P2@.105	Ea	4.01	2.59	6.60
1½"	P2@.110	Ea	4.23	2.71	6.94
2"	P2@.115	Ea	4.45	2.83	7.28
2½"	P2@.120	Ea	4.69	2.96	7.65
3"	P2@.120	Ea	5.13	2.96	8.09
4"	P2@.125	Ea	6.45	3.08	9.53

Copper, Type L with Brazed Joints

Type L hard-drawn copper pipe with wrought copper fittings and brazed joints is used in a wide variety of plumbing and HVAC systems such as heating hot water, chilled water, potable water, compressed air, refrigerant and A.C. condensate.

Brazed joints are those made with silver or other alloy filler metals having melting points at, or above, 1,000 degrees F. Maximum working pressure/temperature relationships for brazed joints are approximately as follows:

Maximum Working Pressure (PSIG)*			
Water Temperature (degrees F.)	Nominal Pipe Size (inches)		
	Up to 1	1¼ to 2	2½ to 4
Up to 350	270	190	150

*For copper pipe and solder-type fittings using brazing alloys melting at, or above, 1,000 degrees F.

This section has been arranged to save the estimator's time by including all normally-used system components such as pipe, fittings, valves, hanger assemblies, riser clamps and miscellaneous items under one heading. Additional items can be found under "Plumbing and Piping Specialties." The cost estimates in this section are based on the conditions, limitations and wage rates described in the section "How to Use This Book" beginning on page 5.

Description	Craft@Hrs	Unit	Material $	Labor $	Total $

Type L copper pipe with brazed joints

Description	Craft@Hrs	Unit	Material $	Labor $	Total $
1/2"	P2@.032	LF	.97	.79	1.76
3/4"	P2@.035	LF	1.54	.86	2.40
1"	P2@.038	LF	2.14	.94	3.08
1¼"	P2@.042	LF	3.00	1.03	4.03
1½"	P2@.046	LF	3.83	1.13	4.96
2"	P2@.053	LF	6.03	1.31	7.34
2½"	P2@.060	LF	8.82	1.48	10.30
3"	P2@.066	LF	16.50	1.63	18.13
4"	P2@.080	LF	18.90	1.97	20.87

Type L copper 45 degree ell with brazed joints C x C

Description	Craft@Hrs	Unit	Material $	Labor $	Total $
1/2"	P2@.129	Ea	.72	3.18	3.90
3/4"	P2@.181	Ea	1.22	4.46	5.68
1"	P2@.232	Ea	3.08	5.72	8.80
1¼"	P2@.284	Ea	4.20	7.00	11.20
1½"	P2@.335	Ea	5.04	8.25	13.29
2"	P2@.447	Ea	8.44	11.00	19.44
2½"	P2@.550	Ea	18.00	13.60	31.60
3"	P2@.654	Ea	26.70	16.10	42.80
4"	P2@.860	Ea	55.60	21.20	76.80

Description	Craft@Hrs	Unit	Material $	Labor $	Total $

Type L copper 90 degree ell with brazed joints C x C

Description	Craft@Hrs	Unit	Material $	Labor $	Total $
1/2"	P2@.129	Ea	.43	3.18	3.61
3/4"	P2@.181	Ea	.91	4.46	5.37
1"	P2@.232	Ea	2.11	5.72	7.83
1¼"	P2@.284	Ea	3.21	7.00	10.21
1½"	P2@.335	Ea	5.00	8.25	13.25
2"	P2@.447	Ea	9.12	11.00	20.12
2½"	P2@.550	Ea	17.40	13.60	31.00
3"	P2@.654	Ea	24.30	16.10	40.40
4"	P2@.860	Ea	56.00	21.20	77.20

Type L copper 90 degree ell with brazed joint Ftg. x C

Description	Craft@Hrs	Unit	Material $	Labor $	Total $
1/2"	P2@.129	Ea	.60	3.18	3.78
3/4"	P2@.181	Ea	1.33	4.46	5.79
1"	P2@.232	Ea	3.20	5.72	8.92
1¼"	P2@.284	Ea	4.88	7.00	11.88
1½"	P2@.335	Ea	6.40	8.25	14.65
2"	P2@.447	Ea	13.90	11.00	24.90
2½"	P2@.550	Ea	29.10	13.60	42.70
3"	P2@.654	Ea	34.70	16.10	50.80
4"	P2@.860	Ea	80.00	21.20	101.20

Type L copper tee with brazed joints C x C x C

Description	Craft@Hrs	Unit	Material $	Labor $	Total $
1/2"	P2@.156	Ea	.70	3.84	4.54
3/4"	P2@.204	Ea	1.70	5.03	6.73
1"	P2@.263	Ea	4.92	6.48	11.40
1¼"	P2@.322	Ea	7.44	7.93	15.37
1½"	P2@.380	Ea	10.30	9.36	19.66
2"	P2@.507	Ea	16.10	12.50	28.60
2½"	P2@.624	Ea	32.50	15.40	47.90
3"	P2@.741	Ea	49.60	18.30	67.90
4"	P2@.975	Ea	107.00	24.00	131.00

Copper, Type L with Brazed Joints

Description	Craft@Hrs	Unit	Material $	Labor $	Total $
Type L copper branch reducing tee with brazed joints C x C x C					
1/2 x 3/8"	P2@.129	Ea	3.30	3.18	6.48
3/4 x 1/2"	P2@.182	Ea	1.54	4.48	6.02
1 x 1/2"	P2@.209	Ea	4.92	5.15	10.07
1 x 3/4"	P2@.234	Ea	4.92	5.77	10.69
1¼ x 1/2"	P2@.241	Ea	7.24	5.94	13.18
1¼ x 3/4"	P2@.264	Ea	7.24	6.50	13.74
1¼ x 1"	P2@.287	Ea	7.24	7.07	14.31
1½ x 1/2"	P2@.291	Ea	7.68	7.17	14.85
1½ x 3/4"	P2@.307	Ea	7.68	7.56	15.24
1½ x 1"	P2@.323	Ea	7.68	7.96	15.64
1½ x 1¼"	P2@.339	Ea	7.68	8.35	16.03
2 x 1/2"	P2@.348	Ea	12.10	8.57	20.67
2 x 3/4"	P2@.373	Ea	12.10	9.19	21.29
2 x 1"	P2@.399	Ea	12.10	9.83	21.93
2 x 1¼"	P2@.426	Ea	12.10	10.50	22.60
2 x 1½"	P2@.452	Ea	12.10	11.10	23.20
2½ x 1/2"	P2@.468	Ea	37.00	11.50	48.50
2½ x 3/4"	P2@.485	Ea	37.00	12.00	49.00
2½ x 1"	P2@.502	Ea	37.00	12.40	49.40
2½ x 1½"	P2@.537	Ea	37.00	13.20	50.20
2½ x 2"	P2@.555	Ea	37.00	13.70	50.70
3 x 1¼"	P2@.562	Ea	40.80	13.80	54.60
3 x 2"	P2@.626	Ea	40.80	15.40	56.20
3 x 2½"	P2@.660	Ea	40.80	16.30	57.10
4 x 1¼"	P2@.669	Ea	73.20	16.50	89.70
4 x 1½"	P2@.718	Ea	73.20	17.70	90.90
4 x 2"	P2@.768	Ea	73.20	18.90	92.10
4 x 2½"	P2@.819	Ea	73.20	20.20	93.40
4 x 3"	P2@.868	Ea	73.20	21.40	94.60

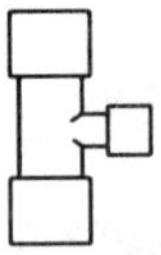

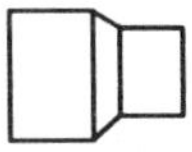

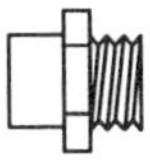

Description	Craft@Hrs	Unit	Material $	Labor $	Total $
Type L copper reducer with brazed joints					
1/2 x 3/8"	P2@.148	Ea	.71	3.65	4.36
3/4 x 1/2"	P2@.155	Ea	1.06	3.82	4.88
1 x 3/4"	P2@.206	Ea	1.72	5.08	6.80
1¼ x 1/2"	P2@.206	Ea	2.96	5.08	8.04
1¼ x 3/4"	P2@.232	Ea	2.96	5.72	8.68
1¼ x 1"	P2@.258	Ea	2.96	6.36	9.32
1½ x 1/2"	P2@.232	Ea	3.90	5.72	9.62
1½ x 3/4"	P2@.257	Ea	3.90	6.33	10.23
1½ x 1"	P2@.283	Ea	3.90	6.97	10.87
1½ x 1¼"	P2@.308	Ea	3.90	7.59	11.49
2 x 1/2"	P2@.287	Ea	7.33	7.07	14.40
2 x 3/4"	P2@.277	Ea	7.33	6.83	14.16
2 x 1"	P2@.338	Ea	7.33	8.33	15.66
2 x 1¼"	P2@.365	Ea	7.33	8.99	16.32
2 x 1½"	P2@.390	Ea	7.33	9.61	16.94
2½ x 1"	P2@.390	Ea	14.60	9.61	24.21
2½ x 1¼"	P2@.416	Ea	14.60	10.30	24.90
2½ x 1½"	P2@.442	Ea	14.60	10.90	25.50
2½ x 2"	P2@.497	Ea	14.60	12.20	26.80
3 x 1¼"	P2@.468	Ea	18.30	11.50	29.80
3 x 1½"	P2@.493	Ea	18.30	12.10	30.40
3 x 2"	P2@.548	Ea	18.30	13.50	31.80
3 x 2½"	P2@.600	Ea	18.30	14.80	33.10
4 x 2"	P2@.652	Ea	36.60	16.10	52.70
4 x 2½"	P2@.704	Ea	36.60	17.30	53.90
4 x 3"	P2@.755	Ea	36.60	18.60	55.20
Type L copper adapter with brazed joint C x MPT					
1/2"	P2@.090	Ea	.87	2.22	3.09
3/4"	P2@.126	Ea	1.45	3.10	4.55
1"	P2@.162	Ea	3.54	3.99	7.53
1¼"	P2@.198	Ea	5.22	4.88	10.10
1½"	P2@.234	Ea	5.99	5.77	11.76
2"	P2@.312	Ea	10.20	7.69	17.89
2½"	P2@.384	Ea	31.20	9.46	40.66
3"	P2@.456	Ea	37.30	11.20	48.50
4"	P2@.600	Ea	63.60	14.80	78.40
Type L copper adapter with brazed joint C x FPT					
1/2"	P2@.090	Ea	1.30	2.22	3.52
3/4"	P2@.126	Ea	1.79	3.10	4.89
1"	P2@.162	Ea	3.66	3.99	7.65
1¼"	P2@.198	Ea	5.99	4.88	10.87
1½"	P2@.234	Ea	9.40	5.77	15.17
2"	P2@.312	Ea	12.80	7.69	20.49
2½"	P2@.384	Ea	35.40	9.46	44.86
3"	P2@.456	Ea	51.20	11.20	62.40
4"	P2@.600	Ea	87.10	14.80	101.90

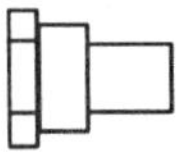

Copper, Type L with Brazed Joints

Description	Craft@Hrs	Unit	Material $	Labor $	Total $
Type L copper flush bushing with brazed joints					
1/2 x 3/8"	P2@.148	Ea	.82	3.65	4.47
3/4 x 1/2"	P2@.155	Ea	1.50	3.82	5.32
1 x 3/4"	P2@.206	Ea	2.32	5.08	7.40
1 x 1/2"	P2@.206	Ea	2.32	5.08	7.40
1¼ x 1"	P2@.258	Ea	2.81	6.36	9.17
1½ x 1¼"	P2@.308	Ea	3.55	7.59	11.14
2 x 1½"	P2@.390	Ea	10.20	9.61	19.81
Type L copper union with brazed joint					
1/2"	P2@.146	Ea	3.82	3.60	7.42
3/4"	P2@.205	Ea	4.80	5.05	9.85
1"	P2@.263	Ea	8.52	6.48	15.00
1¼"	P2@.322	Ea	14.80	7.93	22.73
1½"	P2@.380	Ea	19.40	9.36	28.76
2"	P2@.507	Ea	33.10	12.50	45.60
Type L copper dielectric union with brazed joint					
1/2"	P2@.146	Ea	3.39	3.60	6.99
3/4"	P2@.205	Ea	3.39	5.05	8.44
1"	P2@.263	Ea	7.76	6.48	14.24
1¼"	P2@.322	Ea	12.70	7.93	20.63
1½"	P2@.380	Ea	18.60	9.36	27.96
2"	P2@.507	Ea	26.30	12.50	38.80
Type L copper cap with brazed joint					
1/2"	P2@.083	Ea	.29	2.05	2.34
3/4"	P2@.116	Ea	.50	2.86	3.36
1"	P2@.149	Ea	1.18	3.67	4.85
1¼"	P2@.182	Ea	1.68	4.48	6.16
1½"	P2@.215	Ea	2.46	5.30	7.76
2"	P2@.286	Ea	4.50	7.05	11.55
2½"	P2@.352	Ea	13.90	8.67	22.57
3"	P2@.418	Ea	19.00	10.30	29.30
4"	P2@.550	Ea	37.30	13.60	50.90
Type L copper coupling with brazed joint					
1/2"	P2@.129	Ea	.32	3.18	3.50
3/4"	P2@.181	Ea	.61	4.46	5.07
1"	P2@.232	Ea	1.60	5.72	7.32
1¼"	P2@.284	Ea	2.11	7.00	9.11
1½"	P2@.335	Ea	2.79	8.25	11.04
2"	P2@.447	Ea	4.62	11.00	15.62
2½"	P2@.550	Ea	9.84	13.60	23.44
3"	P2@.654	Ea	15.10	16.10	31.20
4"	P2@.860	Ea	29.60	21.20	50.80

Description	Craft@Hrs	Unit	Material $	Labor $	Total $

Class 125 bronze body gate valve, solder ends

	Craft@Hrs	Unit	Material $	Labor $	Total $
1/2"	P2@.240	Ea	12.20	5.91	18.11
3/4"	P2@.300	Ea	14.70	7.39	22.09
1"	P2@.360	Ea	18.30	8.87	27.17
1¼"	P2@.480	Ea	31.30	11.80	43.10
1½"	P2@.540	Ea	34.50	13.30	47.80
2"	P2@.600	Ea	48.40	14.80	63.20
2½"	P2@1.00	Ea	113.00	24.60	137.60
3"	P2@1.50	Ea	161.00	37.00	198.00

Class 125 iron body gate valve, flanged ends

	Craft@Hrs	Unit	Material $	Labor $	Total $
2"	P2@.500	Ea	164.00	12.30	176.30
2½"	P2@.600	Ea	168.00	14.80	182.80
3"	P2@.750	Ea	188.00	18.50	206.50
4"	P2@1.35	Ea	270.00	33.30	303.30
5"	P2@2.00	Ea	460.00	49.30	509.30
6"	P2@2.50	Ea	460.00	61.60	521.60

Class 125 bronze body globe valve, solder ends

	Craft@Hrs	Unit	Material $	Labor $	Total $
1/2"	P2@.240	Ea	18.30	5.91	24.21
3/4"	P2@.300	Ea	23.30	7.39	30.69
1"	P2@.360	Ea	33.80	8.87	42.67
1¼"	P2@.480	Ea	41.30	11.80	53.10
1½"	P2@.540	Ea	53.30	13.30	66.60
2"	P2@.600	Ea	81.30	14.80	96.10

Class 125 iron body globe valve, flanged ends

	Craft@Hrs	Unit	Material $	Labor $	Total $
2½"	P2@.600	Ea	195.00	14.80	209.80
3"	P2@.750	Ea	229.00	18.50	247.50
4"	P2@1.35	Ea	328.00	33.30	361.30

200 PSI iron body butterfly valve, lug type

	Craft@Hrs	Unit	Material $	Labor $	Total $
2"	P2@.200	Ea	77.60	4.93	82.53
2½"	P2@.300	Ea	81.00	7.39	88.39
3"	P2@.400	Ea	89.80	9.86	99.66
4"	P2@.500	Ea	120.00	12.30	132.30

200 PSI iron body butterfly valve, wafer type

	Craft@Hrs	Unit	Material $	Labor $	Total $
2"	P2@.200	Ea	66.80	4.93	71.73
2½"	P2@.300	Ea	69.50	7.39	76.89
3"	P2@.400	Ea	76.30	9.86	86.16
4"	P2@.500	Ea	95.20	12.30	107.50

Copper, Type L with Brazed Joints

Description	Craft@Hrs	Unit	Material $	Labor $	Total $
Class 125 bronze body 2-piece ball valve, solder ends					
1/2"	P2@.240	Ea	6.08	5.91	11.99
3/4"	P2@.300	Ea	9.70	7.39	17.09
1"	P2@.360	Ea	12.40	8.87	21.27
1¼"	P2@.480	Ea	21.70	11.80	33.50
1½"	P2@.540	Ea	27.10	13.30	40.40
2"	P2@.600	Ea	34.60	14.80	49.40
Class 125 bronze body swing check valve, solder ends					
1/2"	P2@.240	Ea	20.50	5.91	26.41
3/4"	P2@.300	Ea	24.90	7.39	32.29
1"	P2@.360	Ea	34.30	8.87	43.17
1¼"	P2@.480	Ea	47.40	11.80	59.20
1½"	P2@.540	Ea	56.30	13.30	69.60
2"	P2@.600	Ea	82.20	14.80	97.00
2½"	P2@1.00	Ea	203.00	24.60	227.60
3"	P2@1.50	Ea	300.00	37.00	337.00
Class 125 iron body swing check valve, flanged ends					
2"	P2@.500	Ea	174.00	12.30	186.30
2½"	P2@.600	Ea	221.00	14.80	235.80
3"	P2@.750	Ea	240.00	18.50	258.50
4"	P2@1.35	Ea	378.00	33.30	411.30
Class 125 iron body silent check valve, wafer type					
2"	P2@.500	Ea	97.20	12.30	109.50
2½"	P2@.600	Ea	112.00	14.80	126.80
3"	P2@.750	Ea	124.00	18.50	142.50
4"	P2@1.35	Ea	167.00	33.30	200.30
Class 150 bronze body strainer, threaded ends					
1/2"	P2@.230	Ea	21.30	5.67	26.97
3/4"	P2@.260	Ea	23.20	6.41	29.61
1"	P2@.330	Ea	34.90	8.13	43.03
1¼"	P2@.440	Ea	56.10	10.80	66.90
1½"	P2@.495	Ea	75.20	12.20	87.40
2"	P2@.550	Ea	100.00	13.60	113.60
Class 125 iron body strainer, flanged ends					
2"	P2@.500	Ea	180.00	12.30	192.30
2½"	P2@.600	Ea	183.00	14.80	197.80
3"	P2@.750	Ea	212.00	18.50	230.50
4"	P2@1.35	Ea	382.00	33.30	415.30

Description	Craft@Hrs	Unit	Material $	Labor $	Total $
Installation of 2-way control valve					
1/2"	P2@.210	Ea	--	5.17	5.17
3/4"	P2@.275	Ea	--	6.78	6.78
1"	P2@.350	Ea	--	8.62	8.62
1¼"	P2@.430	Ea	--	10.60	10.60
1½"	P2@.505	Ea	--	12.40	12.40
2"	P2@.675	Ea	--	16.60	16.60
2½"	P2@.830	Ea	--	20.50	20.50
3"	P2@.990	Ea	--	24.40	24.40
4"	P2@1.30	Ea	--	32.00	32.00
Installation of 3-way control valve					
1/2"	P2@.260	Ea	--	6.41	6.41
3/4"	P2@.365	Ea	--	8.99	8.99
1"	P2@.475	Ea	--	11.70	11.70
1¼"	P2@.575	Ea	--	14.20	14.20
1½"	P2@.680	Ea	--	16.80	16.80
2"	P2@.910	Ea	--	22.40	22.40
2½"	P2@1.12	Ea	--	27.60	27.60
3"	P2@1.33	Ea	--	32.80	32.80
4"	P2@2.00	Ea	--	49.30	49.30
Companion flange					
2"	P2@.290	Ea	47.50	7.15	54.65
2½"	P2@.380	Ea	61.80	9.36	71.16
3"	P2@.460	Ea	61.40	11.30	72.70
4"	P2@.600	Ea	91.70	14.80	106.50
Bolt and gasket sets					
2"	P2@.500	Ea	3.85	12.30	16.15
2½"	P2@.650	Ea	4.08	16.00	20.08
3"	P2@.750	Ea	4.28	18.50	22.78
4"	P2@1.00	Ea	7.75	24.60	32.35
Thermometer with well					
7"	P2@.250	Ea	14.30	6.16	20.46
9"	P2@.250	Ea	14.80	6.16	20.96
Pressure gauge					
2½"	P2@.200	Ea	15.60	4.93	20.53
3½"	P2@.200	Ea	27.10	4.93	32.03

Copper, Type L with Brazed Joints

Description	Craft@Hrs	Unit	Material $	Labor $	Total $
Hanger with swivel assembly					
1/2"	P2@.450	Ea	.56	11.10	11.66
3/4"	P2@.450	Ea	.56	11.10	11.66
1"	P2@.450	Ea	.56	11.10	11.66
1¼"	P2@.450	Ea	.56	11.10	11.66
1½"	P2@.450	Ea	.62	11.10	11.72
2"	P2@.450	Ea	.65	11.10	11.75
2½"	P2@.450	Ea	3.95	11.10	15.05
3"	P2@.450	Ea	4.32	11.10	15.42
4"	P2@.450	Ea	5.87	11.10	16.97

Description	Craft@Hrs	Unit	Material $	Labor $	Total $
Riser clamp					
1/2"	P2@.100	Ea	3.27	2.46	5.73
3/4"	P2@.100	Ea	3.27	2.46	5.73
1"	P2@.100	Ea	3.30	2.46	5.76
1¼"	P2@.105	Ea	4.01	2.59	6.60
1½"	P2@.110	Ea	4.23	2.71	6.94
2"	P2@.115	Ea	4.45	2.83	7.28
2½"	P2@.120	Ea	4.69	2.96	7.65
3"	P2@.120	Ea	5.13	2.96	8.09
4"	P2@.125	Ea	6.45	3.08	9.53

Copper, Type L with Soft-Soldered Joints

Type L hard-drawn copper pipe with wrought copper fittings and soft-soldered joints is used in a wide variety of plumbing and HVAC systems such as heating hot water, chilled water, potable water and A.C. condensate.

Soft-soldered joints are those made with solders having melting points in the 350 degree F. to 500 degree F. range. Maximum working temperature/pressure relationships for soft-soldered joints are approximately as follows:

Maximum Working Pressures (PSIG)*				
Soft-solder Type	Water Temperature (degrees F.)	Nominal Pipe Size (inches)		
		Up to 1	1¼ to 2	2½ x 4
50-50 tin-lead**	100	200	175	150
	150	150	125	100
	200	100	90	75
	250	85	75	50
95-5 tin-antimony	100	500	400	300
	150	400	350	275
	200	300	250	200
	250	200	175	150

*For copper pipe and solder-type fittings using soft-solders melting at approximately 350 degrees F. to 500 degrees F.

The use of any solder containing lead is **not allowed in potable water systems.

This section has been arranged to save the estimator's time by including all normally-used system components such as pipe, fittings, valves, hanger assemblies, riser clamps and miscellaneous items under one heading. Additional items can be found under "Plumbing and Piping Specialties." The cost estimates in this section are based on the conditions, limitations and wage rates described in the section "How to Use This Book" beginning on page 5.

Description	Craft@Hrs	Unit	Material $	Labor $	Total $

Type L copper pipe with soft soldered joints

Description	Craft@Hrs	Unit	Material $	Labor $	Total $
1/2"	P2@.032	LF	.97	.79	1.76
3/4"	P2@.035	LF	1.54	.86	2.40
1"	P2@.038	LF	2.14	.94	3.08
1¼"	P2@.042	LF	3.00	1.03	4.03
1½"	P2@.046	LF	3.83	1.13	4.96
2"	P2@.053	LF	6.03	1.31	7.34
2½"	P2@.060	LF	8.82	1.48	10.30
3"	P2@.066	LF	16.50	1.63	18.13
4"	P2@.080	LF	18.90	1.97	20.87

Copper, Type L with Soft-Soldered Joints

Description	Craft@Hrs	Unit	Material $	Labor $	Total $

Type L copper 45 degree ell with soft soldered joints C x C

Description	Craft@Hrs	Unit	Material $	Labor $	Total $
1/2"	P2@.107	Ea	.72	2.64	3.36
3/4"	P2@.150	Ea	1.22	3.70	4.92
1"	P2@.193	Ea	3.08	4.76	7.84
1¼"	P2@.236	Ea	4.20	5.82	10.02
1½"	P2@.278	Ea	5.04	6.85	11.89
2"	P2@.371	Ea	8.44	9.14	17.58
2½"	P2@.457	Ea	18.00	11.30	29.30
3"	P2@.543	Ea	26.70	13.40	40.10
4"	P2@.714	Ea	55.60	17.60	73.20

Type L copper 90 degree ell with soft soldered joints C x C

Description	Craft@Hrs	Unit	Material $	Labor $	Total $
1/2"	P2@.107	Ea	.43	2.64	3.07
3/4"	P2@.150	Ea	.91	3.70	4.61
1"	P2@.193	Ea	2.11	4.76	6.87
1¼"	P2@.236	Ea	3.21	5.82	9.03
1½"	P2@.278	Ea	5.00	6.85	11.85
2"	P2@.371	Ea	9.12	9.14	18.26
2½"	P2@.457	Ea	17.40	11.30	28.70
3"	P2@.543	Ea	24.30	13.40	37.70
4"	P2@.714	Ea	56.00	17.60	73.60

Type L copper 90 degree ell with soft soldered joint Ftg. x C

Description	Craft@Hrs	Unit	Material $	Labor $	Total $
1/2"	P2@.107	Ea	.60	2.64	3.24
3/4"	P2@.150	Ea	1.33	3.70	5.03
1"	P2@.193	Ea	3.20	4.76	7.96
1¼"	P2@.236	Ea	4.88	5.82	10.70
1½"	P2@.278	Ea	6.40	6.85	13.25
2"	P2@.371	Ea	13.90	9.14	23.04
2½"	P2@.457	Ea	29.10	11.30	40.40
3"	P2@.543	Ea	34.20	13.40	47.60
4"	P2@.714	Ea	80.00	17.60	97.60

Type L copper tee with soft soldered joints C x C x C

Description	Craft@Hrs	Unit	Material $	Labor $	Total $
1/2"	P2@.129	Ea	.70	3.18	3.88
3/4"	P2@.181	Ea	1.70	4.46	6.16
1"	P2@.233	Ea	4.92	5.74	10.66
1¼"	P2@.285	Ea	7.44	7.02	14.46
1½"	P2@.337	Ea	10.20	8.30	18.50
2"	P2@.449	Ea	16.10	11.10	27.20
2½"	P2@.552	Ea	32.50	13.60	46.10
3"	P2@.656	Ea	49.60	16.20	65.80
4"	P2@.863	Ea	107.00	21.30	128.30

Description	Craft@Hrs	Unit	Material $	Labor $	Total $

Type L copper branch reducing tee with soft soldered joints CxCxC

Description	Craft@Hrs	Unit	Material $	Labor $	Total $
1/2 x 3/8"	P2@.121	Ea	3.30	2.98	6.28
3/4 x 1/2"	P2@.170	Ea	1.54	4.19	5.73
1 x 1/2"	P2@.195	Ea	4.92	4.80	9.72
1 x 3/4"	P2@.219	Ea	4.92	5.40	10.32
1¼ x 1/2"	P2@.225	Ea	7.24	5.54	12.78
1¼ x 3/4"	P2@.247	Ea	7.24	6.09	13.33
1¼ x 1"	P2@.268	Ea	7.24	6.60	13.84
1½ x 1/2"	P2@.272	Ea	7.68	6.70	14.38
1½ x 3/4"	P2@.287	Ea	7.68	7.07	14.75
1½ x 1"	P2@.302	Ea	7.68	7.44	15.12
1½ x 1¼"	P2@.317	Ea	7.68	7.81	15.49
2 x 1/2"	P2@.325	Ea	12.10	8.01	20.11
2 x 3/4"	P2@.348	Ea	12.10	8.57	20.67
2 x 1"	P2@.373	Ea	12.10	9.19	21.29
2 x 1¼"	P2@.398	Ea	12.10	9.81	21.91
2 x 1½"	P2@.422	Ea	12.10	10.40	22.50
2½ x 1/2"	P2@.437	Ea	37.00	10.80	47.80
2½ x 3/4"	P2@.453	Ea	37.00	11.20	48.20
2½ x 1"	P2@.469	Ea	37.00	11.60	48.60
2½ x 1½"	P2@.502	Ea	37.00	12.40	49.40
2½ x 2"	P2@.519	Ea	37.00	12.80	49.80
3 x 1¼"	P2@.525	Ea	40.80	12.90	53.70
3 x 2"	P2@.585	Ea	40.80	14.40	55.20
3 x 2½"	P2@.617	Ea	40.80	15.20	56.00
4 x 1¼"	P2@.625	Ea	73.20	15.40	88.60
4 x 1½"	P2@.671	Ea	73.20	16.50	89.70
4 x 2"	P2@.718	Ea	73.20	17.70	90.90
4 x 2½"	P2@.765	Ea	73.20	18.80	92.00
4 x 3"	P2@.811	Ea	73.20	20.00	93.20

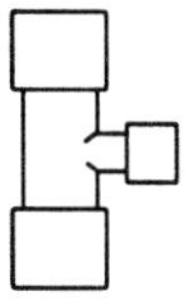

Copper, Type L with Soft-Soldered Joints

Description	Craft@Hrs	Unit	Material $	Labor $	Total $
Type L copper reducer with soft soldered joints					
1/2 x 3/8"	P2@.124	Ea	.71	3.06	3.77
3/4 x 1/2"	P2@.129	Ea	1.06	3.18	4.24
1 x 3/4"	P2@.172	Ea	1.72	4.24	5.96
1¼ x 1/2"	P2@.172	Ea	2.96	4.24	7.20
1¼ x 3/4"	P2@.193	Ea	2.96	4.76	7.72
1¼ x 1"	P2@.215	Ea	2.96	5.30	8.26
1½ x 1/2"	P2@.193	Ea	3.90	4.76	8.66
1½ x 3/4"	P2@.214	Ea	3.90	5.27	9.17
1½ x 1"	P2@.236	Ea	3.90	5.82	9.72
1½ x 1¼"	P2@.257	Ea	3.90	6.33	10.23
2 x 1/2"	P2@.239	Ea	7.33	5.89	13.22
2 x 3/4"	P2@.261	Ea	7.33	6.43	13.76
2 x 1"	P2@.282	Ea	7.33	6.95	14.28
2 x 1¼"	P2@.304	Ea	7.33	7.49	14.82
2 x 1½"	P2@.325	Ea	7.33	8.01	15.34
2½ x 1"	P2@.325	Ea	14.60	8.01	22.61
2½ x 1¼"	P2@.347	Ea	14.60	8.55	23.15
2½ x 1½"	P2@.368	Ea	14.60	9.07	23.67
2½ x 2"	P2@.414	Ea	14.60	10.20	24.80
3 x 1¼"	P2@.390	Ea	18.30	9.61	27.91
3 x 1½"	P2@.411	Ea	18.30	10.10	28.40
3 x 2"	P2@.457	Ea	18.30	11.30	29.60
3 x 2½"	P2@.500	Ea	18.30	12.30	30.60
4 x 2"	P2@.543	Ea	36.60	13.40	50.00
4 x 2½"	P2@.586	Ea	36.60	14.40	51.00
4 x 3"	P2@.629	Ea	36.60	15.50	52.10
Type L copper adapter with soft soldered joint C x MPT					
1/2"	P2@.075	Ea	.87	1.85	2.72
3/4"	P2@.105	Ea	1.45	2.59	4.04
1"	P2@.134	Ea	3.54	3.30	6.84
1¼"	P2@.164	Ea	5.22	4.04	9.26
1½"	P2@.194	Ea	5.99	4.78	10.77
2"	P2@.259	Ea	10.20	6.38	16.58
2½"	P2@.319	Ea	31.20	7.86	39.06
3"	P2@.378	Ea	37.30	9.31	46.61
4"	P2@.498	Ea	63.60	12.30	75.90
Type L copper adapter with soft soldered joint C x FPT					
1/2"	P2@.075	Ea	1.30	1.85	3.15
3/4"	P2@.105	Ea	1.79	2.59	4.38
1"	P2@.134	Ea	3.66	3.30	6.96
1¼"	P2@.164	Ea	5.99	4.04	10.03
1½"	P2@.194	Ea	9.40	4.78	14.18
2"	P2@.259	Ea	12.80	6.38	19.18
2½"	P2@.319	Ea	35.40	7.86	43.26
3"	P2@.378	Ea	51.20	9.31	60.51
4"	P2@.498	Ea	87.10	12.30	99.40

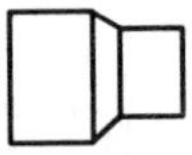

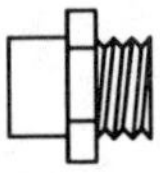

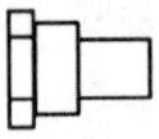

Description	Craft@Hrs	Unit	Material $	Labor $	Total $

Type L copper flush bushing with soft soldered joints

Description	Craft@Hrs	Unit	Material $	Labor $	Total $
1/2 x 3/8"	P2@.124	Ea	.82	3.06	3.88
3/4 x 1/2"	P2@.129	Ea	1.50	3.18	4.68
1 x 3/4"	P2@.172	Ea	2.32	4.24	6.56
1 x 1/2"	P2@.172	Ea	2.32	4.24	6.56
1¼ x 1"	P2@.215	Ea	2.81	5.30	8.11
1½ x 1¼"	P2@.257	Ea	3.55	6.33	9.88
2 x 1½"	P2@.325	Ea	10.20	8.01	18.21

Type L copper union with soft soldered joint

Description	Craft@Hrs	Unit	Material $	Labor $	Total $
1/2"	P2@.121	Ea	3.82	2.98	6.80
3/4"	P2@.170	Ea	4.80	4.19	8.99
1"	P2@.218	Ea	8.52	5.37	13.89
1¼"	P2@.267	Ea	14.80	6.58	21.38
1½"	P2@.315	Ea	19.40	7.76	27.16
2"	P2@.421	Ea	33.10	10.40	43.50

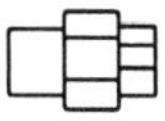

Type L copper dielectric union with soft soldered joint

Description	Craft@Hrs	Unit	Material $	Labor $	Total $
1/2"	P2@.121	Ea	3.39	2.98	6.37
3/4"	P2@.170	Ea	3.39	4.19	7.58
1"	P2@.218	Ea	7.76	5.37	13.13
1¼"	P2@.267	Ea	12.70	6.58	19.28
1½"	P2@.315	Ea	18.60	7.76	26.36
2"	P2@.421	Ea	26.30	10.40	36.70

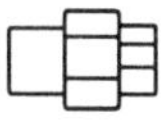

Type L copper cap with soft soldered joint

Description	Craft@Hrs	Unit	Material $	Labor $	Total $
1/2"	P2@.069	Ea	.29	1.70	1.99
3/4"	P2@.096	Ea	.50	2.37	2.87
1"	P2@.124	Ea	1.18	3.06	4.24
1¼"	P2@.151	Ea	1.68	3.72	5.40
1½"	P2@.178	Ea	2.46	4.39	6.85
2"	P2@.237	Ea	4.50	5.84	10.34
2½"	P2@.292	Ea	13.90	7.19	21.09
3"	P2@.347	Ea	19.00	8.55	27.55
4"	P2@.457	Ea	37.30	11.30	48.60

Type L copper coupling with soft soldered joints

Description	Craft@Hrs	Unit	Material $	Labor $	Total $
1/2"	P2@.107	Ea	.32	2.64	2.96
3/4"	P2@.150	Ea	.61	3.70	4.31
1"	P2@.193	Ea	1.60	4.76	6.36
1¼"	P2@.236	Ea	2.11	5.82	7.93
1½"	P2@.278	Ea	2.79	6.85	9.64
2"	P2@.371	Ea	4.62	9.14	13.76
2½"	P2@.457	Ea	9.84	11.30	21.14
3"	P2@.543	Ea	15.10	13.40	28.50
4"	P2@.714	Ea	29.60	17.60	47.20

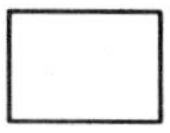

Copper, Type L with Soft-Soldered Joints

Description	Craft@Hrs	Unit	Material $	Labor $	Total $
Class 125 bronze body gate valve, solder ends					
1/2"	P2@.200	Ea	12.20	4.93	17.13
3/4"	P2@.249	Ea	14.70	6.14	20.84
1"	P2@.299	Ea	18.30	7.37	25.67
1¼"	P2@.398	Ea	31.30	9.81	41.11
1½"	P2@.448	Ea	34.50	11.00	45.50
2"	P2@.498	Ea	48.40	12.30	60.70
2½"	P2@.830	Ea	113.00	20.50	133.50
3"	P2@1.24	Ea	161.00	30.60	191.60
Class 125 iron body gate valve, flanged ends					
2"	P2@.500	Ea	164.00	12.30	176.30
2½"	P2@.600	Ea	168.00	14.80	182.80
3"	P2@.750	Ea	188.00	18.50	206.50
4"	P2@1.35	Ea	270.00	33.30	303.30
5"	P2@2.00	Ea	460.00	49.30	509.30
6"	P2@2.50	Ea	460.00	61.60	521.60
Class 125 bronze body globe valve, solder ends					
1/2"	P2@.200	Ea	18.30	4.93	23.23
3/4"	P2@.249	Ea	23.30	6.14	29.44
1"	P2@.299	Ea	33.80	7.37	41.17
1¼"	P2@.398	Ea	41.30	9.81	51.11
1½"	P2@.448	Ea	53.30	11.00	64.30
2"	P2@.498	Ea	81.30	12.30	93.60
Class 125 iron body globe valve, flanged ends					
2½"	P2@.600	Ea	195.00	14.80	209.80
3"	P2@.750	Ea	229.00	18.50	247.50
4"	P2@1.35	Ea	328.00	33.30	361.30
200 PSI iron body butterfly valve, lug type					
2"	P2@.200	Ea	77.60	4.93	82.53
2½"	P2@.300	Ea	81.00	7.39	88.39
3"	P2@.400	Ea	89.80	9.86	99.66
4"	P2@.500	Ea	120.00	12.30	132.30
200 PSI iron body butterfly valve, wafer type					
2"	P2@.200	Ea	66.80	4.93	71.73
2½"	P2@.300	Ea	69.50	7.39	76.89
3"	P2@.400	Ea	76.30	9.86	86.16
4"	P2@.500	Ea	95.20	12.30	107.50

Description	Craft@Hrs	Unit	Material $	Labor $	Total $

Class 125 bronze body 2-piece ball valve, solder ends

Description	Craft@Hrs	Unit	Material $	Labor $	Total $
1/2"	P2@.200	Ea	6.08	4.93	11.01
3/4"	P2@.249	Ea	9.70	6.14	15.84
1"	P2@.299	Ea	12.40	7.37	19.77
1¼"	P2@.398	Ea	21.70	9.81	31.51
1½"	P2@.448	Ea	27.10	11.00	38.10
2"	P2@.498	Ea	34.60	12.30	46.90

Class 125 bronze body swing check valve, solder ends

Description	Craft@Hrs	Unit	Material $	Labor $	Total $
1/2"	P2@.200	Ea	20.50	4.93	25.43
3/4"	P2@.249	Ea	24.90	6.14	31.04
1"	P2@.299	Ea	34.30	7.37	41.67
1¼"	P2@.398	Ea	47.40	9.81	57.21
1½"	P2@.448	Ea	56.30	11.00	67.30
2"	P2@.498	Ea	82.20	12.30	94.50
2½"	P2@.830	Ea	203.00	20.50	223.50
3"	P2@1.24	Ea	300.00	30.60	330.60

Class 125 iron body swing check valve, flanged ends

Description	Craft@Hrs	Unit	Material $	Labor $	Total $
2"	P2@.500	Ea	174.00	12.30	186.30
2½"	P2@.600	Ea	221.00	14.80	235.80
3"	P2@.750	Ea	240.00	18.50	258.50
4"	P2@1.35	Ea	378.00	33.30	411.30

Class 125 iron body silent check valve, wafer type

Description	Craft@Hrs	Unit	Material $	Labor $	Total $
2"	P2@.500	Ea	97.20	12.30	109.50
2½"	P2@.600	Ea	112.00	14.80	126.80
3"	P2@.750	Ea	124.00	18.50	142.50
4"	P2@1.35	Ea	169.00	33.30	202.30

Class 150 bronze body strainer, threaded ends

Description	Craft@Hrs	Unit	Material $	Labor $	Total $
1/2"	P2@.230	Ea	21.30	5.67	26.97
3/4"	P2@.260	Ea	23.20	6.41	29.61
1"	P2@.330	Ea	34.90	8.13	43.03
1¼"	P2@.440	Ea	56.10	10.80	66.90
1½"	P2@.495	Ea	75.20	12.20	87.40
2"	P2@.550	Ea	100.00	13.60	113.60

Class 125 iron body strainer, flanged ends

Description	Craft@Hrs	Unit	Material $	Labor $	Total $
2"	P2@.500	Ea	180.00	12.30	192.30
2½"	P2@.600	Ea	183.00	14.80	197.80
3"	P2@.750	Ea	212.00	18.50	230.50
4"	P2@1.35	Ea	382.00	33.30	415.30

Copper, Type L with Soft-Soldered Joints

Description	Craft@Hrs	Unit	Material $	Labor $	Total $
Installation of 2-way control valve					
1/2"	P2@.210	Ea	--	5.17	5.17
3/4"	P2@.275	Ea	--	6.78	6.78
1"	P2@.350	Ea	--	8.62	8.62
1¼"	P2@.430	Ea	--	10.60	10.60
1½"	P2@.505	Ea	--	12.40	12.40
2"	P2@.675	Ea	--	16.60	16.60
2½"	P2@.830	Ea	--	20.50	20.50
3"	P2@.990	Ea	--	24.40	24.40
4"	P2@1.30	Ea	--	32.00	32.00
Installation of 3-way control valve					
1/2"	P2@.260	Ea	--	6.41	6.41
3/4"	P2@.365	Ea	--	8.99	8.99
1"	P2@.475	Ea	--	11.70	11.70
1¼"	P2@.575	Ea	--	14.20	14.20
1½"	P2@.680	Ea	--	16.80	16.80
2"	P2@.910	Ea	--	22.40	22.40
2½"	P2@1.12	Ea	--	27.60	27.60
3"	P2@1.33	Ea	--	32.80	32.80
4"	P2@2.00	Ea	--	49.30	49.30
Companion flange					
2"	P2@.290	Ea	47.50	7.15	54.65
2½"	P2@.380	Ea	61.80	9.36	71.16
3"	P2@.460	Ea	61.40	11.30	72.70
4"	P2@.600	Ea	91.70	14.80	106.50
Bolt and gasket sets					
2"	P2@.500	Ea	3.85	12.30	16.15
2½"	P2@.650	Ea	4.08	16.00	20.08
3"	P2@.750	Ea	4.28	18.50	22.78
4"	P2@1.00	Ea	7.75	24.60	32.35
Thermometer with well					
7"	P2@.250	Ea	14.30	6.16	20.46
9"	P2@.250	Ea	14.80	6.16	20.96
Pressure gauge					
2½"	P2@.200	Ea	15.60	4.93	20.53
3½"	P2@.200	Ea	27.10	4.93	32.03

Description	Craft@Hrs	Unit	Material $	Labor $	Total $

Hanger with swivel assembly

Description	Craft@Hrs	Unit	Material $	Labor $	Total $
1/2"	P2@.450	Ea	.56	11.10	11.66
3/4"	P2@.450	Ea	.56	11.10	11.66
1"	P2@.450	Ea	.56	11.10	11.66
1¼"	P2@.450	Ea	.56	11.10	11.66
1½"	P2@.450	Ea	.62	11.10	11.72
2"	P2@.450	Ea	.65	11.10	11.75
2½"	P2@.450	Ea	3.95	11.10	15.05
3"	P2@.450	Ea	4.32	11.10	15.42
4"	P2@.450	Ea	5.87	11.10	16.97

Riser clamp

Description	Craft@Hrs	Unit	Material $	Labor $	Total $
1/2"	P2@.100	Ea	3.27	2.46	5.73
3/4"	P2@.100	Ea	3.27	2.46	5.73
1"	P2@.100	Ea	3.30	2.46	5.76
1¼"	P2@.105	Ea	4.01	2.59	6.60
1½"	P2@.110	Ea	4.23	2.71	6.94
2"	P2@.115	Ea	4.45	2.83	7.28
2½"	P2@.120	Ea	4.69	2.96	7.65
3"	P2@.120	Ea	5.13	2.96	8.09
4"	P2@.125	Ea	6.45	3.08	9.53

Copper, Type M with Brazed Joints

Type M hard-drawn copper pipe with wrought copper fittings and brazed joints is used in a wide variety of plumbing and HVAC systems such as heating hot water, chilled water and potable water.

Brazed joints are those made with silver or other alloy filler metals having melting points at, or above, 1,000 degrees F. Maximum working pressure/temperature relationships for brazed joints are approximately as follows:

Maximum Working Pressure (PSIG)*			
Water Temperature	Nominal Pipe Size (inches)		
(degrees F.)	Up to 1	1¼ to 2	2½ to 4
Up to 350	270	190	150

For copper pipe and solder-type fittings using brazing alloys melting at, or above, 1,000 degrees F.

This section has been arranged to save the estimator's time by including all normally-used system components such as pipe, fittings, valves, hanger assemblies, riser clamps and miscellaneous items under one heading. Additional items can be found under "Plumbing and Piping Specialties." The cost estimates in this section are based on the conditions, limitations and wage rates described in the section "How to Use This Book" beginning on page 5.

Description	Craft@Hrs	Unit	Material $	Labor $	Total $

Type M copper pipe with brazed joints

Description	Craft@Hrs	Unit	Material $	Labor $	Total $
1/2"	P2@.032	LF	.70	.79	1.49
3/4"	P2@.035	LF	1.10	.86	1.96
1"	P2@.038	LF	1.58	.94	2.52
1¼"	P2@.042	LF	2.38	1.03	3.41
1½"	P2@.046	LF	3.34	1.13	4.47
2"	P2@.053	LF	5.54	1.31	6.85
2½"	P2@.060	LF	7.66	1.48	9.14
3"	P2@.066	LF	9.74	1.63	11.37
4"	P2@.080	LF	16.50	1.97	18.47

Type M copper 45 degree ell with brazed joints C x C

Description	Craft@Hrs	Unit	Material $	Labor $	Total $
1/2"	P2@.129	Ea	.72	3.18	3.90
3/4"	P2@.181	Ea	1.22	4.46	5.68
1"	P2@.232	Ea	3.08	5.72	8.80
1¼"	P2@.284	Ea	4.20	7.00	11.20
1½"	P2@.335	Ea	5.04	8.25	13.29
2"	P2@.447	Ea	8.49	11.00	19.49
2½"	P2@.550	Ea	18.00	13.60	31.60
3"	P2@.654	Ea	26.70	16.10	42.80
4"	P2@.860	Ea	55.60	21.20	76.80

Description	Craft@Hrs	Unit	Material $	Labor $	Total $

Type M copper 90 degree ell with brazed joints C x C

Description	Craft@Hrs	Unit	Material $	Labor $	Total $
1/2"	P2@.129	Ea	.43	3.18	3.61
3/4"	P2@.181	Ea	.91	4.46	5.37
1"	P2@.232	Ea	2.11	5.72	7.83
1¼"	P2@.284	Ea	3.21	7.00	10.21
1½"	P2@.335	Ea	5.00	8.25	13.25
2"	P2@.447	Ea	9.12	11.00	20.12
2½"	P2@.550	Ea	17.40	13.60	31.00
3"	P2@.654	Ea	24.30	16.10	40.40
4"	P2@.860	Ea	56.00	21.20	77.20

Type M copper 90 degree ell with brazed joints Ftg. x C

Description	Craft@Hrs	Unit	Material $	Labor $	Total $
1/2"	P2@.129	Ea	.60	3.18	3.78
3/4"	P2@.181	Ea	1.33	4.46	5.79
1"	P2@.232	Ea	3.20	5.72	8.92
1¼"	P2@.284	Ea	4.88	7.00	11.88
1½"	P2@.335	Ea	6.40	8.25	14.65
2"	P2@.447	Ea	13.90	11.00	24.90
2½"	P2@.550	Ea	29.10	13.60	42.70
3"	P2@.654	Ea	34.70	16.10	50.80
4"	P2@.860	Ea	80.00	21.20	101.20

Type M copper tee with brazed joints C x C x C

Description	Craft@Hrs	Unit	Material $	Labor $	Total $
1/2"	P2@.156	Ea	.70	3.84	4.54
3/4"	P2@.204	Ea	1.70	5.03	6.73
1"	P2@.263	Ea	4.92	6.48	11.40
1¼"	P2@.322	Ea	7.44	7.93	15.37
1½"	P2@.380	Ea	10.30	9.36	19.66
2"	P2@.507	Ea	16.10	12.50	28.60
2½"	P2@.624	Ea	32.50	15.40	47.90
3"	P2@.741	Ea	49.60	18.30	67.90
4"	P2@.975	Ea	107.00	24.00	131.00

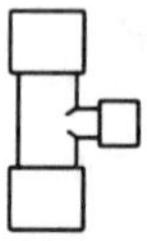

Copper, Type M with Brazed Joints

Description	Craft@Hrs	Unit	Material $	Labor $	Total $
Type M copper reducing branch tee with brazed joints C x C x C					
1/2 x 3/8"	P2@.129	Ea	3.30	3.18	6.48
3/4 x 1/2"	P2@.182	Ea	1.54	4.48	6.02
1 x 1/2"	P2@.209	Ea	4.92	5.15	10.07
1 x 3/4"	P2@.234	Ea	4.92	5.77	10.69
1¼ x 1/2"	P2@.241	Ea	7.24	5.94	13.18
1¼ x 3/4"	P2@.264	Ea	7.24	6.50	13.74
1¼ x 1"	P2@.287	Ea	7.24	7.07	14.31
1½ x 1/2"	P2@.291	Ea	7.68	7.17	14.85
1½ x 3/4"	P2@.307	Ea	7.68	7.56	15.24
1½ x 1"	P2@.323	Ea	7.68	7.96	15.64
1½ x 1¼"	P2@.339	Ea	7.68	8.35	16.03
2 x 1/2"	P2@.348	Ea	12.10	8.57	20.67
2 x 3/4"	P2@.373	Ea	12.10	9.19	21.29
2 x 1"	P2@.399	Ea	12.10	9.83	21.93
2 x 1¼"	P2@.426	Ea	12.10	10.50	22.60
2 x 1½"	P2@.452	Ea	12.10	11.10	23.20
2½ x 1/2"	P2@.468	Ea	37.00	11.50	48.50
2½ x 3/4"	P2@.485	Ea	37.00	12.00	49.00
2½ x 1"	P2@.502	Ea	37.00	12.40	49.40
2½ x 1½"	P2@.537	Ea	37.00	13.20	50.20
2½ x 2"	P2@.555	Ea	37.00	13.70	50.70
3 x 1¼"	P2@.562	Ea	40.80	13.80	54.60
3 x 2"	P2@.626	Ea	40.80	15.40	56.20
3 x 2½"	P2@.660	Ea	40.80	16.30	57.10
4 x 1¼"	P2@.669	Ea	73.20	16.50	89.70
4 x 1½"	P2@.718	Ea	73.20	17.70	90.90
4 x 2"	P2@.768	Ea	73.20	18.90	92.10
4 x 2½"	P2@.819	Ea	73.20	20.20	93.40
4 x 3"	P2@.868	Ea	73.20	21.40	94.60

Description	Craft@Hrs	Unit	Material $	Labor $	Total $

Type M copper reducer with brazed joints

Description	Craft@Hrs	Unit	Material $	Labor $	Total $
1/2 x 3/8"	P2@.148	Ea	.71	3.65	4.36
3/4 x 1/2"	P2@.155	Ea	1.06	3.82	4.88
1 x 3/4"	P2@.206	Ea	1.72	5.08	6.80
1¼ x 1/2"	P2@.206	Ea	2.96	5.08	8.04
1¼ x 3/4"	P2@.232	Ea	2.96	5.72	8.68
1¼ x 1"	P2@.258	Ea	2.96	6.36	9.32
1½ x 1/2"	P2@.232	Ea	3.90	5.72	9.62
1½ x 3/4"	P2@.257	Ea	3.90	6.33	10.23
1½ x 1"	P2@.283	Ea	3.90	6.97	10.87
1½ x 1¼"	P2@.308	Ea	3.90	7.59	11.49
2 x 1/2"	P2@.287	Ea	7.73	7.07	14.80
2 x 3/4"	P2@.277	Ea	7.73	6.83	14.56
2 x 1"	P2@.338	Ea	7.73	8.33	16.06
2 x 1¼"	P2@.365	Ea	7.73	8.99	16.72
2 x 1½"	P2@.390	Ea	7.73	9.61	17.34
2½ x 1"	P2@.390	Ea	14.60	9.61	24.21
2½ x 1¼"	P2@.416	Ea	14.60	10.30	24.90
2½ x 1½"	P2@.442	Ea	14.60	10.90	25.50
2½ x 2"	P2@.497	Ea	14.60	12.20	26.80
3 x 1½"	P2@.493	Ea	18.30	12.10	30.40
3 x 2"	P2@.548	Ea	18.30	13.50	31.80
3 x 2½"	P2@.600	Ea	18.30	14.80	33.10
4 x 2"	P2@.652	Ea	36.60	16.10	52.70
4 x 2½"	P2@.704	Ea	36.60	17.30	53.90
4 x 3"	P2@.755	Ea	36.60	18.60	55.20

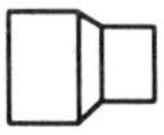

Type M copper adapter with brazed joint C x MPT

Description	Craft@Hrs	Unit	Material $	Labor $	Total $
1/2"	P2@.090	Ea	.87	2.22	3.09
3/4"	P2@.126	Ea	1.45	3.10	4.55
1"	P2@.162	Ea	3.54	3.99	7.53
1¼"	P2@.198	Ea	5.22	4.88	10.10
1½"	P2@.234	Ea	5.99	5.77	11.76
2"	P2@.312	Ea	10.20	7.69	17.89
2½"	P2@.384	Ea	31.20	9.46	40.66
3"	P2@.456	Ea	37.30	11.20	48.50
4"	P2@.600	Ea	63.60	14.80	78.40

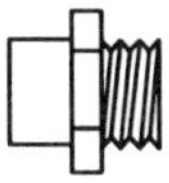

Type M copper adapter with brazed joint C x FPT

Description	Craft@Hrs	Unit	Material $	Labor $	Total $
1/2"	P2@.090	Ea	1.30	2.22	3.52
3/4"	P2@.126	Ea	1.79	3.10	4.89
1"	P2@.162	Ea	3.66	3.99	7.65
1¼"	P2@.198	Ea	5.99	4.88	10.87
1½"	P2@.234	Ea	9.40	5.77	15.17
2"	P2@.312	Ea	12.80	7.69	20.49
2½"	P2@.384	Ea	35.40	9.46	44.86
3"	P2@.456	Ea	51.20	11.20	62.40
4"	P2@.600	Ea	87.10	14.80	101.90

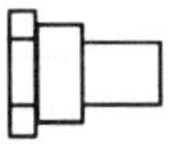

Copper, Type M with Brazed Joints

Description	Craft@Hrs	Unit	Material $	Labor $	Total $
Type M copper flush bushing with brazed joints					
1/2 x 3/8"	P2@.148	Ea	.82	3.65	4.47
3/4 x 1/2"	P2@.155	Ea	1.50	3.82	5.32
1 x 3/4"	P2@.206	Ea	2.32	5.08	7.40
1 x 1/2"	P2@.206	Ea	2.32	5.08	7.40
1¼ x 1"	P2@.258	Ea	2.81	6.36	9.17
1½ x 1¼"	P2@.308	Ea	3.55	7.59	11.14
2 x 1½"	P2@.390	Ea	10.20	9.61	19.81
Type M copper union with brazed joint					
1/2"	P2@.146	Ea	3.82	3.60	7.42
3/4"	P2@.205	Ea	4.80	5.05	9.85
1"	P2@.263	Ea	8.52	6.48	15.00
1¼"	P2@.322	Ea	14.80	7.93	22.73
1½"	P2@.380	Ea	19.40	9.36	28.76
2"	P2@.507	Ea	33.10	12.50	45.60
Type M copper dielectric union with brazed joint					
1/2"	P2@.146	Ea	3.55	3.60	7.15
3/4"	P2@.205	Ea	3.55	5.05	8.60
1"	P2@.263	Ea	6.21	6.48	12.69
1¼"	P2@.322	Ea	10.20	7.93	18.13
1½"	P2@.380	Ea	15.00	9.36	24.36
2"	P2@.507	Ea	21.00	12.50	33.50
Type M copper cap with brazed joint					
1/2"	P2@.083	Ea	.29	2.05	2.34
3/4"	P2@.116	Ea	.50	2.86	3.36
1"	P2@.149	Ea	1.18	3.67	4.85
1¼"	P2@.182	Ea	1.68	4.48	6.16
1½"	P2@.215	Ea	2.46	5.30	7.76
2"	P2@.286	Ea	4.50	7.05	11.55
2½"	P2@.352	Ea	13.90	8.67	22.57
3"	P2@.418	Ea	19.00	10.30	29.30
4"	P2@.550	Ea	37.30	13.60	50.90
Type M copper coupling with brazed joints					
1/2"	P2@.129	Ea	.32	3.18	3.50
3/4"	P2@.181	Ea	.61	4.46	5.07
1"	P2@.232	Ea	1.60	5.72	7.32
1¼"	P2@.284	Ea	2.11	7.00	9.11
1½"	P2@.335	Ea	2.79	8.25	11.04
2"	P2@.447	Ea	4.62	11.00	15.62
2½"	P2@.550	Ea	9.84	13.60	23.44
3"	P2@.654	Ea	15.10	16.10	31.20
4"	P2@.860	Ea	29.60	21.20	50.80

Description	Craft@Hrs	Unit	Material $	Labor $	Total $

Class 125 bronze body gate valve, solder ends

Description	Craft@Hrs	Unit	Material $	Labor $	Total $
1/2"	P2@.240	Ea	12.20	5.91	18.11
3/4"	P2@.300	Ea	14.70	7.39	22.09
1"	P2@.360	Ea	18.30	8.87	27.17
1¼"	P2@.480	Ea	31.30	11.80	43.10
1½"	P2@.540	Ea	34.50	13.30	47.80
2"	P2@.600	Ea	48.40	14.80	63.20
2½"	P2@1.00	Ea	113.00	24.60	137.60
3"	P2@1.50	Ea	161.00	37.00	198.00

Class 125 iron body gate valve, flanged ends

Description	Craft@Hrs	Unit	Material $	Labor $	Total $
2"	P2@.500	Ea	164.00	12.30	176.30
2½"	P2@.600	Ea	168.00	14.80	182.80
3"	P2@.750	Ea	188.00	18.50	206.50
4"	P2@1.35	Ea	270.00	33.30	303.30
5"	P2@2.00	Ea	460.00	49.30	509.30
6"	P2@2.50	Ea	460.00	61.60	521.60

Class 125 bronze body globe valve, solder ends

Description	Craft@Hrs	Unit	Material $	Labor $	Total $
1/2"	P2@.240	Ea	18.30	5.91	24.21
3/4"	P2@.300	Ea	23.30	7.39	30.69
1"	P2@.360	Ea	33.80	8.87	42.67
1¼"	P2@.480	Ea	41.30	11.80	53.10
1½"	P2@.540	Ea	53.30	13.30	66.60
2"	P2@.600	Ea	81.30	14.80	96.10

Class 125 iron body globe valve, flanged ends

Description	Craft@Hrs	Unit	Material $	Labor $	Total $
2½"	P2@.600	Ea	195.00	14.80	209.80
3"	P2@.750	Ea	229.00	18.50	247.50
4"	P2@1.35	Ea	328.00	33.30	361.30

200 PSI iron body butterfly valve, lug type

Description	Craft@Hrs	Unit	Material $	Labor $	Total $
2"	P2@.200	Ea	77.60	4.93	82.53
2½"	P2@.300	Ea	81.00	7.39	88.39
3"	P2@.400	Ea	89.80	9.86	99.66
4"	P2@.500	Ea	120.00	12.30	132.30

200 PSI iron body butterfly valve, wafer type

Description	Craft@Hrs	Unit	Material $	Labor $	Total $
2"	P2@.200	Ea	66.80	4.93	71.73
2½"	P2@.300	Ea	69.50	7.39	76.89
3"	P2@.400	Ea	76.30	9.86	86.16
4"	P2@.500	Ea	95.20	12.30	107.50

Copper, Type M with Brazed Joints

Description	Craft@Hrs	Unit	Material $	Labor $	Total $
Class 125 bronze body 2-piece ball valve, solder					
1/2"	P2@.240	Ea	6.08	5.91	11.99
3/4"	P2@.300	Ea	9.70	7.39	17.09
1"	P2@.360	Ea	12.40	8.87	21.27
1¼"	P2@.480	Ea	21.70	11.80	33.50
1½"	P2@.540	Ea	27.10	13.30	40.40
2"	P2@.600	Ea	34.60	14.80	49.40
Class 125 bronze body swing check valve, solder ends					
1/2"	P2@.240	Ea	20.50	5.91	26.41
3/4"	P2@.300	Ea	24.90	7.39	32.29
1"	P2@.360	Ea	34.30	8.87	43.17
1¼"	P2@.480	Ea	47.40	11.80	59.20
1½"	P2@.540	Ea	56.30	13.30	69.60
2"	P2@.600	Ea	82.20	14.80	97.00
2½"	P2@1.00	Ea	203.00	24.60	227.60
3"	P2@1.50	Ea	300.00	37.00	337.00
Class 125 iron body swing check valve, flanged ends					
2"	P2@.500	Ea	174.00	12.30	186.30
2½"	P2@.600	Ea	221.00	14.80	235.80
3"	P2@.750	Ea	240.00	18.50	258.50
4"	P2@1.35	Ea	378.00	33.30	411.30
Class 125 iron body silent check valve, wafer type					
2"	P2@.500	Ea	97.20	12.30	109.50
2½"	P2@.600	Ea	112.00	14.80	126.80
3"	P2@.750	Ea	124.00	18.50	142.50
4"	P2@1.35	Ea	167.00	33.30	200.30
Class 150 bronze body strainer, threaded ends					
1/2"	P2@.230	Ea	21.30	5.67	26.97
3/4"	P2@.260	Ea	23.20	6.41	29.61
1"	P2@.330	Ea	34.90	8.13	43.03
1¼"	P2@.440	Ea	56.10	10.80	66.90
1½"	P2@.495	Ea	75.20	12.20	87.40
2"	P2@.550	Ea	100.00	13.60	113.60
Class 125 iron body strainer, flanged ends					
2"	P2@.500	Ea	180.00	12.30	192.30
2½"	P2@.600	Ea	183.00	14.80	197.80
3"	P2@.750	Ea	212.00	18.50	230.50
4"	P2@1.35	Ea	382.00	33.30	415.30

Description	Craft@Hrs	Unit	Material $	Labor $	Total $

Installation of 2-way control valve

Description	Craft@Hrs	Unit	Material $	Labor $	Total $
1/2"	P2@.210	Ea	--	5.17	5.17
3/4"	P2@.275	Ea	--	6.78	6.78
1"	P2@.350	Ea	--	8.62	8.62
1¼"	P2@.430	Ea	--	10.60	10.60
1½"	P2@.505	Ea	--	12.40	12.40
2"	P2@.675	Ea	--	16.60	16.60
2½"	P2@.830	Ea	--	20.50	20.50
3"	P2@.990	Ea	--	24.40	24.40
4"	P2@1.30	Ea	--	32.00	32.00

Installation of 3-way control valve

Description	Craft@Hrs	Unit	Material $	Labor $	Total $
1/2"	P2@.260	Ea	--	6.41	6.41
3/4"	P2@.365	Ea	--	8.99	8.99
1"	P2@.475	Ea	--	11.70	11.70
1¼"	P2@.575	Ea	--	14.20	14.20
1½"	P2@.680	Ea	--	16.80	16.80
2"	P2@.910	Ea	--	22.40	22.40
2½"	P2@1.12	Ea	--	27.60	27.60
3"	P2@1.33	Ea	--	32.80	32.80
4"	P2@2.00	Ea	--	49.30	49.30

Companion flange

Description	Craft@Hrs	Unit	Material $	Labor $	Total $
2"	P2@.290	Ea	47.50	7.15	54.65
2½"	P2@.380	Ea	61.80	9.36	71.16
3"	P2@.460	Ea	61.40	11.30	72.70
4"	P2@.600	Ea	91.70	14.80	106.50

Bolt and gasket sets

Description	Craft@Hrs	Unit	Material $	Labor $	Total $
2"	P2@.500	Ea	3.85	12.30	16.15
2½"	P2@.650	Ea	4.08	16.00	20.08
3"	P2@.750	Ea	4.28	18.50	22.78
4"	P2@1.00	Ea	7.75	24.60	32.35

Thermometer with well

Description	Craft@Hrs	Unit	Material $	Labor $	Total $
7"	P2@.250	Ea	14.30	6.16	20.46
9"	P2@.250	Ea	14.80	6.16	20.96

Pressure gauge

Description	Craft@Hrs	Unit	Material $	Labor $	Total $
2½"	P2@.200	Ea	15.60	4.93	20.53
3½"	P2@.200	Ea	27.10	4.93	32.03

Copper, Type M with Brazed Joints

Description	Craft@Hrs	Unit	Material $	Labor $	Total $
Hanger with swivel assembly					
1/2"	P2@.450	Ea	.56	11.10	11.66
3/4"	P2@.450	Ea	.56	11.10	11.66
1"	P2@.450	Ea	.56	11.10	11.66
1¼"	P2@.450	Ea	.56	11.10	11.66
1½"	P2@.450	Ea	.62	11.10	11.72
2"	P2@.450	Ea	.65	11.10	11.75
2½"	P2@.450	Ea	3.95	11.10	15.05
3"	P2@.450	Ea	4.32	11.10	15.42
4"	P2@.450	Ea	5.87	11.10	16.97
Riser clamp					
1/2"	P2@.100	Ea	3.27	2.46	5.73
3/4"	P2@.100	Ea	3.27	2.46	5.73
1"	P2@.100	Ea	3.30	2.46	5.76
1¼"	P2@.105	Ea	4.01	2.59	6.60
1½"	P2@.110	Ea	4.23	2.71	6.94
2"	P2@.115	Ea	4.45	2.83	7.28
2½"	P2@.120	Ea	4.69	2.96	7.65
3"	P2@.120	Ea	5.13	2.96	8.09
4"	P2@.125	Ea	6.45	3.08	9.53

Type M hard-drawn copper pipe with wrought copper fittings and soft-soldered joints is used in a wide variety of plumbing and HVAC systems such as heating hot water, chilled water, potable water and A.C. condensate.

Soft-soldered joints are those made with solders having melting points in the 350 degree F. to 500 degree F. range. Maximum working pressure/temperature relationships for soft-soldered joints are approximately as follows:

Maximum Working Pressures (PSIG)*				
Soft-solder Type	Water Temperature (degrees F.)	Nominal Pipe Size (inches)		
		Up to 1	1¼ to 2	2½ x 4
50-50 tin-lead**	100	200	175	150
	150	150	125	100
	200	100	90	75
	250	85	75	50
95-5 tin-antimony	100	500	400	300
	150	400	350	275
	200	300	250	200
	250	200	175	150

*For copper pipe and solder-type fittings using soft-solders melting at approximately 350 degrees F. to 500 degrees F.

The use of any solder containing lead is **not allowed in potable water systems.

This section has been arranged to save the estimator's time by including all normally-used system components such as pipe, fittings, valves, hanger assemblies, riser clamps and miscellaneous items under one heading. Additional items can be found under "Plumbing and Piping Specialties." The cost estimates in this section are based on the conditions, limitations and wage rates described in the section "How to Use This Book" beginning on page 5.

Description	Craft@Hrs	Unit	Material $	Labor $	Total $

Type M copper pipe with soft soldered joints

Description	Craft@Hrs	Unit	Material $	Labor $	Total $
1/2"	P2@.032	LF	.70	.79	1.49
3/4"	P2@.035	LF	1.10	.86	1.96
1"	P2@.038	LF	1.58	.94	2.52
1¼"	P2@.042	LF	2.38	1.03	3.41
1½"	P2@.046	LF	3.34	1.13	4.47
2"	P2@.053	LF	5.54	1.31	6.85
2½"	P2@.060	LF	7.76	1.48	9.24
3"	P2@.066	LF	9.74	1.63	11.37
4"	P2@.080	LF	16.50	1.97	18.47

Copper, Type M with Soft-Soldered Joints

Description	Craft@Hrs	Unit	Material $	Labor $	Total $

Type M copper 45 degree ell with soft soldered joints C x C

Description	Craft@Hrs	Unit	Material $	Labor $	Total $
1/2"	P2@.107	Ea	.72	2.64	3.36
3/4"	P2@.150	Ea	1.22	3.70	4.92
1"	P2@.193	Ea	3.08	4.76	7.84
1¼"	P2@.236	Ea	4.20	5.82	10.02
1½"	P2@.278	Ea	5.04	6.85	11.89
2"	P2@.371	Ea	8.44	9.14	17.58
2½"	P2@.457	Ea	18.00	11.30	29.30
3"	P2@.543	Ea	26.70	13.40	40.10
4"	P2@.714	Ea	55.60	17.60	73.20

Type M copper 90 degree ell with soft soldered joints C x C

Description	Craft@Hrs	Unit	Material $	Labor $	Total $
1/2"	P2@.107	Ea	.43	2.64	3.07
3/4"	P2@.150	Ea	.91	3.70	4.61
1"	P2@.193	Ea	2.11	4.76	6.87
1¼"	P2@.236	Ea	3.21	5.82	9.03
1½"	P2@.278	Ea	5.00	6.85	11.85
2"	P2@.371	Ea	9.12	9.14	18.26
2½"	P2@.457	Ea	17.40	11.30	28.70
3"	P2@.543	Ea	24.30	13.40	37.70
4"	P2@.714	Ea	56.00	17.60	73.60

Type M copper 90 degree ell with soft soldered joint Ftg. x C

Description	Craft@Hrs	Unit	Material $	Labor $	Total $
1/2"	P2@.107	Ea	.60	2.64	3.24
3/4"	P2@.150	Ea	1.33	3.70	5.03
1"	P2@.193	Ea	3.20	4.76	7.96
1¼"	P2@.236	Ea	4.88	5.82	10.70
1½"	P2@.278	Ea	6.40	6.85	13.25
2"	P2@.371	Ea	13.90	9.14	23.04
2½"	P2@.457	Ea	29.10	11.30	40.40
3"	P2@.543	Ea	34.70	13.40	48.10
4"	P2@.714	Ea	80.00	17.60	97.60

Type M copper tee with soft soldered joints C x C x C

Description	Craft@Hrs	Unit	Material $	Labor $	Total $
1/2"	P2@.129	Ea	.70	3.18	3.88
3/4"	P2@.181	Ea	1.70	4.46	6.16
1"	P2@.233	Ea	4.92	5.74	10.66
1¼"	P2@.285	Ea	7.44	7.02	14.46
1½"	P2@.337	Ea	10.30	8.30	18.60
2"	P2@.449	Ea	16.10	11.10	27.20
2½"	P2@.552	Ea	32.50	13.60	46.10
3"	P2@.656	Ea	49.60	16.20	65.80
4"	P2@.863	Ea	107.00	21.30	128.30

Description	Craft@Hrs	Unit	Material $	Labor $	Total $

Type M copper branch reducing tee with soft soldered joints CxCxC

Description	Craft@Hrs	Unit	Material $	Labor $	Total $
1/2 x 3/8"	P2@.121	Ea	3.30	2.98	6.28
3/4 x 1/2"	P2@.170	Ea	1.54	4.19	5.73
1 x 1/2"	P2@.195	Ea	4.92	4.80	9.72
1 x 3/4"	P2@.219	Ea	4.92	5.40	10.32
1¼ x 1/2"	P2@.225	Ea	7.24	5.54	12.78
1¼ x 3/4"	P2@.247	Ea	7.24	6.09	13.33
1¼ x 1"	P2@.268	Ea	7.24	6.60	13.84
1½ x 1/2"	P2@.272	Ea	7.68	6.70	14.38
1½ x 3/4"	P2@.287	Ea	7.68	7.07	14.75
1½ x 1"	P2@.302	Ea	7.68	7.44	15.12
1½ x 1¼"	P2@.317	Ea	7.68	7.81	15.49
2 x 1/2"	P2@.325	Ea	12.10	8.01	20.11
2 x 3/4"	P2@.348	Ea	12.10	8.57	20.67
2 x 1"	P2@.373	Ea	12.10	9.19	21.29
2 x 1¼"	P2@.398	Ea	12.10	9.81	21.91
2 x 1½"	P2@.422	Ea	12.10	10.40	22.50
2½ x 1/2"	P2@.437	Ea	37.00	10.80	47.80
2½ x 3/4"	P2@.453	Ea	37.00	11.20	48.20
2½ x 1"	P2@.469	Ea	37.00	11.60	48.60
2½ x 1½"	P2@.502	Ea	37.00	12.40	49.40
2½ x 2"	P2@.519	Ea	37.00	12.80	49.80
3 x 1¼"	P2@.525	Ea	40.80	12.90	53.70
3 x 2"	P2@.585	Ea	40.80	14.40	55.20
3 x 2½"	P2@.617	Ea	40.80	15.20	56.00
4 x 1¼"	P2@.625	Ea	73.20	15.40	88.60
4 x 1½"	P2@.671	Ea	73.20	16.50	89.70
4 x 2"	P2@.718	Ea	73.20	17.70	90.90
4 x 2½"	P2@.765	Ea	73.20	18.80	92.00
4 x 3"	P2@.811	Ea	73.20	20.00	93.20

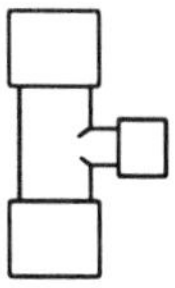

Copper, Type M with Soft-Soldered Joints

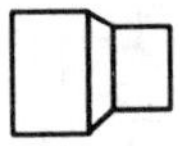

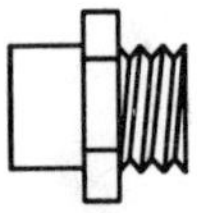

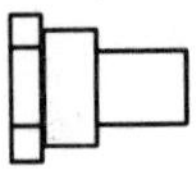

Description	Craft@Hrs	Unit	Material $	Labor $	Total $
Type M copper reducer with soft soldered joints					
1/2 x 3/8"	P2@.124	Ea	.71	3.06	3.77
3/4 x 1/2"	P2@.129	Ea	1.06	3.18	4.24
1 x 3/4"	P2@.172	Ea	1.72	4.24	5.96
1¼ x 1/2"	P2@.172	Ea	2.96	4.24	7.20
1¼ x 3/4"	P2@.193	Ea	2.96	4.76	7.72
1¼ x 1"	P2@.215	Ea	2.96	5.30	8.26
1½ x 1/2"	P2@.193	Ea	3.90	4.76	8.66
1½ x 3/4"	P2@.214	Ea	3.90	5.27	9.17
1½ x 1"	P2@.236	Ea	3.90	5.82	9.72
1½ x 1¼"	P2@.257	Ea	3.90	6.33	10.23
2 x 1/2"	P2@.239	Ea	7.33	5.89	13.22
2 x 3/4"	P2@.261	Ea	7.33	6.43	13.76
2 x 1"	P2@.282	Ea	7.33	6.95	14.28
2 x 1¼"	P2@.304	Ea	7.33	7.49	14.82
2 x 1½"	P2@.325	Ea	7.33	8.01	15.34
2½ x 1"	P2@.325	Ea	14.60	8.01	22.61
2½ x 1¼"	P2@.347	Ea	14.60	8.55	23.15
2½ x 1½"	P2@.368	Ea	14.60	9.07	23.67
2½ x 2"	P2@.414	Ea	14.60	10.20	24.80
3 x 1¼"	P2@.390	Ea	18.30	9.61	27.91
3 x 1½"	P2@.411	Ea	18.30	10.10	28.40
3 x 2"	P2@.457	Ea	18.30	11.30	29.60
3 x 2½"	P2@.500	Ea	18.30	12.30	30.60
4 x 2"	P2@.543	Ea	36.60	13.40	50.00
4 x 2½"	P2@.586	Ea	36.60	14.40	51.00
4 x 3"	P2@.629	Ea	36.60	15.50	52.10
Type M copper adapter with soft soldered joint C x MPT					
1/2"	P2@.075	Ea	.87	1.85	2.72
3/4"	P2@.105	Ea	1.45	2.59	4.04
1"	P2@.134	Ea	3.54	3.30	6.84
1¼"	P2@.164	Ea	5.22	4.04	9.26
1½"	P2@.194	Ea	5.99	4.78	10.77
2"	P2@.259	Ea	10.20	6.38	16.58
2½"	P2@.319	Ea	31.20	7.86	39.06
3"	P2@.378	Ea	37.30	9.31	46.61
4"	P2@.498	Ea	63.60	12.30	75.90
Type M copper adapter with soft soldered joint C x FPT					
1/2"	P2@.075	Ea	1.30	1.85	3.15
3/4"	P2@.105	Ea	1.79	2.59	4.38
1"	P2@.134	Ea	3.66	3.30	6.96
1¼"	P2@.164	Ea	5.99	4.04	10.03
1½"	P2@.194	Ea	9.40	4.78	14.18
2"	P2@.259	Ea	12.80	6.38	19.18
2½"	P2@.319	Ea	35.40	7.86	43.26
3"	P2@.378	Ea	51.20	9.31	60.51
4"	P2@.498	Ea	87.10	12.30	99.40

Description	Craft@Hrs	Unit	Material $	Labor $	Total $

Type M copper flush bushing with soft soldered joints

Description	Craft@Hrs	Unit	Material $	Labor $	Total $
1/2 x 3/8"	P2@.124	Ea	.82	3.06	3.88
3/4 x 1/2"	P2@.129	Ea	1.50	3.18	4.68
1　x 3/4"	P2@.172	Ea	2.32	4.24	6.56
1　x 1/2"	P2@.172	Ea	2.32	4.24	6.56
1¼ x 1"	P2@.215	Ea	2.81	5.30	8.11
1½ x 1¼"	P2@.257	Ea	3.55	6.33	9.88
2　x 1½"	P2@.325	Ea	10.20	8.01	18.21

Type M copper union with soft soldered joint

Description	Craft@Hrs	Unit	Material $	Labor $	Total $
1/2"	P2@.121	Ea	3.82	2.98	6.80
3/4"	P2@.170	Ea	4.80	4.19	8.99
1"	P2@.218	Ea	8.52	5.37	13.89
1¼"	P2@.267	Ea	14.80	6.58	21.38
1½"	P2@.315	Ea	19.40	7.76	27.16
2"	P2@.421	Ea	33.10	10.40	43.50

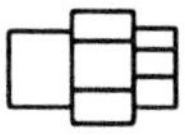

Type M copper dielectric union with soft soldered joint

Description	Craft@Hrs	Unit	Material $	Labor $	Total $
1/2"	P2@.121	Ea	3.39	2.98	6.37
3/4"	P2@.170	Ea	3.39	4.19	7.58
1"	P2@.218	Ea	7.76	5.37	13.13
1¼"	P2@.267	Ea	12.70	6.58	19.28
1½"	P2@.315	Ea	18.60	7.76	26.36
2"	P2@.421	Ea	26.30	10.40	36.70

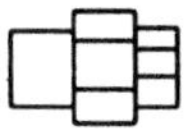

Type M copper cap with soft soldered joint

Description	Craft@Hrs	Unit	Material $	Labor $	Total $
1/2"	P2@.069	Ea	.29	1.70	1.99
3/4"	P2@.096	Ea	.50	2.37	2.87
1"	P2@.124	Ea	1.18	3.06	4.24
1¼"	P2@.151	Ea	1.68	3.72	5.40
1½"	P2@.178	Ea	2.46	4.39	6.85
2"	P2@.237	Ea	4.50	5.84	10.34
2½"	P2@.292	Ea	13.90	7.19	21.09
3"	P2@.347	Ea	19.00	8.55	27.55
4"	P2@.457	Ea	37.30	11.30	48.60

Type M copper coupling with soft soldered joints

Description	Craft@Hrs	Unit	Material $	Labor $	Total $
1/2"	P2@.107	Ea	.32	2.64	2.96
3/4"	P2@.150	Ea	.61	3.70	4.31
1"	P2@.193	Ea	1.60	4.76	6.36
1¼"	P2@.236	Ea	2.11	5.82	7.93
1½"	P2@.278	Ea	2.79	6.85	9.64
2"	P2@.371	Ea	4.62	9.14	13.76
2½"	P2@.457	Ea	9.84	11.30	21.14
3"	P2@.543	Ea	15.10	13.40	28.50
4"	P2@.714	Ea	29.60	17.60	47.20

Copper, Type M with Soft-Soldered Joints

Description	Craft@Hrs	Unit	Material $	Labor $	Total $
Class 125 bronze body gate valve, solder ends					
1/2"	P2@.200	Ea	12.20	4.93	17.03
3/4"	P2@.249	Ea	14.70	6.14	20.84
1"	P2@.299	Ea	18.30	7.37	25.67
1¼"	P2@.398	Ea	31.30	9.81	41.11
1½"	P2@.448	Ea	34.50	11.00	45.50
2"	P2@.498	Ea	48.60	12.30	60.90
2½"	P2@.830	Ea	113.00	20.50	133.50
3"	P2@1.24	Ea	161.00	30.60	191.60
Class 125 iron body gate valve, flanged ends					
2"	P2@.500	Ea	164.00	12.30	176.30
2½"	P2@.600	Ea	168.00	14.80	182.80
3"	P2@.750	Ea	188.00	18.50	206.50
4"	P2@1.35	Ea	270.00	33.30	303.30
5"	P2@2.00	Ea	460.00	49.30	509.30
6"	P2@2.50	Ea	460.00	61.60	521.60
Class 125 bronze body globe valve, solder ends					
1/2"	P2@.200	Ea	18.30	4.93	23.23
3/4"	P2@.249	Ea	23.30	6.14	29.44
1"	P2@.299	Ea	33.30	7.37	40.67
1¼"	P2@.398	Ea	41.30	9.81	51.11
1½"	P2@.448	Ea	53.30	11.00	64.30
2"	P2@.498	Ea	81.30	12.30	93.60
Class 125 iron body globe valve, flanged ends					
2½"	P2@.600	Ea	195.00	14.80	209.80
3"	P2@.750	Ea	229.00	18.50	247.50
4"	P2@1.35	Ea	328.00	33.30	361.30
200 PSI iron body butterfly valve, lug type					
2"	P2@.200	Ea	77.60	4.93	82.53
2½"	P2@.300	Ea	81.00	7.39	88.39
3"	P2@.400	Ea	89.80	9.86	99.66
4"	P2@.500	Ea	120.00	12.30	132.30
200 PSI iron body butterfly valve, wafer type					
2"	P2@.200	Ea	66.80	4.93	71.73
2½"	P2@.300	Ea	69.50	7.39	76.89
3"	P2@.400	Ea	76.30	9.86	86.16
4"	P2@.500	Ea	95.20	12.30	107.50

Description	Craft@Hrs	Unit	Material $	Labor $	Total $

Class 125 bronze body 2-piece ball valve, solder ends

1/2"	P2@.200	Ea	6.08	4.93	11.01
3/4"	P2@.249	Ea	9.70	6.14	15.84
1"	P2@.299	Ea	12.40	7.37	19.77
1¼"	P2@.398	Ea	21.70	9.81	31.51
1½"	P2@.448	Ea	27.10	11.00	38.10
2"	P2@.498	Ea	34.60	12.30	46.90

Class 125 bronze body swing check valve, solder ends

1/2"	P2@.200	Ea	20.50	4.93	25.43
3/4"	P2@.249	Ea	24.90	6.14	31.04
1"	P2@.299	Ea	34.30	7.37	41.67
1¼"	P2@.398	Ea	47.40	9.81	57.21
1½"	P2@.448	Ea	56.30	11.00	67.30
2"	P2@.498	Ea	82.20	12.30	94.50
2½"	P2@.830	Ea	203.00	20.50	223.50
3"	P2@1.24	Ea	300.00	30.60	330.60

Class 125 iron body swing check valve, flanged ends

2"	P2@.500	Ea	174.00	12.30	186.30
2½"	P2@.600	Ea	221.00	14.80	235.80
3"	P2@.750	Ea	240.00	18.50	258.50
4"	P2@1.35	Ea	378.00	33.30	411.30

Class 125 iron body silent check valve, wafer type

2"	P2@.500	Ea	97.20	12.30	109.50
2½"	P2@.600	Ea	112.00	14.80	126.80
3"	P2@.750	Ea	124.00	18.50	142.50
4"	P2@1.35	Ea	167.00	33.30	200.30

Class 150 bronze body strainer, threaded ends

1/2"	P2@.230	Ea	21.30	5.67	26.97
3/4"	P2@.260	Ea	23.20	6.41	29.61
1"	P2@.330	Ea	34.90	8.13	43.03
1¼"	P2@.440	Ea	56.10	10.80	66.90
1½"	P2@.495	Ea	75.20	12.20	87.40
2"	P2@.550	Ea	100.00	13.60	113.60

Class 125 iron body strainer, flanged ends

2"	P2@.500	Ea	180.00	12.30	192.30
2½"	P2@.600	Ea	183.00	14.80	197.80
3"	P2@.750	Ea	212.00	18.50	230.50
4"	P2@1.35	Ea	382.00	33.30	415.30

Copper, Type M with Soft-Soldered Joints

Description	Craft@Hrs	Unit	Material $	Labor $	Total $
Installation of 2-way control valve					
1/2"	P2@.210	Ea	--	5.17	5.17
3/4"	P2@.275	Ea	--	6.78	6.78
1"	P2@.350	Ea	--	8.62	8.62
1¼"	P2@.430	Ea	--	10.60	10.60
1½"	P2@.505	Ea	--	12.40	12.40
2"	P2@.675	Ea	--	16.60	16.60
2½"	P2@.830	Ea	--	20.50	20.50
3"	P2@.990	Ea	--	24.40	24.40
4"	P2@1.30	Ea	--	32.00	32.00
Installation of 3-way control valve					
1/2"	P2@.260	Ea	--	6.41	6.41
3/4"	P2@.365	Ea	--	8.99	8.99
1"	P2@.475	Ea	--	11.70	11.70
1¼"	P2@.575	Ea	--	14.20	14.20
1½"	P2@.680	Ea	--	16.80	16.80
2"	P2@.910	Ea	--	22.40	22.40
2½"	P2@1.12	Ea	--	27.60	27.60
3"	P2@1.33	Ea	--	32.80	32.80
4"	P2@2.00	Ea	--	49.30	49.30
Companion flange					
2"	P2@.290	Ea	47.50	7.15	54.65
2½"	P2@.380	Ea	61.80	9.36	71.16
3"	P2@.460	Ea	61.40	11.30	72.70
4"	P2@.600	Ea	91.70	14.80	106.50
Bolt and gasket sets					
2"	P2@.500	Ea	3.85	12.30	16.15
2½"	P2@.650	Ea	4.08	16.00	20.08
3"	P2@.750	Ea	4.28	18.50	22.78
4"	P2@1.00	Ea	7.75	24.60	32.35
Thermometer with well					
7"	P2@.250	Ea	14.30	6.16	20.46
9"	P2@.250	Ea	14.80	6.16	20.96
Pressure gauge					
2½"	P2@.200	Ea	15.60	4.93	20.53
3½"	P2@.200	Ea	27.10	4.93	32.03

Description	Craft@Hrs	Unit	Material $	Labor $	Total $
Hanger with swivel assembly					
1/2"	P2@.450	Ea	.56	11.10	11.66
3/4"	P2@.450	Ea	.56	11.10	11.66
1"	P2@.450	Ea	.56	11.10	11.66
1¼"	P2@.450	Ea	.56	11.10	11.66
1½"	P2@.450	Ea	.62	11.10	11.72
2"	P2@.450	Ea	.65	11.10	11.75
2½"	P2@.450	Ea	3.95	11.10	15.05
3"	P2@.450	Ea	4.32	11.10	15.42
4"	P2@.450	Ea	5.87	11.10	16.97
Riser clamp					
1/2"	P2@.100	Ea	3.27	2.46	5.73
3/4"	P2@.100	Ea	3.27	2.46	5.73
1"	P2@.100	Ea	3.30	2.46	5.76
1¼"	P2@.105	Ea	4.01	2.59	6.60
1½"	P2@.110	Ea	4.23	2.71	6.94
2"	P2@.115	Ea	4.45	2.83	7.28
2½"	P2@.120	Ea	4.69	2.96	7.65
3"	P2@.120	Ea	5.13	2.96	8.09
4"	P2@.125	Ea	6.45	3.08	9.53

Ductile Iron, Class 153, Cement-Lined with Mechanical Joints

Description	Craft@Hrs	Unit	Material $	Labor $	Total $

Cement-lined Class 153 ductile iron pipe with mechanical joints

Description	Craft@Hrs	Unit	Material $	Labor $	Total $
4"	P2@.140	LF	5.83	3.45	9.28
6"	P2@.180	LF	8.36	4.44	12.80
8"	P2@.200	LF	11.60	4.93	16.53
10"	P2@.230	LF	17.20	5.67	22.87
12"	P2@.270	LF	23.40	6.65	30.05
14"	P2@.310	LF	35.60	7.64	43.24
16"	P2@.350	LF	45.80	8.62	54.42

Cement-lined Class 153 ductile iron 1/16 bend with mechanical joints

Description	Craft@Hrs	Unit	Material $	Labor $	Total $
4"	P2@.900	Ea	23.70	22.20	45.90
6"	P2@1.30	Ea	36.30	32.00	68.30
8"	P2@1.70	Ea	57.20	41.90	99.10
10"	P2@2.15	Ea	78.10	53.00	131.10
12"	P2@2.40	Ea	107.00	59.10	166.10
14"	P2@2.70	Ea	238.00	66.50	304.50
16"	P2@3.00	Ea	268.00	73.90	341.90

Cement-lined Class 153 ductile iron 1/8 bend with mechanical joints

Description	Craft@Hrs	Unit	Material $	Labor $	Total $
4"	P2@.900	Ea	23.70	22.20	45.90
6"	P2@1.30	Ea	35.20	32.00	67.20
8"	P2@1.70	Ea	56.10	41.90	98.00
10"	P2@2.15	Ea	77.00	53.00	130.00
12"	P2@2.40	Ea	106.00	59.10	165.10
14"	P2@2.70	Ea	230.00	66.50	296.50
16"	P2@3.00	Ea	268.00	73.90	341.90

Cement-lined Class 153 ductile iron 1/4 bend with mechanical joints

Description	Craft@Hrs	Unit	Material $	Labor $	Total $
4"	P2@.900	Ea	25.30	22.20	47.50
6"	P2@1.30	Ea	41.30	32.00	73.30
8"	P2@1.70	Ea	86.40	41.90	128.30
10"	P2@2.15	Ea	129.00	53.00	182.00
12"	P2@2.40	Ea	177.00	59.10	236.10
14"	P2@2.70	Ea	455.00	66.50	521.50
16"	P2@3.00	Ea	478.00	73.90	551.90

Cement-lined Class 153 ductile iron tee with mechanical joints

Description	Craft@Hrs	Unit	Material $	Labor $	Total $
4"	P2@1.65	Ea	38.50	40.70	79.20
6"	P2@2.35	Ea	56.10	57.90	114.00
8"	P2@3.15	Ea	86.40	77.60	164.00
10"	P2@4.00	Ea	129.00	98.60	227.60
12"	P2@4.45	Ea	177.00	110.00	287.00
14"	P2@4.90	Ea	455.00	121.00	576.00
16"	P2@5.30	Ea	478.00	131.00	609.00

Ductile Iron, Class 153, Cement-Lined with Mechanical Joints

Description	Craft@Hrs	Unit	Material $	Labor $	Total $
Cement-lined Class 153 ductile iron wye with mechanical joints					
4"	P2@1.65	Ea	86.40	40.70	127.10
6"	P2@2.35	Ea	107.00	57.90	164.90
8"	P2@3.15	Ea	160.00	77.60	237.60
10"	P2@4.00	Ea	301.00	98.60	399.60
12"	P2@4.45	Ea	452.00	110.00	562.00
14"	P2@4.90	Ea	927.00	121.00	1,048.00
16"	P2@5.30	Ea	1,160.00	131.00	1,291.00
Cement-lined Class 153 ductile iron reducer with mechanical joints					
6"	P2@1.45	Ea	28.10	35.70	63.80
8"	P2@2.00	Ea	44.40	49.30	93.70
10"	P2@2.55	Ea	58.90	62.80	121.70
12"	P2@3.05	Ea	75.90	75.20	151.10
14"	P2@3.60	Ea	170.00	88.70	258.70
16"	P2@4.20	Ea	207.00	103.00	310.00

Ductile Iron, Class 153, Double Cement-Lined with Mechanical Joints

Description	Craft@Hrs	Unit	Material $	Labor $	Total $

Double Cement-lined Class 153 ductile iron pipe with mechanical joints

Description	Craft@Hrs	Unit	Material $	Labor $	Total $
4"	P2@.140	LF	6.12	3.45	9.57
6"	P2@.180	LF	8.78	4.44	13.22
8"	P2@.200	LF	12.20	4.93	17.13
10"	P2@.230	LF	18.10	5.67	23.77
12"	P2@.270	LF	24.60	6.65	31.25
14"	P2@.310	LF	37.40	7.64	45.04
16"	P2@.350	LF	48.10	8.62	56.72

Double cement-lined Class 153 ductile iron 1/16 bend with mechanical joints

Description	Craft@Hrs	Unit	Material $	Labor $	Total $
4"	P2@.900	Ea	27.00	22.20	49.20
6"	P2@1.30	Ea	41.80	32.00	73.80
8"	P2@1.70	Ea	66.00	41.90	107.90
10"	P2@2.15	Ea	90.00	53.00	143.00
12"	P2@2.40	Ea	123.00	59.10	182.10
14"	P2@2.70	Ea	273.00	66.50	339.50
16"	P2@3.00	Ea	309.00	73.90	382.90

Double Cement-lined Class 153 ductile iron 1/8 bend with mechanical joints

Description	Craft@Hrs	Unit	Material $	Labor $	Total $
4"	P2@.900	Ea	27.00	22.20	49.20
6"	P2@1.30	Ea	40.70	32.00	72.70
8"	P2@1.70	Ea	64.40	41.90	106.30
10"	P2@2.15	Ea	88.60	53.00	141.60
12"	P2@2.40	Ea	122.00	59.10	181.10
14"	P2@2.70	Ea	265.00	66.50	331.50
16"	P2@3.00	Ea	309.00	73.90	382.90

Double Cement-lined Class 153 ductile iron 1/4 bend with mechanical joints

Description	Craft@Hrs	Unit	Material $	Labor $	Total $
4"	P2@.900	Ea	29.20	22.20	51.40
6"	P2@1.30	Ea	47.30	32.00	79.30
8"	P2@1.70	Ea	73.20	41.90	115.10
10"	P2@2.15	Ea	115.00	53.00	168.00
12"	P2@2.40	Ea	149.00	59.10	208.10
14"	P2@2.70	Ea	353.00	66.50	419.50
16"	P2@3.00	Ea	402.00	73.90	475.90

Ductile Iron, Class 153, Double Cement-Lined with Mechanical Joints

Description	Craft@Hrs	Unit	Material $	Labor $	Total $

Double Cement-lined Class 153 ductile iron tee with mechanical joints

Description	Craft@Hrs	Unit	Material $	Labor $	Total $
4"	P2@1.65	Ea	44.60	40.70	85.30
6"	P2@2.35	Ea	64.40	57.90	122.30
8"	P2@3.15	Ea	99.60	77.60	177.20
10"	P2@4.00	Ea	149.00	98.60	247.60
12"	P2@4.45	Ea	203.00	110.00	313.00
14"	P2@4.90	Ea	523.00	121.00	644.00
16"	P2@5.30	Ea	549.00	131.00	680.00

Double Cement-lined Class 153 ductile iron wye with mechanical joints

Description	Craft@Hrs	Unit	Material $	Labor $	Total $
4"	P2@1.65	Ea	99.60	40.70	140.30
6"	P2@2.35	Ea	127.00	57.90	184.90
8"	P2@3.15	Ea	186.00	77.60	263.60
10"	P2@4.00	Ea	350.00	98.60	448.60
12"	P2@4.45	Ea	520.00	110.00	630.00
14"	P2@4.90	Ea	1,070.00	121.00	1,191.00
16"	P2@5.30	Ea	1,330.00	131.00	1,461.00

Double Cement-lined Class 153 ductile iron reducer with mechanical joints

Description	Craft@Hrs	Unit	Material $	Labor $	Total $
6"	P2@1.45	Ea	32.50	35.70	68.20
8"	P2@2.00	Ea	46.20	49.30	95.50
10"	P2@2.55	Ea	60.00	62.80	122.80
12"	P2@3.05	Ea	87.50	75.20	162.70
14"	P2@3.60	Ea	188.00	88.70	276.70
16"	P2@4.20	Ea	238.00	103.00	341.00

Ductile Iron, Class 110, Cement-Lined with Mechanical Joints

Description	Craft@Hrs	Unit	Material $	Labor $	Total $
Cement-lined Class 110 ductile iron pipe with mechanical joints					
4"	P2@.180	LF	6.65	4.44	11.09
6"	P2@.220	LF	9.53	5.42	14.95
8"	P2@.240	LF	13.20	5.91	19.11
10"	P2@.270	LF	19.60	6.65	26.25
12"	P2@.320	LF	26.60	7.88	34.48
16"	P2@.400	LF	52.20	9.86	62.06
Cement-lined Class 110 ductile iron 1/16 bend with mechanical joints					
4"	P2@1.40	Ea	66.60	34.50	101.10
6"	P2@2.00	Ea	89.70	49.30	139.00
8"	P2@2.50	Ea	124.00	61.60	185.60
10"	P2@3.20	Ea	187.00	78.80	265.80
12"	P2@3.60	Ea	240.00	88.70	328.70
16"	P2@4.50	Ea	620.00	111.00	731.00
Cement-lined Class 110 ductile iron 1/8 bend with mechanical joints					
4"	P2@1.40	Ea	64.90	34.50	99.40
6"	P2@2.00	Ea	88.00	49.30	137.30
8"	P2@2.50	Ea	124.00	61.60	185.60
10"	P2@3.20	Ea	180.00	78.80	258.80
12"	P2@3.60	Ea	239.00	88.70	327.70
16"	P2@4.50	Ea	608.00	111.00	719.00
Cement-lined Class 110 ductile iron 1/4 bend with mechanical joints					
4"	P2@1.40	Ea	72.60	34.50	107.10
6"	P2@2.00	Ea	97.40	49.30	146.70
8"	P2@2.50	Ea	141.00	61.60	202.60
10"	P2@3.20	Ea	216.00	78.80	294.80
12"	P2@3.60	Ea	326.00	88.70	414.70
16"	P2@4.50	Ea	787.00	111.00	898.00
Cement-lined Class 110 ductile iron tee with mechanical joints					
4"	P2@2.50	Ea	109.00	61.60	170.60
6"	P2@3.50	Ea	146.00	86.20	232.20
8"	P2@4.70	Ea	208.00	116.00	324.00
10"	P2@6.00	Ea	358.00	148.00	506.00
12"	P2@6.70	Ea	455.00	165.00	620.00
16"	P2@8.00	Ea	1,190.00	197.00	1,387.00
Cement-lined Class 110 ductile iron reducer with mechanical joints					
6"	P2@2.20	Ea	72.60	54.20	126.80
8"	P2@3.00	Ea	100.00	73.90	173.90
10"	P2@3.80	Ea	164.00	93.60	257.60
12"	P2@4.60	Ea	194.00	113.00	307.00
16"	P2@6.30	Ea	532.00	102.00	634.00

Polypropylene, DWV, Schedule 40, with Heat-Fusioned Joints

Polypropylene is used almost exclusively for acid and laboratory drain, waste and vent (DWV) systems.

Joints are made by applying low-voltage current to electrical resistance coils imbedded in propylene collars, which are slipped over the ends of the pipe, then inserted into the hubs of the fittings. Compression clamps are placed around the assemblies and tightened to compress the joints prior to applying electrical current from the power unit to fuse the joint.

This section has been arranged to save the estimator's time by including all normally-used system components such as pipe, fittings, hanger assemblies and riser clamps under one heading. The cost estimates in this section are based on the conditions, limitations and wage rates described in the section "How to Use This Book" beginning on page 5.

Description	Craft@Hrs	Unit	Material $	Labor $	Total $
Schedule 40 DWV polypropylene pipe with heat-fusioned joints					
1½"	P2@.050	LF	2.10	1.23	3.33
2"	P2@.055	LF	2.84	1.36	4.20
3"	P2@.060	LF	5.13	1.48	6.61
4"	P2@.080	LF	7.28	1.97	9.25
6"	P2@.110	LF	13.10	2.71	15.81
Schedule 40 DWV polypropylene 1/8 bend					
1½"	P2@.350	Ea	9.25	8.62	17.87
2"	P2@.400	Ea	11.30	9.86	21.16
3"	P2@.450	Ea	20.70	11.10	31.80
4"	P2@.500	Ea	23.40	12.30	35.70
6"	P2@.750	Ea	65.00	18.50	83.50
Schedule 40 DWV polypropylene 1/4 bend vent					
1½"	P2@.350	Ea	9.40	8.62	18.02
2"	P2@.400	Ea	11.70	9.86	21.56
3"	P2@.450	Ea	19.80	11.10	30.90
4"	P2@.500	Ea	31.60	12.30	43.90
6"	P2@.750	Ea	77.60	18.50	96.10
Schedule 40 DWV polypropylene long sweep 1/4 bend					
1½"	P2@.350	Ea	9.73	8.62	18.35
2"	P2@.400	Ea	13.30	9.86	23.16
3"	P2@.450	Ea	22.70	11.10	33.80
4"	P2@.500	Ea	32.60	12.30	44.90
Schedule 40 DWV polypropylene P-trap					
1½"	P2@.500	Ea	18.30	12.30	30.60
2"	P2@.560	Ea	25.60	13.80	39.40
3"	P2@.630	Ea	46.50	15.50	62.00
4"	P2@.700	Ea	82.50	17.20	99.70

Polypropylene, DWV, Schedule 40, with Heat-Fusioned Joints

Description	Craft@Hrs	Unit	Material $	Labor $	Total $

Schedule 40 DWV polypropylene sanitary tee

Description	Craft@Hrs	Unit	Material $	Labor $	Total $
1½"	P2@.500	Ea	11.70	12.30	24.00
2"	P2@.560	Ea	14.00	13.80	27.80
3"	P2@.630	Ea	28.00	15.50	43.50
4"	P2@.700	Ea	42.00	17.20	59.20

Schedule 40 DWV polypropylene reducing sanitary tee

Description	Craft@Hrs	Unit	Material $	Labor $	Total $
2 x 1½"	P2@.510	Ea	14.00	12.60	26.60
3 x 1½"	P2@.530	Ea	24.30	13.10	37.40
3 x 2"	P2@.580	Ea	26.20	14.30	40.50
4 x 2"	P2@.650	Ea	37.50	16.00	53.50
4 x 3"	P2@.700	Ea	39.20	17.20	56.40
6 x 4"	P2@.850	Ea	71.00	20.90	91.90

Schedule 40 DWV polypropylene combination

Description	Craft@Hrs	Unit	Material $	Labor $	Total $
1½"	P2@.500	Ea	15.60	12.30	27.90
2"	P2@.560	Ea	20.80	13.80	34.60
3"	P2@.630	Ea	33.10	15.50	48.60
4"	P2@.700	Ea	47.10	17.20	64.30

Schedule 40 DWV polypropylene reducing combination

Description	Craft@Hrs	Unit	Material $	Labor $	Total $
2 x 1½"	P2@.500	Ea	18.70	12.30	31.00
3 x 1½"	P2@.520	Ea	27.60	12.80	40.40
3 x 2"	P2@.570	Ea	29.90	14.00	43.90
4 x 2"	P2@.580	Ea	41.30	14.30	55.60
4 x 3"	P2@.630	Ea	43.30	15.50	58.80

Schedule 40 DWV polypropylene wye

Description	Craft@Hrs	Unit	Material $	Labor $	Total $
1½"	P2@.500	Ea	11.90	12.30	24.20
2"	P2@.560	Ea	17.00	13.80	30.80
3"	P2@.630	Ea	29.80	15.50	45.30
4"	P2@.700	Ea	43.40	17.20	60.60
6"	P2@1.05	Ea	110.00	25.90	135.90

Schedule 40 DWV polypropylene reducing wye

Description	Craft@Hrs	Unit	Material $	Labor $	Total $
2 x 1½"	P2@.510	Ea	16.70	12.60	29.30
3 x 2"	P2@.580	Ea	27.80	14.30	42.10
4 x 2"	P2@.600	Ea	40.00	14.80	54.80
4 x 3"	P2@.650	Ea	41.80	16.00	57.80
6 x 2"	P2@.900	Ea	69.70	22.20	91.90
6 x 3"	P2@.900	Ea	70.80	22.20	93.00
6 x 4"	P2@.950	Ea	71.60	23.40	95.00

Polypropylene, DWV, Schedule 40, with Heat-Fusioned Joints

Description	Craft@Hrs	Unit	Material $	Labor $	Total $

Schedule 40 DWV polypropylene female adapter

Description	Craft@Hrs	Unit	Material $	Labor $	Total $
1½"	P2@.280	Ea	5.79	6.90	12.69
2"	P2@.330	Ea	7.88	8.13	16.01
3"	P2@.350	Ea	13.00	8.62	21.62
4"	P2@.400	Ea	25.00	9.86	34.86

Schedule 40 DWV polypropylene male adapter

Description	Craft@Hrs	Unit	Material $	Labor $	Total $
1½"	P2@.280	Ea	5.72	6.90	12.62
2"	P2@.330	Ea	6.92	8.13	15.05
3"	P2@.350	Ea	11.40	8.62	20.02
4"	P2@.400	Ea	23.70	9.86	33.56

Schedule 40 DWV polypropylene cleanout adapter

Description	Craft@Hrs	Unit	Material $	Labor $	Total $
1½"	P2@.210	Ea	6.09	5.17	11.26
2"	P2@.280	Ea	7.17	6.90	14.07
3"	P2@.430	Ea	13.20	10.60	23.80
4"	P2@.640	Ea	22.80	15.80	38.60
6"	P2@.850	Ea	39.40	20.90	60.30

Schedule 40 DWV polypropylene plug

Description	Craft@Hrs	Unit	Material $	Labor $	Total $
1½"	P2@.060	Ea	2.11	1.48	3.59
2"	P2@.065	Ea	2.30	1.60	3.90
3"	P2@.070	Ea	4.02	1.72	5.74
4"	P2@.090	Ea	4.51	2.22	6.73

Schedule 40 DWV polypropylene coupling

Description	Craft@Hrs	Unit	Material $	Labor $	Total $
1½"	P2@.350	Ea	7.35	8.62	15.97
2"	P2@.400	Ea	9.18	9.86	19.04
3"	P2@.450	Ea	11.70	11.10	22.80
4"	P2@.500	Ea	16.70	12.30	29.00
6"	P2@.750	Ea	26.50	18.50	45.00

Schedule 40 DWV polypropylene reducer

Description	Craft@Hrs	Unit	Material $	Labor $	Total $
2 x 1½"	P2@.360	Ea	7.84	8.87	16.71
3 x 2"	P2@.410	Ea	9.04	10.10	19.14
4 x 2"	P2@.430	Ea	20.00	10.60	30.60
4 x 3"	P2@.450	Ea	20.50	11.10	31.60

Schedule 40 DWV polypropylene flange

Description	Craft@Hrs	Unit	Material $	Labor $	Total $
2"	P2@.350	Ea	25.70	8.62	34.32
3"	P2@.380	Ea	53.00	9.36	62.36
4"	P2@.430	Ea	68.80	10.60	79.40
6"	P2@.630	Ea	91.20	15.50	106.70

Polypropylene, DWV, Schedule 40, with Heat-Fusioned Joints

Description	Craft@Hrs	Unit	Material $	Labor $	Total $
Schedule 40 DWV polypropylene bolt and gasket set					
2"	P2@.500	Ea	3.85	12.30	16.15
2½"	P2@.650	Ea	4.08	16.00	20.08
3"	P2@.750	Ea	4.28	18.50	22.78
4"	P2@1.00	Ea	7.75	24.60	32.35
Hanger with swivel assembly					
1½"	P2@.450	Ea	.62	11.10	11.72
2"	P2@.450	Ea	.65	11.10	11.75
3"	P2@.450	Ea	4.32	11.10	15.42
4"	P2@.450	Ea	5.87	11.10	16.97
6"	P2@.500	Ea	8.78	12.30	21.08
Riser clamp					
1½"	P2@.110	Ea	4.23	2.71	6.94
2"	P2@.115	Ea	4.45	2.83	7.28
3"	P2@.120	Ea	5.13	2.96	8.09
4"	P2@.125	Ea	6.45	3.08	9.53
6"	P2@.200	Ea	8.58	4.93	13.51

PVC, Schedule 40, with Solvent-Weld Joints

PVC (Polyvinyl Chloride) Schedule 40 pipe with Schedule 40 socket-type fittings and solvent-welded joints is widely used in process, chemical, A.C. condensate, potable water and irrigation piping systems. PVC is available in Type I (normal impact) and Type II (high impact). Type I has a maximum temperature rating of 150 degrees F., and Type II is rated at 140 degrees F.

Consult the manufacturers for maximum pressure ratings and recommended joint solvents for specific applications.

Because of the current unreliability of many plastic valves, most engineers specify standard bronze-body and iron-body valves for use in PVC water piping systems. Plastic valves, however, have to be used in systems conveying liquids that may be injurious to metallic valves.

This section has been arranged to save the estimator's time by including all normally-used system components such as pipe, fittings, valves, hanger assemblies, riser clamps and miscellaneous items under one heading. Additional items can be found under "Plumbing and Piping Specialties." The cost estimates in this section are based on the conditions, limitations and wage rates described in the section "How to Use This Book" beginning on page 5.

Description	Craft@Hrs	Unit	Material $	Labor $	Total $

Schedule 40 PVC pipe with solvent-weld joints

Description	Craft@Hrs	Unit	Material $	Labor $	Total $
1/2"	P2@.020	LF	.13	.49	.62
3/4"	P2@.025	LF	.17	.62	.79
1"	P2@.030	LF	.25	.74	.99
1¼"	P2@.035	LF	.33	.86	1.19
1½"	P2@.040	LF	.39	.99	1.38
2"	P2@.045	LF	.55	1.11	1.66
2½"	P2@.050	LF	.87	1.23	2.10
3"	P2@.055	LF	1.13	1.36	2.49
4"	P2@.070	LF	1.62	1.72	3.34
5"	P2@.095	LF	2.21	2.34	4.55
6"	P2@.120	LF	2.84	2.96	5.80
8"	P2@.160	LF	4.30	3.94	8.24

Schedule 40 PVC 45 degree ell S x S

Description	Craft@Hrs	Unit	Material $	Labor $	Total $
1/2"	P2@.100	Ea	.44	2.46	2.90
3/4"	P2@.115	Ea	.68	2.83	3.51
1"	P2@.120	Ea	.82	2.96	3.78
1¼"	P2@.160	Ea	1.16	3.94	5.10
1½"	P2@.180	Ea	1.44	4.44	5.88
2"	P2@.200	Ea	1.88	4.93	6.81
2½"	P2@.250	Ea	4.89	6.16	11.05
3"	P2@.300	Ea	7.60	7.39	14.99
4"	P2@.500	Ea	13.60	12.30	25.90
5"	P2@.550	Ea	27.10	13.60	40.70
6"	P2@.600	Ea	33.70	14.80	48.50

PVC, Schedule 40, with Solvent-Weld Joints

Description	Craft@Hrs	Unit	Material $	Labor $	Total $

Schedule 40 PVC 90 degree ell S x S

Description	Craft@Hrs	Unit	Material $	Labor $	Total $
1/2"	P2@.100	Ea	.27	2.46	2.73
3/4"	P2@.115	Ea	.30	2.83	3.13
1"	P2@.120	Ea	.54	2.96	3.50
1¼"	P2@.160	Ea	.96	3.94	4.90
1½"	P2@.180	Ea	1.02	4.44	5.46
2"	P2@.200	Ea	1.61	4.93	6.54
2½"	P2@.250	Ea	4.89	6.16	11.05
3"	P2@.300	Ea	5.84	7.39	13.23
4"	P2@.500	Ea	10.50	12.30	22.80
5"	P2@.550	Ea	27.10	13.60	40.70
6"	P2@.600	Ea	33.30	14.80	48.10

Schedule 40 PVC tee S x S x S

Description	Craft@Hrs	Unit	Material $	Labor $	Total $
1/2"	P2@.130	Ea	.33	3.20	3.53
3/4"	P2@.140	Ea	.38	3.45	3.83
1"	P2@.170	Ea	.72	4.19	4.91
1¼"	P2@.220	Ea	1.13	5.42	6.55
1½"	P2@.250	Ea	1.37	6.16	7.53
2"	P2@.280	Ea	1.98	6.90	8.88
2½"	P2@.350	Ea	6.54	8.62	15.16
3"	P2@.420	Ea	8.58	10.30	18.88
4"	P2@.560	Ea	15.50	13.80	29.30
5"	P2@.710	Ea	37.50	17.50	55.00
6"	P2@.840	Ea	52.30	20.70	73.00

Description	Craft@Hrs	Unit	Material $	Labor $	Total $
Schedule 40 PVC reducing tee S x S x S					
3/4 x 1/2"	P2@.140	Ea	.42	3.45	3.87
1 x 1/2"	P2@.170	Ea	.76	4.19	4.95
1 x 3/4"	P2@.170	Ea	.82	4.19	5.01
1¼ x 1/2"	P2@.220	Ea	1.23	5.42	6.65
1¼ x 3/4"	P2@.220	Ea	1.23	5.42	6.65
1¼ x 1"	P2@.220	Ea	1.23	5.42	6.65
1½ x 1/2"	P2@.250	Ea	2.16	6.16	8.32
1½ x 3/4"	P2@.250	Ea	2.16	6.16	8.32
1½ x 1"	P2@.250	Ea	2.16	6.16	8.32
1½ x 1¼"	P2@.250	Ea	2.16	6.16	8.32
2 x 1/2"	P2@.280	Ea	2.12	6.90	9.02
2 x 3/4"	P2@.280	Ea	2.12	6.90	9.02
2 x 1"	P2@.280	Ea	2.12	6.90	9.02
2 x 1¼"	P2@.280	Ea	2.12	6.90	9.02
2 x 1½"	P2@.280	Ea	2.12	6.90	9.02
2½ x 1"	P2@.350	Ea	6.52	8.62	15.14
2½ x 1¼"	P2@.350	Ea	6.52	8.62	15.14
2½ x 1½"	P2@.350	Ea	6.52	8.62	15.14
2½ x 2"	P2@.350	Ea	6.52	8.62	15.14
3 x 2"	P2@.420	Ea	9.34	10.30	19.64
3 x 2½"	P2@.420	Ea	9.34	10.30	19.64
4 x 2"	P2@.560	Ea	15.50	13.80	29.30
4 x 2½"	P2@.560	Ea	15.50	13.80	29.30
4 x 3"	P2@.560	Ea	15.50	13.80	29.30
5 x 3"	P2@.710	Ea	36.40	17.50	53.90
5 x 4"	P2@.710	Ea	36.40	17.50	53.90
6 x 3"	P2@.840	Ea	52.30	20.70	73.00
6 x 4"	P2@.840	Ea	52.30	20.70	73.00
6 x 5"	P2@.840	Ea	52.30	20.70	73.00

PVC, Schedule 40, with Solvent-Weld Joints

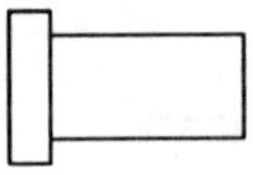

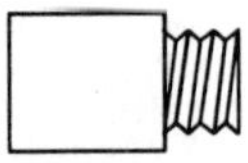

Description	Craft@Hrs	Unit	Material $	Labor $	Total $
Schedule 40 PVC reducing bushing SPIG x S					
3/4 x 1/2"	P2@.050	Ea	.28	1.23	1.51
1 x 1/2"	P2@.060	Ea	.52	1.48	2.00
1 x 3/4"	P2@.060	Ea	.52	1.48	2.00
1¼ x 1/2"	P2@.080	Ea	.68	1.97	2.65
1¼ x 3/4"	P2@.080	Ea	.68	1.97	2.65
1¼ x 1"	P2@.080	Ea	.68	1.97	2.65
1½ x 1/2"	P2@.100	Ea	.72	2.46	3.18
1½ x 3/4"	P2@.100	Ea	.72	2.46	3.18
1½ x 1"	P2@.100	Ea	.72	2.46	3.18
1½ x 1¼"	P2@.100	Ea	.72	2.46	3.18
2 x 1"	P2@.110	Ea	1.19	2.71	3.90
2 x 1¼"	P2@.110	Ea	1.19	2.71	3.90
2 x 1½"	P2@.110	Ea	1.19	2.71	3.90
2½ x 1¼"	P2@.130	Ea	1.92	3.20	5.12
2½ x 1½"	P2@.130	Ea	1.92	3.20	5.12
2½ x 2"	P2@.130	Ea	1.92	3.20	5.12
3 x 2"	P2@.150	Ea	2.85	3.70	6.55
3 x 2½"	P2@.150	Ea	2.85	3.70	6.55
4 x 2½"	P2@.200	Ea	6.36	4.93	11.29
4 x 3"	P2@.200	Ea	6.36	4.93	11.29
5 x 3"	P2@.250	Ea	8.93	6.16	15.09
5 x 4"	P2@.250	Ea	8.93	6.16	15.09
6 x 4"	P2@.300	Ea	15.70	7.39	23.09
6 x 5"	P2@.300	Ea	15.70	7.39	23.09
Schedule 40 PVC adapter MPT x S					
1/2"	P2@.060	Ea	.24	1.48	1.72
3/4"	P2@.065	Ea	.27	1.60	1.87
1"	P2@.070	Ea	.48	1.72	2.20
1¼"	P2@.100	Ea	.58	2.46	3.04
1½"	P2@.110	Ea	.79	2.71	3.50
2"	P2@.120	Ea	1.03	2.96	3.99
2½"	P2@.150	Ea	3.04	3.70	6.74
3"	P2@.180	Ea	4.45	4.44	8.89
4"	P2@.240	Ea	5.67	5.91	11.58
5"	P2@.370	Ea	9.36	9.12	18.48
6"	P2@.420	Ea	14.80	10.30	25.10

Description	Craft@Hrs	Unit	Material $	Labor $	Total $

Schedule 40 PVC adapter FPT x S

Description	Craft@Hrs	Unit	Material $	Labor $	Total $
1/2"	P2@.110	Ea	.30	2.71	3.01
3/4"	P2@.120	Ea	.38	2.96	3.34
1"	P2@.130	Ea	.44	3.20	3.64
1¼"	P2@.170	Ea	.68	4.19	4.87
1½"	P2@.210	Ea	.79	5.17	5.96
2"	P2@.290	Ea	1.06	7.15	8.21
2½"	P2@.330	Ea	2.67	8.13	10.80
3"	P2@.470	Ea	3.59	11.60	15.19
4"	P2@.630	Ea	5.95	15.50	21.45
5"	P2@.890	Ea	15.40	21.90	37.30
6"	P2@.970	Ea	21.90	23.90	45.80

Schedule 40 PVC cap S

Description	Craft@Hrs	Unit	Material $	Labor $	Total $
1/2"	P2@.060	Ea	.24	1.48	1.72
3/4"	P2@.065	Ea	.28	1.60	1.88
1"	P2@.070	Ea	.45	1.72	2.17
1¼"	P2@.100	Ea	.62	2.46	3.08
1½"	P2@.110	Ea	.68	2.71	3.39
2"	P2@.120	Ea	.82	2.96	3.78
2½"	P2@.150	Ea	2.63	3.70	6.33
3"	P2@.180	Ea	2.88	4.44	7.32
4"	P2@.240	Ea	6.54	5.91	12.45
5"	P2@.370	Ea	10.90	9.12	20.02
6"	P2@.420	Ea	15.70	10.30	26.00

Schedule 40 PVC plug MPT

Description	Craft@Hrs	Unit	Material $	Labor $	Total $
1/2"	P2@.150	Ea	.72	3.70	4.42
3/4"	P2@.160	Ea	.78	3.94	4.72
1"	P2@.180	Ea	1.27	4.44	5.71
1¼"	P2@.210	Ea	1.33	5.17	6.50
1½"	P2@.230	Ea	1.44	5.67	7.11
2"	P2@.240	Ea	1.85	5.91	7.76
2½"	P2@.260	Ea	2.57	6.41	8.98
3"	P2@.300	Ea	3.90	7.39	11.29
4"	P2@.400	Ea	8.79	9.86	18.65

PVC, Schedule 40, with Solvent-Weld Joints

Description	Craft@Hrs	Unit	Material $	Labor $	Total $
Schedule 40 PVC coupling S x S					
1/2"	P2@.100	Ea	.18	2.46	2.64
3/4"	P2@.115	Ea	.24	2.83	3.07
1"	P2@.120	Ea	.42	2.96	3.38
1¼"	P2@.160	Ea	.58	3.94	4.52
1½"	P2@.180	Ea	.62	4.44	5.06
2"	P2@.200	Ea	.97	4.93	5.90
2½"	P2@.250	Ea	2.12	6.16	8.28
3"	P2@.300	Ea	3.33	7.39	10.72
4"	P2@.500	Ea	4.80	12.30	17.10
5"	P2@.550	Ea	8.80	13.60	22.40
6"	P2@.600	Ea	15.20	14.80	30.00
Schedule 40 PVC union S x S					
1/2"	P2@.125	Ea	2.63	3.08	5.71
3/4"	P2@.140	Ea	3.00	3.45	6.45
1"	P2@.160	Ea	3.08	3.94	7.02
1¼"	P2@.200	Ea	9.93	4.93	14.86
1½"	P2@.225	Ea	10.40	5.54	15.94
2"	P2@.250	Ea	13.90	6.16	20.06
Class 125 bronze body gate valve, threaded ends					
1/2"	P2@.210	Ea	15.60	5.17	20.77
3/4"	P2@.250	Ea	19.20	6.16	25.36
1"	P2@.300	Ea	26.00	7.39	33.39
1¼"	P2@.400	Ea	37.10	9.86	46.96
1½"	P2@.450	Ea	45.00	11.10	56.10
2"	P2@.500	Ea	56.20	12.30	68.50
2½"	P2@.750	Ea	150.00	18.50	168.50
3"	P2@.950	Ea	207.00	23.40	230.40
Class 125 iron body gate valve, flanged ends					
2"	P2@.500	Ea	164.00	12.30	176.30
2½"	P2@.600	Ea	168.00	14.80	182.80
3"	P2@.750	Ea	188.00	18.50	206.50
4"	P2@1.35	Ea	270.00	33.30	303.30
5"	P2@2.00	Ea	460.00	49.30	509.30
6"	P2@2.50	Ea	460.00	61.60	521.60
8"	P2@3.00	Ea	789.00	73.90	862.90

Description	Craft@Hrs	Unit	Material $	Labor $	Total $

Class 125 bronze body globe valve, threaded ends

Description	Craft@Hrs	Unit	Material $	Labor $	Total $
1/2"	P2@.210	Ea	19.40	5.17	24.57
3/4"	P2@.250	Ea	24.60	6.16	30.76
1"	P2@.300	Ea	32.90	7.39	40.29
1¼"	P2@.400	Ea	43.60	9.86	53.46
1½"	P2@.450	Ea	56.30	11.10	67.40
2"	P2@.500	Ea	85.70	12.30	98.00

200 PSI iron body butterfly valve, lug type

Description	Craft@Hrs	Unit	Material $	Labor $	Total $
2"	P2@.200	Ea	73.40	4.93	78.33
2½"	P2@.300	Ea	74.20	7.39	81.59
3"	P2@.400	Ea	84.20	9.86	94.06
4"	P2@.500	Ea	100.00	12.30	112.30
5"	P2@.650	Ea	143.00	16.00	159.00
6"	P2@.750	Ea	163.00	18.50	181.50
8"	P2@.950	Ea	213.00	23.40	236.40

200 PSI iron body butterfly valve, wafer type

Description	Craft@Hrs	Unit	Material $	Labor $	Total $
2"	P2@.200	Ea	73.40	4.93	78.33
2½"	P2@.300	Ea	73.40	7.39	80.79
3"	P2@.400	Ea	74.20	9.86	84.06
4"	P2@.500	Ea	89.30	12.30	101.60
5"	P2@.650	Ea	122.00	16.00	138.00
6"	P2@.750	Ea	148.00	18.50	166.50
8"	P2@.950	Ea	210.00	23.40	233.40

Class 125 bronze body 2-piece ball valve, threaded ends

Description	Craft@Hrs	Unit	Material $	Labor $	Total $
1/2"	P2@.210	Ea	5.17	5.17	10.34
3/4"	P2@.250	Ea	8.53	6.16	14.69
1"	P2@.300	Ea	10.70	7.39	18.09
1¼"	P2@.400	Ea	18.00	9.86	27.86
1½"	P2@.450	Ea	23.00	11.10	34.10
2"	P2@.500	Ea	28.80	12.30	41.10

Class 125 bronze body swing check valve, threaded

Description	Craft@Hrs	Unit	Material $	Labor $	Total $
1/2"	P2@.210	Ea	20.80	5.17	25.97
3/4"	P2@.250	Ea	25.10	6.16	31.26
1"	P2@.300	Ea	34.60	7.39	41.99
1¼"	P2@.400	Ea	47.90	9.86	57.76
1½"	P2@.450	Ea	57.10	11.10	68.20
2"	P2@.500	Ea	83.30	12.30	95.60

PVC, Schedule 40, with Solvent-Weld Joints

Description	Craft@Hrs	Unit	Material $	Labor $	Total $
Class 125 iron body swing check valve, flanged ends					
2"	P2@.500	Ea	185.00	12.30	197.30
2½"	P2@.600	Ea	235.00	14.80	249.80
3"	P2@.750	Ea	256.00	18.50	274.50
4"	P2@1.35	Ea	403.00	33.30	436.30
5"	P2@2.00	Ea	547.00	49.30	596.30
6"	P2@2.50	Ea	691.00	61.60	752.60
8"	P2@3.00	Ea	1,310.00	73.90	1,383.90
Class 125 iron body silent check valve, wafer type					
2"	P2@.500	Ea	97.20	12.30	109.50
2½"	P2@.600	Ea	112.00	14.80	126.80
3"	P2@.750	Ea	124.00	18.50	142.50
4"	P2@1.35	Ea	167.00	33.30	200.30
5"	P2@2.00	Ea	232.00	49.30	281.30
6"	P2@2.50	Ea	315.00	61.60	376.60
8"	P2@3.00	Ea	542.00	73.90	615.90
Class 150 bronze body strainer, threaded ends					
1/2"	P2@.210	Ea	15.10	5.17	20.27
3/4"	P2@.250	Ea	19.60	6.16	25.76
1"	P2@.300	Ea	24.80	7.39	32.19
1¼"	P2@.400	Ea	31.00	9.86	40.86
1½"	P2@.450	Ea	40.20	11.10	51.30
2"	P2@.500	Ea	65.20	12.30	77.50
Class 125 iron body strainer, flanged ends					
2"	P2@.500	Ea	83.00	12.30	95.30
2½"	P2@.600	Ea	91.80	14.80	106.60
3"	P2@.750	Ea	104.00	18.50	122.50
4"	P2@1.35	Ea	186.00	33.30	219.30
5"	P2@2.00	Ea	309.00	49.30	358.30
6"	P2@2.50	Ea	378.00	61.60	439.60
8"	P2@3.00	Ea	634.00	73.90	707.90

Description	Craft@Hrs	Unit	Material $	Labor $	Total $

Installation of 2-way control valve

Description	Craft@Hrs	Unit	Material $	Labor $	Total $
1/2"	P2@.210	Ea	--	5.17	5.17
3/4"	P2@.275	Ea	--	6.78	6.78
1"	P2@.350	Ea	--	8.62	8.62
1¼"	P2@.430	Ea	--	10.60	10.60
1½"	P2@.505	Ea	--	12.40	12.40
2"	P2@.675	Ea	--	16.60	16.60
2½"	P2@.830	Ea	--	20.50	20.50
3"	P2@.990	Ea	--	24.40	24.40
4"	P2@1.30	Ea	--	32.00	32.00
6"	P2@2.50	Ea	--	61.60	61.60
8"	P2@3.00	Ea	--	73.90	73.90

Installation of 3-way control valve

Description	Craft@Hrs	Unit	Material $	Labor $	Total $
1/2"	P2@.260	Ea	--	6.41	6.41
3/4"	P2@.365	Ea	--	8.99	8.99
1"	P2@.475	Ea	--	11.70	11.70
1¼"	P2@.575	Ea	--	14.20	14.20
1½"	P2@.680	Ea	--	16.80	16.80
2"	P2@.910	Ea	--	22.40	22.40
2½"	P2@1.12	Ea	--	27.60	27.60
3"	P2@1.33	Ea	--	32.80	32.80
4"	P2@2.00	Ea	--	49.30	49.30
6"	P2@3.00	Ea	--	73.90	73.90
8"	P2@4.00	Ea	--	98.60	98.60

Companion flange

Description	Craft@Hrs	Unit	Material $	Labor $	Total $
2"	P2@.140	Ea	12.10	3.45	15.55
2½"	P2@.180	Ea	18.70	4.44	23.14
3"	P2@.210	Ea	20.60	5.17	25.77
4"	P2@.280	Ea	26.10	6.90	33.00
5"	P2@.310	Ea	41.00	7.64	48.64
6"	P2@.420	Ea	41.00	10.30	51.30

Bolt and gasket set

Description	Craft@Hrs	Unit	Material $	Labor $	Total $
2"	P2@.500	Ea	3.85	12.30	16.15
2½"	P2@.650	Ea	4.08	16.00	20.08
3"	P2@.750	Ea	4.28	18.50	22.78
4"	P2@1.00	Ea	7.75	24.60	32.35
5"	P2@1.10	Ea	11.50	27.10	38.60
6"	P2@1.20	Ea	12.10	29.60	41.70
8"	P2@1.25	Ea	12.90	30.80	43.70

Thermometer with well

Description	Craft@Hrs	Unit	Material $	Labor $	Total $
7"	P2@.250	Ea	14.30	6.16	20.46
9"	P2@.250	Ea	14.80	6.16	20.96

PVC, Schedule 40, with Solvent-Weld Joints

Description	Craft@Hrs	Unit	Material $	Labor $	Total $
Pressure gauge					
2½"	P2@.200	Ea	15.60	4.93	20.53
3½"	P2@.200	Ea	27.10	4.93	32.03
Hanger with swivel assembly					
1/2"	P2@.450	Ea	.56	11.10	11.66
3/4"	P2@.450	Ea	.56	11.10	11.66
1"	P2@.450	Ea	.56	11.10	11.66
1¼"	P2@.450	Ea	.56	11.10	11.66
1½"	P2@.450	Ea	.62	11.10	11.72
2"	P2@.450	Ea	.65	11.10	11.75
2½"	P2@.450	Ea	3.95	11.10	15.05
3"	P2@.450	Ea	4.32	11.10	15.42
4"	P2@.450	Ea	5.87	11.10	16.97
5"	P2@.500	Ea	8.79	12.30	21.09
6"	P2@.500	Ea	11.70	12.30	24.00
8"	P2@.550	Ea	17.30	13.60	30.90
Riser clamp					
1/2"	P2@.100	Ea	3.27	2.46	5.73
3/4"	P2@.100	Ea	3.27	2.46	5.73
1"	P2@.100	Ea	3.30	2.46	5.76
1¼"	P2@.105	Ea	4.01	2.59	6.60
1½"	P2@.110	Ea	4.23	2.71	6.94
2"	P2@.115	Ea	4.45	2.83	7.28
2½"	P2@.120	Ea	4.69	2.96	7.65
3"	P2@.120	Ea	5.13	2.96	8.09
4"	P2@.125	Ea	6.45	3.08	9.53
5"	P2@.180	Ea	8.58	4.44	13.02
6"	P2@.200	Ea	10.70	4.93	15.63
8"	P2@.200	Ea	17.40	4.93	22.33

PVC (Polyvinyl Chloride) Schedule 80 pipe with Schedule 80 socket-type fittings and solvent-welded joints is widely used in process, chemical, A.C. condensate, potable water and irrigation piping systems. PVC is available in Type I (normal impact) and Type II (high impact). Type I has a maximum temperature rating of 150 degrees F., and Type II is rated at 140 degrees F.

Schedule 80 PVC can be threaded, but it is more economical to use solvent-welded joints.

Manufacturers should be consulted for maximum pressure ratings and recommended joints solvents for specific applications.

Because of the current unreliability of many plastic valves, most engineers specify standard bronze-body and iron-body valves for use in PVC water piping systems. Plastic valves, however, have to be used in systems conveying liquids that may be injurious to metallic valves.

This section has been arranged to save the estimator's time by including all normally-used system components such as pipe, fittings, valves, hanger assemblies, riser clamps and miscellaneous items under one heading. Additional items can be found under "Plumbing and Piping Specialties." The cost estimates in this section are based on the conditions, limitations and wage rates described in the section "How to Use This Book" beginning on page 5.

Description	Craft@Hrs	Unit	Material $	Labor $	Total $

Schedule 80 PVC pipe with solvent-weld joints

Description	Craft@Hrs	Unit	Material $	Labor $	Total $
1/2"	P2@.021	LF	.17	.52	.69
3/4"	P2@.026	LF	.24	.64	.88
1"	P2@.032	LF	.35	.79	1.14
1¼"	P2@.037	LF	.48	.91	1.39
1½"	P2@.042	LF	.58	1.03	1.61
2"	P2@.047	LF	.82	1.16	1.98
2½"	P2@.052	LF	1.22	1.28	2.50
3"	P2@.057	LF	1.63	1.40	3.03
4"	P2@.074	LF	2.39	1.82	4.21
6"	P2@.160	LF	4.56	3.94	8.50
8"	P2@.190	LF	6.95	4.68	11.63

Schedule 80 PVC 45 degree ell S x S

Description	Craft@Hrs	Unit	Material $	Labor $	Total $
1/2"	P2@.105	Ea	2.51	2.59	5.10
3/4"	P2@.120	Ea	3.80	2.96	6.76
1"	P2@.125	Ea	5.71	3.08	8.79
1¼"	P2@.170	Ea	7.28	4.19	11.47
1½"	P2@.190	Ea	8.61	4.68	13.29
2"	P2@.210	Ea	11.20	5.17	16.37
2½"	P2@.262	Ea	23.40	6.46	29.86
3"	P2@.315	Ea	28.50	7.76	36.26
4"	P2@.520	Ea	51.30	12.80	64.10
6"	P2@.630	Ea	64.60	15.50	80.10

PVC, Schedule 80, with Solvent-Weld Joints

Description	Craft@Hrs	Unit	Material $	Labor $	Total $
Schedule 80 PVC 90 degree ell S x S					
1/2"	P2@.105	Ea	1.32	2.59	3.91
3/4"	P2@.120	Ea	1.70	2.96	4.66
1"	P2@.125	Ea	2.73	3.08	5.81
1¼"	P2@.170	Ea	3.64	4.19	7.83
1½"	P2@.190	Ea	3.91	4.68	8.59
2"	P2@.210	Ea	4.72	5.17	9.89
2½"	P2@.262	Ea	11.00	6.46	17.46
3"	P2@.315	Ea	12.40	7.76	20.16
4"	P2@.520	Ea	18.90	12.80	31.70
6"	P2@.630	Ea	53.80	15.50	69.30
Schedule 80 PVC tee S x S x S					
1/2"	P2@.135	Ea	3.74	3.33	7.07
3/4"	P2@.145	Ea	3.91	3.57	7.48
1"	P2@.180	Ea	4.89	4.44	9.33
1¼"	P2@.230	Ea	13.40	5.67	19.07
1½"	P2@.265	Ea	13.40	6.53	19.93
2"	P2@.295	Ea	16.80	7.27	24.07
2½"	P2@.368	Ea	18.30	9.07	27.37
3"	P2@.440	Ea	22.90	10.80	33.70
4"	P2@.585	Ea	26.50	14.40	40.90
6"	P2@.880	Ea	90.40	21.70	112.10
Schedule 80 PVC reducing tee S x S x S					
3/4 x 1/2"	P2@.145	Ea	2.86	3.57	6.43
1 x 1/2"	P2@.175	Ea	4.14	4.31	8.45
1 x 3/4"	P2@.175	Ea	4.14	4.31	8.45
1¼ x 1/2"	P2@.230	Ea	6.83	5.67	12.50
1¼ x 3/4"	P2@.230	Ea	6.83	5.67	12.50
1¼ x 1"	P2@.230	Ea	8.97	5.67	14.64
1½ x 3/4"	P2@.260	Ea	8.97	6.41	15.38
1½ x 1"	P2@.260	Ea	8.97	6.41	15.38
1½ x 1¼"	P2@.260	Ea	8.97	6.41	15.38
2 x 1/2"	P2@.290	Ea	11.40	7.15	18.55
2 x 3/4"	P2@.290	Ea	11.40	7.15	18.55
2 x 1"	P2@.290	Ea	11.40	7.15	18.55
2 x 1¼"	P2@.290	Ea	11.40	7.15	18.55
2 x 1½"	P2@.290	Ea	11.40	7.15	18.55
3 x 2"	P2@.430	Ea	20.90	10.60	31.50
3 x 2½"	P2@.430	Ea	20.90	10.60	31.50
4 x 2"	P2@.570	Ea	33.70	14.00	47.70
4 x 2½"	P2@.570	Ea	33.70	14.00	47.70
4 x 3"	P2@.570	Ea	33.70	14.00	47.70
6 x 3"	P2@.850	Ea	90.40	20.90	111.30
6 x 4"	P2@.850	Ea	90.40	20.90	111.30
6 x 5"	P2@.850	Ea	90.40	20.90	111.30

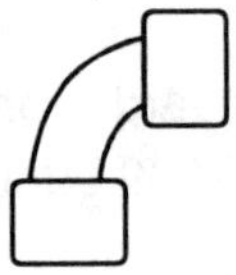

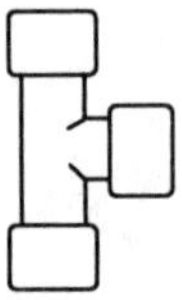

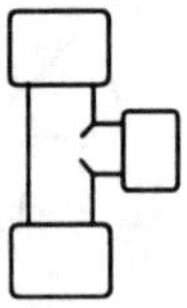

Description	Craft@Hrs	Unit	Material $	Labor $	Total $

Schedule 80 PVC reducing bushing SPIG x S

Description	Craft@Hrs	Unit	Material $	Labor $	Total $
3/4 x 1/2"	P2@.053	Ea	.77	1.31	2.08
1 x 1/2"	P2@.063	Ea	2.21	1.55	3.76
1 x 3/4"	P2@.063	Ea	2.21	1.55	3.76
1¼ x 1/2"	P2@.084	Ea	3.47	2.07	5.54
1¼ x 3/4"	P2@.084	Ea	3.47	2.07	5.54
1¼ x 1"	P2@.084	Ea	3.47	2.07	5.54
1½ x 1/2"	P2@.105	Ea	4.72	2.59	7.31
1½ x 3/4"	P2@.105	Ea	4.72	2.59	7.31
1½ x 1"	P2@.105	Ea	4.72	2.59	7.31
1½ x 1¼"	P2@.105	Ea	4.72	2.59	7.31
2 x 1"	P2@.116	Ea	6.74	2.86	9.60
2 x 1¼"	P2@.116	Ea	6.74	2.86	9.60
2 x 1½"	P2@.116	Ea	6.74	2.86	9.60
2½ x 1¼"	P2@.137	Ea	11.60	3.38	14.98
2½ x 1½"	P2@.137	Ea	11.60	3.38	14.98
2½ x 2"	P2@.137	Ea	11.60	3.38	14.98
3 x 2"	P2@.158	Ea	18.60	3.89	22.49
3 x 2½"	P2@.158	Ea	18.60	3.89	22.49
4 x 2½"	P2@.210	Ea	25.70	5.17	30.87
4 x 3"	P2@.210	Ea	25.70	5.17	30.87
6 x 4"	P2@.315	Ea	35.70	7.76	43.46
6 x 5"	P2@.315	Ea	35.70	7.76	43.46

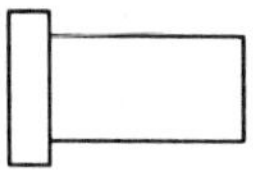

Schedule 80 PVC adapter MPT x S

Description	Craft@Hrs	Unit	Material $	Labor $	Total $
1/2"	P2@.063	Ea	2.80	1.55	4.35
3/4"	P2@.068	Ea	3.09	1.68	4.77
1"	P2@.074	Ea	5.35	1.82	7.17
1¼"	P2@.105	Ea	6.25	2.59	8.84
1½"	P2@.116	Ea	8.98	2.86	11.84
2"	P2@.126	Ea	13.00	3.10	16.10
2½"	P2@.158	Ea	14.80	3.89	18.69
3"	P2@.189	Ea	16.40	4.66	21.06
4"	P2@.252	Ea	29.10	6.21	35.31

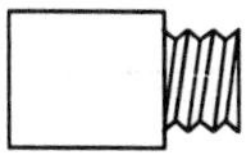

Schedule 80 PVC adapter FPT x S

Description	Craft@Hrs	Unit	Material $	Labor $	Total $
1/2"	P2@.116	Ea	2.25	2.86	5.11
3/4"	P2@.126	Ea	3.33	3.10	6.43
1"	P2@.137	Ea	4.90	3.38	8.28
1¼"	P2@.179	Ea	7.95	4.41	12.36
1½"	P2@.221	Ea	9.76	5.45	15.21
2"	P2@.305	Ea	17.00	7.52	24.52
2½"	P2@.345	Ea	26.90	8.50	35.40
3"	P2@.490	Ea	30.20	12.10	42.30
4"	P2@.655	Ea	52.00	16.10	68.10

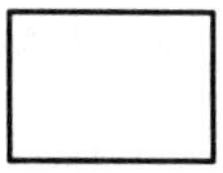

PVC, Schedule 80, with Solvent-Weld Joints

Description	Craft@Hrs	Unit	Material $	Labor $	Total $
Schedule 80 PVC cap S					
1/2"	P2@.063	Ea	2.35	1.55	3.90
3/4"	P2@.070	Ea	2.46	1.72	4.18
1"	P2@.075	Ea	4.38	1.85	6.23
1¼"	P2@.105	Ea	5.30	2.59	7.89
1½"	P2@.115	Ea	5.30	2.83	8.13
2"	P2@.125	Ea	10.50	3.08	13.58
2½"	P2@.158	Ea	21.40	3.89	25.29
3"	P2@.189	Ea	25.00	4.66	29.66
4"	P2@.250	Ea	42.10	6.16	48.26
6"	P2@.300	Ea	63.80	7.39	71.19
Schedule 80 PVC plug MPT					
1/2"	P2@.155	Ea	2.26	3.82	6.08
3/4"	P2@.165	Ea	2.35	4.07	6.42
1"	P2@.190	Ea	2.89	4.68	7.57
1¼"	P2@.220	Ea	4.23	5.42	9.65
1½"	P2@.240	Ea	5.15	5.91	11.06
2"	P2@.255	Ea	5.27	6.28	11.55
2½"	P2@.275	Ea	14.80	6.78	21.58
3"	P2@.320	Ea	17.20	7.88	25.08
4"	P2@.425	Ea	33.50	10.50	44.00
Schedule 80 PVC coupling S x S					
1/2"	P2@.105	Ea	2.40	2.59	4.99
3/4"	P2@.120	Ea	3.24	2.96	6.20
1"	P2@.125	Ea	3.33	3.08	6.41
1¼"	P2@.170	Ea	5.08	4.19	9.27
1½"	P2@.190	Ea	5.47	4.68	10.15
2"	P2@.210	Ea	5.87	5.17	11.04
2½"	P2@.262	Ea	14.50	6.46	20.96
3"	P2@.315	Ea	16.60	7.76	24.36
4"	P2@.520	Ea	20.80	12.80	33.60
6"	P2@.630	Ea	44.80	15.50	60.30
Schedule 80 PVC union S x S					
1/2"	P2@.125	Ea	4.91	3.08	7.99
3/4"	P2@.140	Ea	6.24	3.45	9.69
1"	P2@.160	Ea	7.14	3.94	11.08
1¼"	P2@.200	Ea	14.20	4.93	19.13
1½"	P2@.225	Ea	16.00	5.54	21.54
2"	P2@.250	Ea	21.70	6.16	27.86
3"	P2@.275	Ea	40.40	6.78	47.18

Description	Craft@Hrs	Unit	Material $	Labor $	Total $

Class 125 bronze body gate valve, threaded ends

Description	Craft@Hrs	Unit	Material $	Labor $	Total $
1/2"	P2@.210	Ea	15.60	5.17	20.77
3/4"	P2@.250	Ea	19.20	6.16	25.36
1"	P2@.300	Ea	26.00	7.39	33.39
1¼"	P2@.400	Ea	37.10	9.86	46.96
1½"	P2@.450	Ea	45.00	11.10	56.10
2"	P2@.500	Ea	56.20	12.30	68.50
2½"	P2@.750	Ea	150.00	18.50	168.50
3"	P2@.950	Ea	207.00	23.40	230.40

125 iron body gate valve, flanged ends

Description	Craft@Hrs	Unit	Material $	Labor $	Total $
2"	P2@.500	Ea	164.00	12.30	176.30
2½"	P2@.600	Ea	168.00	14.80	182.80
3"	P2@.750	Ea	188.00	18.50	206.50
4"	P2@1.35	Ea	270.00	33.30	303.30
5"	P2@2.00	Ea	460.00	49.30	509.30
6"	P2@2.50	Ea	460.00	61.60	521.60
8"	P2@3.00	Ea	789.00	73.90	862.90

125 bronze body globe valve, threaded ends

Description	Craft@Hrs	Unit	Material $	Labor $	Total $
1/2"	P2@.210	Ea	19.40	5.17	24.57
3/4"	P2@.250	Ea	24.60	6.16	30.76
1"	P2@.300	Ea	32.90	7.39	40.29
1¼"	P2@.400	Ea	43.60	9.86	53.46
1½"	P2@.450	Ea	56.30	11.10	67.40
2"	P2@.500	Ea	85.70	12.30	98.00

200 PSI iron body butterfly valve, lug type

Description	Craft@Hrs	Unit	Material $	Labor $	Total $
2"	P2@.200	Ea	73.40	4.93	78.33
2½"	P2@.300	Ea	74.20	7.39	81.59
3"	P2@.400	Ea	84.20	9.86	94.06
4"	P2@.500	Ea	100.00	12.30	112.30
5"	P2@.650	Ea	143.00	16.00	159.00
6"	P2@.750	Ea	163.00	18.50	181.50
8"	P2@.950	Ea	213.00	23.40	236.40

200 PSI iron body butterfly valve, wafer type

Description	Craft@Hrs	Unit	Material $	Labor $	Total $
2"	P2@.200	Ea	73.40	4.93	78.33
2½"	P2@.300	Ea	73.40	7.39	80.79
3"	P2@.400	Ea	74.20	9.86	84.06
4"	P2@.500	Ea	89.20	12.30	101.50
5"	P2@.650	Ea	122.00	16.00	138.00
6"	P2@.750	Ea	148.00	18.50	166.50
8"	P2@.950	Ea	210.00	23.40	233.40

PVC, Schedule 80, with Solvent-Weld Joints

Description	Craft@Hrs	Unit	Material $	Labor $	Total $
Class 125 bronze body 2-piece ball valve, threaded ends					
1/2"	P2@.210	Ea	5.17	5.17	10.34
3/4"	P2@.250	Ea	8.53	6.16	14.69
1"	P2@.300	Ea	10.70	7.39	18.09
1¼"	P2@.400	Ea	18.00	9.86	27.86
1½"	P2@.450	Ea	23.00	11.10	34.10
2"	P2@.500	Ea	28.80	12.30	41.10
Class 125 bronze body swing check valve, threaded ends					
1/2"	P2@.210	Ea	20.80	5.17	25.97
3/4"	P2@.250	Ea	25.10	6.16	31.26
1"	P2@.300	Ea	34.60	7.39	41.99
1¼"	P2@.400	Ea	47.90	9.86	57.76
1½"	P2@.450	Ea	57.10	11.10	68.20
2"	P2@.500	Ea	83.30	12.30	95.60
Class 125 iron body swing check valve, flanged ends					
2"	P2@.500	Ea	185.00	12.30	197.30
2½"	P2@.600	Ea	235.00	14.80	249.80
3"	P2@.750	Ea	256.00	18.50	274.50
4"	P2@1.35	Ea	403.00	33.30	436.30
5"	P2@2.00	Ea	547.00	49.30	596.30
6"	P2@2.50	Ea	691.00	61.60	752.60
8"	P2@3.00	Ea	1,310.00	73.90	1,383.90
Class 125 iron body silent check valve, wafer type					
2"	P2@.500	Ea	97.20	12.30	109.50
2½"	P2@.600	Ea	112.00	14.80	126.80
3"	P2@.750	Ea	124.00	18.50	142.50
4"	P2@1.35	Ea	167.00	33.30	200.30
5"	P2@2.00	Ea	232.00	49.30	281.30
6"	P2@2.50	Ea	315.00	61.60	376.60
8"	P2@3.00	Ea	542.00	73.90	615.90
Class 150 bronze body strainer, threaded ends					
1/2"	P2@.210	Ea	15.10	5.17	20.27
3/4"	P2@.250	Ea	19.60	6.16	25.76
1"	P2@.300	Ea	24.80	7.39	32.19
1¼"	P2@.400	Ea	31.00	9.86	40.86
1½"	P2@.450	Ea	40.20	11.10	51.30
2"	P2@.500	Ea	65.20	12.30	77.50

Description	Craft@Hrs	Unit	Material $	Labor $	Total $

Class 125 iron body strainer, flanged ends

Description	Craft@Hrs	Unit	Material $	Labor $	Total $
2"	P2@.500	Ea	83.00	12.30	95.30
2½"	P2@.600	Ea	91.80	14.80	106.60
3"	P2@.750	Ea	109.00	18.50	127.50
4"	P2@1.35	Ea	186.00	33.30	219.30
5"	P2@2.00	Ea	309.00	49.30	358.30
6"	P2@2.50	Ea	378.00	61.60	439.60
8"	P2@3.00	Ea	634.00	73.90	707.90

Installation of 2-way control valve

Description	Craft@Hrs	Unit	Material $	Labor $	Total $
1/2"	P2@.210	Ea	--	5.17	5.17
3/4"	P2@.275	Ea	--	6.78	6.78
1"	P2@.350	Ea	--	8.62	8.62
1¼"	P2@.430	Ea	--	10.60	10.60
1½"	P2@.505	Ea	--	12.40	12.40
2"	P2@.675	Ea	--	16.60	16.60
2½"	P2@.830	Ea	--	20.50	20.50
3"	P2@.990	Ea	--	24.40	24.40
4"	P2@1.30	Ea	--	32.00	32.00
6"	P2@2.50	Ea	--	61.60	61.60
8"	P2@3.00	Ea	--	73.90	73.90

Installation of 3-way control valve

Description	Craft@Hrs	Unit	Material $	Labor $	Total $
1/2"	P2@.260	Ea	--	6.41	6.41
3/4"	P2@.365	Ea	--	8.99	8.99
1"	P2@.475	Ea	--	11.70	11.70
1¼"	P2@.575	Ea	--	14.20	14.20
1½"	P2@.680	Ea	--	16.80	16.80
2"	P2@.910	Ea	--	22.40	22.40
2½"	P2@1.12	Ea	--	27.60	27.60
3"	P2@1.33	Ea	--	32.80	32.80
4"	P2@2.00	Ea	--	49.30	49.30
6"	P2@3.00	Ea	--	73.90	73.90
8"	P2@4.00	Ea	--	98.60	98.60

Companion flange

Description	Craft@Hrs	Unit	Material $	Labor $	Total $
2"	P2@.140	Ea	12.10	3.45	15.55
2½"	P2@.180	Ea	18.70	4.44	23.14
3"	P2@.210	Ea	20.70	5.17	25.87
4"	P2@.280	Ea	26.10	6.90	33.00
5"	P2@.310	Ea	41.00	7.64	48.64
6"	P2@.420	Ea	41.00	10.30	51.30

PVC, Schedule 80, with Solvent-Weld Joints

Description	Craft@Hrs	Unit	Material $	Labor $	Total $
Bolt and gasket set					
2"	P2@.500	Ea	3.85	12.30	16.15
2½"	P2@.650	Ea	4.08	16.00	20.08
3"	P2@.750	Ea	4.28	18.50	22.78
4"	P2@1.00	Ea	7.75	24.60	32.35
5"	P2@1.10	Ea	11.50	27.10	38.60
6"	P2@1.20	Ea	12.10	29.60	41.70
8"	P2@1.25	Ea	12.90	30.80	43.70
Thermometer with well					
7"	P2@.250	Ea	13.60	6.16	19.76
9"	P2@.250	Ea	14.10	6.16	20.26
Pressure gauge					
2½"	P2@.200	Ea	14.40	4.93	19.33
3½"	P2@.200	Ea	24.00	4.93	28.93
Hanger with swivel assembly					
1/2"	P2@.450	Ea	.56	11.10	11.66
3/4"	P2@.450	Ea	.56	11.10	11.66
1"	P2@.450	Ea	.56	11.10	11.66
1¼"	P2@.450	Ea	.56	11.10	11.66
1½"	P2@.450	Ea	.62	11.10	11.72
2"	P2@.450	Ea	.65	11.10	11.75
2½"	P2@.450	Ea	3.95	11.10	15.05
3"	P2@.450	Ea	4.32	11.10	15.42
4"	P2@.450	Ea	5.87	11.10	16.97
5"	P2@.500	Ea	8.79	12.30	21.09
6"	P2@.500	Ea	11.70	12.30	24.00
8"	P2@.550	Ea	17.30	13.60	30.90
Riser clamp					
1/2"	P2@.100	Ea	3.27	2.46	5.73
3/4"	P2@.100	Ea	3.27	2.46	5.73
1"	P2@.100	Ea	3.30	2.46	5.76
1¼"	P2@.105	Ea	4.01	2.59	6.60
1½"	P2@.110	Ea	4.23	2.71	6.94
2"	P2@.115	Ea	4.45	2.83	7.28
2½"	P2@.120	Ea	4.69	2.96	7.65
3"	P2@.120	Ea	5.13	2.96	8.09
4"	P2@.125	Ea	6.45	3.08	9.53
5"	P2@.180	Ea	8.58	4.44	13.02
6"	P2@.200	Ea	10.70	4.93	15.63
8"	P2@.200	Ea	17.40	4.93	22.33

PVC (Polyvinyl Chloride) is polymerized vinyl chloride produced from acetylene and anhydrous hydrochloric acid.

PVC sewer pipe (ASTM D-3034) 10' lay length, is for underground installation and is used primarily to convey sewage and rain water. It is of the bell & spigot type with joints made water-tight by means of an elastomeric gasket factory-installed in the hub end of the pipe and fittings.

This section includes pipe and all commonly-used fittings. Additional items can be found under "Plumbing and Piping Specialties." The cost estimates in this section are based on the conditions, limitations and wage rates described in the section "How to Use This Book" beginning on page 5.

Description	Craft@Hrs	Unit	Material $	Labor $	Total $

PVC sewer pipe with bell and spigot gasketed joints

Description	Craft@Hrs	Unit	Material $	Labor $	Total $
4"	P2@.050	LF	.83	1.23	2.06
6"	P2@.070	LF	1.83	1.72	3.55
8"	P2@.090	LF	3.26	2.22	5.48
10"	P2@.110	LF	5.13	2.71	7.84
12"	P2@.140	LF	7.34	3.45	10.79
15"	P2@.180	LF	11.00	4.44	15.44
18"	P2@.230	LF	16.90	5.67	22.57

PVC sewer pipe 1/16 bend B x B

Description	Craft@Hrs	Unit	Material $	Labor $	Total $
4"	P2@.320	Ea	1.97	7.88	9.85
6"	P2@.400	Ea	3.89	9.86	13.75
8"	P2@.480	Ea	11.30	11.80	23.10
10"	P2@.800	Ea	29.20	19.70	48.90
12"	P2@1.06	Ea	38.10	26.10	64.20
15"	P2@1.34	Ea	101.00	33.00	134.00
18"	P2@1.60	Ea	147.00	39.40	186.40

PVC sewer pipe 1/16 bend B x S

Description	Craft@Hrs	Unit	Material $	Labor $	Total $
4"	P2@.320	Ea	1.88	7.88	9.76
6"	P2@.400	Ea	3.69	9.86	13.55
8"	P2@.480	Ea	11.80	11.80	23.60
10"	P2@.800	Ea	28.40	19.70	48.10
12"	P2@1.06	Ea	37.20	26.10	63.30
15"	P2@1.34	Ea	80.40	33.00	113.40
18"	P2@1.60	Ea	123.00	39.40	162.40

PVC pipe sewer 1/8 bend B x B

Description	Craft@Hrs	Unit	Material $	Labor $	Total $
4"	P2@.320	Ea	1.98	7.88	9.86
6"	P2@.400	Ea	4.01	9.86	13.87
8"	P2@.480	Ea	11.20	11.80	23.00
10"	P2@.800	Ea	27.40	19.70	47.10
12"	P2@1.06	Ea	39.80	26.10	65.90
15"	P2@1.34	Ea	89.20	33.00	122.20
18"	P2@1.60	Ea	143.00	39.40	182.40

PVC, Sewer, with Bell & Spigot Gasketed Joints

Description	Craft@Hrs	Unit	Material $	Labor $	Total $
PVC sewer pipe 1/8 bend B x S					
4"	P2@.320	Ea	1.78	7.88	9.66
6"	P2@.400	Ea	3.55	9.86	13.41
8"	P2@.480	Ea	10.70	11.80	22.50
10"	P2@.800	Ea	27.00	19.70	46.70
12"	P2@1.06	Ea	38.60	26.10	64.70
15"	P2@1.34	Ea	69.80	33.00	102.80
18"	P2@1.60	Ea	118.00	39.40	157.40
PVC sewer pipe 1/4 bend B x B					
4"	P2@.320	Ea	2.55	7.88	10.43
6"	P2@.400	Ea	4.56	9.86	14.42
8"	P2@.480	Ea	12.60	11.80	24.40
10"	P2@.800	Ea	40.00	19.70	59.70
12"	P2@1.06	Ea	51.80	26.10	77.90
15"	P2@1.34	Ea	109.00	33.00	142.00
18"	P2@1.60	Ea	185.00	39.40	224.40
PVC sewer pipe 1/4 bend B x S					
4"	P2@.320	Ea	2.61	7.88	10.49
6"	P2@.400	Ea	4.83	9.86	14.69
8"	P2@.480	Ea	13.00	11.80	24.80
10"	P2@.800	Ea	39.90	19.70	59.60
12"	P2@1.06	Ea	50.40	26.10	76.50
15"	P2@1.34	Ea	105.00	33.00	138.00
18"	P2@1.60	Ea	182.00	39.40	221.40
PVC sewer pipe coupling B x B					
4"	P2@.320	Ea	2.43	7.88	10.31
6"	P2@.400	Ea	4.88	9.86	14.74
8"	P2@.480	Ea	8.27	11.80	20.07
10"	P2@.800	Ea	18.40	19.70	38.10
12"	P2@1.06	Ea	26.70	26.10	52.80
15"	P2@1.34	Ea	55.20	33.00	88.20
18"	P2@1.60	Ea	107.00	39.40	146.40
PVC sewer pipe wye B x B x B					
4"	P2@.480	Ea	3.40	11.80	15.20
6"	P2@.600	Ea	7.82	14.80	22.62
8"	P2@.720	Ea	21.40	17.70	39.10
10"	P2@1.20	Ea	59.60	29.60	89.20
12"	P2@1.59	Ea	83.30	39.20	122.50
15"	P2@2.01	Ea	145.00	49.50	194.50
18"	P2@2.40	Ea	259.00	59.10	318.10

Description	Craft@Hrs	Unit	Material $	Labor $	Total $

PVC sewer pipe wye B x S x B

Description	Craft@Hrs	Unit	Material $	Labor $	Total $
10"	P2@1.20	Ea	58.70	29.60	88.30
12"	P2@1.59	Ea	82.50	39.20	121.70
15"	P2@2.01	Ea	134.00	49.50	183.50
18"	P2@2.40	Ea	238.00	59.10	297.10

PVC sewer pipe reducing wye B x B x B

Description	Craft@Hrs	Unit	Material $	Labor $	Total $
6 x 4"	P2@.560	Ea	6.90	13.80	20.70
8 x 4"	P2@.640	Ea	10.30	15.80	26.10
8 x 6"	P2@.680	Ea	12.30	16.80	29.10
10 x 4"	P2@.960	Ea	29.50	23.70	53.20
10 x 6"	P2@1.00	Ea	30.00	24.60	54.60
10 x 8"	P2@1.04	Ea	47.40	25.60	73.00
12 x 4"	P2@1.22	Ea	42.50	30.10	72.60
12 x 6"	P2@1.26	Ea	43.80	31.00	74.80
12 x 8"	P2@1.30	Ea	65.50	32.00	97.50
12 x 10"	P2@1.46	Ea	80.00	36.00	116.00
15 x 4"	P2@1.50	Ea	64.10	37.00	101.10
15 x 6"	P2@1.54	Ea	76.80	37.90	114.70
15 x 8"	P2@1.58	Ea	85.60	38.90	124.50
15 x 10"	P2@1.74	Ea	103.00	42.90	145.90
15 x 12"	P2@1.87	Ea	115.00	46.10	161.10
18 x 4"	P2@1.76	Ea	147.00	43.40	190.40
18 x 6"	P2@1.80	Ea	149.00	44.40	193.40
18 x 8"	P2@1.84	Ea	154.00	45.30	199.30
18 x 10"	P2@2.00	Ea	174.00	49.30	223.30
18 x 12"	P2@2.13	Ea	186.00	52.50	238.50
18 x 15"	P2@2.27	Ea	200.00	55.90	255.90

PVC sewer pipe reducing wye B x S x B

Description	Craft@Hrs	Unit	Material $	Labor $	Total $
10 x 4"	P2@.960	Ea	28.70	23.70	52.40
10 x 6"	P2@1.00	Ea	30.20	24.60	54.80
10 x 8"	P2@1.04	Ea	48.00	25.60	73.60
12 x 4"	P2@1.22	Ea	43.10	30.10	73.20
12 x 6"	P2@1.26	Ea	44.20	31.00	75.20
12 x 8"	P2@1.30	Ea	64.60	32.00	96.60
12 x 10"	P2@1.46	Ea	79.20	36.00	115.20
15 x 4"	P2@1.50	Ea	66.50	37.00	103.50
15 x 6"	P2@1.54	Ea	74.90	37.90	112.80
15 x 8"	P2@1.58	Ea	87.10	38.90	126.00
15 x 10"	P2@1.74	Ea	101.00	42.90	143.90
15 x 12"	P2@1.87	Ea	111.00	46.10	157.10
18 x 4"	P2@1.76	Ea	138.00	43.40	181.40
18 x 6"	P2@1.80	Ea	149.00	44.40	193.40
18 x 8"	P2@1.84	Ea	154.00	45.30	199.30
18 x 10"	P2@2.00	Ea	174.00	49.30	223.30
18 x 12"	P2@2.13	Ea	186.00	52.50	238.50
18 x 15"	P2@2.27	Ea	200.00	55.90	255.90

PVC, Sewer, with Bell & Spigot Gasketed Joints

Description	Craft@Hrs	Unit	Material $	Labor $	Total $
PVC sewer pipe tee-wye B x B x B					
4"	P2@.480	Ea	3.20	11.80	15.00
6"	P2@.600	Ea	6.85	14.80	21.65
6 x 4"	P2@.560	Ea	7.00	13.80	20.80
8 x 4"	P2@.640	Ea	10.40	15.80	26.20
8 x 6"	P2@.680	Ea	10.70	16.80	27.50
PVC sewer pipe tee B x S x B					
10"	P2@1.20	Ea	49.70	29.60	79.30
12"	P2@1.59	Ea	68.90	39.20	108.10
15"	P2@2.01	Ea	91.90	49.50	141.40
18"	P2@2.40	Ea	184.00	59.10	243.10
PVC sewer pipe reducing tee B x S x B					
10 x 4"	P2@.960	Ea	31.40	23.70	55.10
10 x 6"	P2@1.00	Ea	32.50	24.60	57.10
10 x 8"	P2@1.04	Ea	50.80	25.60	76.40
12 x 4"	P2@1.22	Ea	38.40	30.10	68.50
12 x 6"	P2@1.26	Ea	39.10	31.00	70.10
12 x 8"	P2@1.30	Ea	54.00	32.00	86.00
12 x 10"	P2@1.46	Ea	64.90	36.00	100.90
15 x 4"	P2@1.50	Ea	62.00	37.00	99.00
15 x 6"	P2@1.54	Ea	65.90	37.90	103.80
15 x 8"	P2@1.58	Ea	67.10	38.90	106.00
15 x 10"	P2@1.74	Ea	79.30	42.90	122.20
15 x 12"	P2@1.87	Ea	85.80	46.10	131.90
18 x 4"	P2@1.76	Ea	157.00	43.40	200.40
18 x 6"	P2@1.80	Ea	162.00	44.40	206.40
18 x 8"	P2@1.84	Ea	170.00	45.30	215.30
18 x 10"	P2@2.00	Ea	176.00	49.30	225.30
18 x 12"	P2@2.13	Ea	181.00	52.50	233.50
18 x 15"	P2@2.27	Ea	193.00	55.90	248.90

Description	Craft@Hrs	Unit	Material $	Labor $	Total $
PVC sewer pipe reducer S x B					
6 x 4"	P2@.360	Ea	3.55	8.87	12.42
8 x 4"	P2@.400	Ea	9.70	9.86	19.56
8 x 6"	P2@.440	Ea	10.90	10.80	21.70
10 x 4"	P2@.560	Ea	23.90	13.80	37.70
10 x 6"	P2@.600	Ea	24.90	14.80	39.70
10 x 8"	P2@.640	Ea	29.60	15.80	45.40
12 x 4"	P2@.690	Ea	32.00	17.00	49.00
12 x 6"	P2@.730	Ea	32.70	18.00	50.70
12 x 8"	P2@.770	Ea	37.40	19.00	56.40
12 x 10"	P2@.930	Ea	41.50	22.90	64.40
15 x 4"	P2@.830	Ea	53.20	20.50	73.70
15 x 6"	P2@.870	Ea	55.20	21.40	76.60
15 x 8"	P2@.910	Ea	62.50	22.40	84.90
15 x 10"	P2@1.07	Ea	63.00	26.40	89.40
15 x 12"	P2@1.20	Ea	72.20	29.60	101.80
18 x 8"	P2@1.04	Ea	77.20	25.60	102.80
18 x 10"	P2@1.20	Ea	78.30	29.60	107.90
18 x 12"	P2@1.33	Ea	79.40	32.80	112.20
18 x 15"	P2@1.47	Ea	80.00	36.20	116.20

Description	Craft@Hrs	Unit	Material $	Labor $	Total $
PVC sewer pipe adapter B x S					
4"	P2@.320	Ea	2.08	7.88	9.96
6"	P2@.400	Ea	3.86	9.86	13.72
8"	P2@.480	Ea	9.85	11.80	21.65
10"	P2@.800	Ea	12.30	19.70	32.00
12"	P2@1.06	Ea	18.70	26.10	44.80
15"	P2@1.34	Ea	28.30	33.00	61.30
18"	P2@1.60	Ea	62.20	39.40	101.60

Description	Craft@Hrs	Unit	Material $	Labor $	Total $
PVC sewer pipe cleanout adapter, thread x S					
4"	P2@.160	Ea	.95	3.94	4.89
6"	P2@.180	Ea	4.43	4.44	8.87

Description	Craft@Hrs	Unit	Material $	Labor $	Total $
PVC sewer pipe cap					
4"	P2@.160	Ea	1.26	3.94	5.20
6"	P2@.200	Ea	2.38	4.93	7.31
8"	P2@.240	Ea	6.45	5.91	12.36
10"	P2@.400	Ea	19.00	9.86	28.86
12"	P2@.530	Ea	28.50	13.10	41.60
15"	P2@.670	Ea	46.20	16.50	62.70
18"	P2@.800	Ea	67.10	19.70	86.80

PVC, Sewer, with Bell & Spigot Gasketed Joints

Description	Craft@Hrs	Unit	Material $	Labor $	Total $
PVC sewer pipe test plug					
4"	P2@.160	Ea	.96	3.94	4.90
6"	P2@.200	Ea	1.47	4.93	6.40
8"	P2@.240	Ea	5.23	5.91	11.14
10"	P2@.400	Ea	16.40	9.86	26.26
12"	P2@.530	Ea	19.60	13.10	32.70
15"	P2@.670	Ea	38.60	16.50	55.10
18"	P2@.800	Ea	54.40	19.70	74.10

Description	Craft@Hrs	Unit	Material $	Labor $	Total $

Primed steel cam lock non-fire rated access doors

Description	Craft@Hrs	Unit	Material $	Labor $	Total $
8 x 8"	P2@.400	Ea	28.40	9.86	38.26
12 x 12"	P2@.500	Ea	32.90	12.30	45.20
16 x 16"	P2@.600	Ea	39.70	14.80	54.50
18 x 18"	P2@.800	Ea	46.40	19.70	66.10
24 x 24"	P2@1.25	Ea	68.60	30.80	99.40
36 x 36"	P2@1.60	Ea	127.00	39.40	166.40
Add for Allen key lock	--		4.50	--	4.50
Add for cylinder lock	--		11.10	--	11.10

Primed steel cam-lock 1½ hour fire rated access doors

Description	Craft@Hrs	Unit	Material $	Labor $	Total $
8 x 8"	P2@.400	Ea	105.00	9.86	114.86
12 x 12"	P2@.500	Ea	110.00	12.30	122.30
16 x 16"	P2@.600	Ea	130.00	14.80	144.80
18 x 18"	P2@.800	Ea	133.00	19.70	152.70
24 x 24"	P2@1.25	Ea	185.00	30.80	215.80
24 x 36"	P2@1.50	Ea	240.00	37.00	277.00
48 x 48"	P2@2.00	Ea	370.00	49.30	419.30
Add for cylinder lock	--		11.10	--	11.10

Threaded automatic air vent

Description	Craft@Hrs	Unit	Material $	Labor $	Total $
35 PSIG	P2@.150	Ea	35.90	3.70	39.60
150 PSIG	P2@.150	Ea	59.10	3.70	62.80

Threaded manual air vent

Description	Craft@Hrs	Unit	Material $	Labor $	Total $
175 PSIG	P2@.150	Ea	17.20	3.70	20.90

Threaded angle stop

Description	Craft@Hrs	Unit	Material $	Labor $	Total $
12"	P2@.320	Ea	11.10	7.88	18.98
15"	P2@.320	Ea	11.80	7.88	19.68
18"	P2@.320	Ea	13.00	7.88	20.88
24"	P2@.320	Ea	20.05	7.88	27.93

150 pound forged steel slip-on companion flange, welding type

Description	Craft@Hrs	Unit	Material $	Labor $	Total $
2½"	P2@.610	Ea	41.60	15.00	56.60
3"	P2@.730	Ea	43.60	18.00	61.60
4"	P2@.980	Ea	52.00	24.10	76.10
6"	P2@1.47	Ea	80.00	36.20	116.20
8"	P2@1.77	Ea	140.00	43.60	183.60
10"	P2@2.20	Ea	229.00	54.20	283.20
12"	P2@2.64	Ea	335.00	65.00	400.00

Plumbing and Piping Specialties

Description	Craft@Hrs	Unit	Material $	Labor $	Total $

150 pound companion flange, threaded

2"	P2@.290	Ea	38.40	7.15	45.55
2½"	P2@.380	Ea	44.90	9.36	54.26
3"	P2@.460	Ea	44.90	11.30	56.20
4"	P2@.600	Ea	52.70	14.80	67.50
6"	P2@.680	Ea	86.50	16.80	103.30
8"	P2@.760	Ea	147.00	18.70	165.70

300 pound forged steel slip-on companion flange, welding type

2½"	P2@.810	Ea	55.30	20.00	75.30
3"	P2@.980	Ea	55.30	24.10	79.40
4"	P2@1.30	Ea	91.00	32.00	123.00
6"	P2@1.95	Ea	133.00	48.00	181.00
8"	P2@2.35	Ea	228.00	57.90	285.90
10"	P2@2.90	Ea	452.00	71.50	523.50
12"	P2@3.50	Ea	579.00	86.20	665.20

PVC companion flange

2"	P2@.140	Ea	12.10	3.45	15.55
2½"	P2@.180	Ea	18.70	4.44	23.14
3"	P2@.210	Ea	20.60	5.17	25.77
4"	P2@.280	Ea	26.10	6.90	33.00
5"	P2@.310	Ea	41.00	7.64	48.64
6"	P2@.420	Ea	41.00	10.30	51.30

Bolt and gasket set

2"	P2@.500	Ea	3.85	12.30	16.15
2½"	P2@.650	Ea	4.08	16.00	20.08
3"	P2@.750	Ea	4.28	18.50	22.78
4"	P2@1.00	Ea	7.75	24.60	32.35
5"	P2@1.10	Ea	11.50	27.10	38.60
6"	P2@1.20	Ea	12.10	29.60	41.70
8"	P2@1.25	Ea	12.90	30.80	43.70
10"	P2@1.70	Ea	26.70	41.90	68.60
12"	P2@2.20	Ea	28.60	54.20	82.80

Dielectric union with solder ends and brazed joints

1/2"	P2@.146	Ea	3.39	3.60	6.99
3/4"	P2@.205	Ea	3.39	5.05	8.44
1"	P2@.263	Ea	7.76	6.48	14.24
1¼"	P2@.322	Ea	12.70	7.93	20.63
1½"	P2@.380	Ea	18.60	9.36	27.96
2"	P2@.507	Ea	26.30	12.50	38.80

Description	Craft@Hrs	Unit	Material $	Labor $	Total $

125/150# Galvanized ASME expansion tank

Description	Craft@Hrs	Unit	Material $	Labor $	Total $
15 gallon	P2@1.75	Ea	438.00	43.10	481.10
30 gallon	P2@2.00	Ea	545.00	49.30	594.30
40 gallon	P2@2.35	Ea	628.00	57.90	685.90
60 gallon	P2@2.65	Ea	718.00	65.30	783.30
80 gallon	P2@2.95	Ea	818.00	72.70	890.70
100 gallon	P2@3.25	Ea	1,000.00	80.10	1,080.10
120 gallon	P2@3.50	Ea	1,160.00	86.20	1,246.20

Expansion tank fitting

Description	Craft@Hrs	Unit	Material $	Labor $	Total $
3/4"	P2@.300	Ea	32.40	7.39	39.79

Fire hydrant with spool

Description	Craft@Hrs	Unit	Material $	Labor $	Total $
6"	P2@5.00	Ea	795.00	123.00	918.00

Indicator post

Description	Craft@Hrs	Unit	Material $	Labor $	Total $
6"	P2@6.50	Ea	480.00	160.00	640.00

Lead pipe roof flashing

Description	Craft@Hrs	Unit	Material $	Labor $	Total $
2"	P2@.250	Ea	18.40	6.16	24.56
3"	P2@.250	Ea	21.30	6.16	27.46
4"	P2@.250	Ea	23.50	6.16	29.66
6"	P2@.350	Ea	30.00	8.62	38.62
8"	P2@.400	Ea	38.10	9.86	47.96

Braided stainless steel pipe connector, solder ends, brazed joints

Description	Craft@Hrs	Unit	Material $	Labor $	Total $
1/2"	P2@.240	Ea	19.50	5.91	25.41
3/4"	P2@.300	Ea	26.10	7.39	33.49
1"	P2@.360	Ea	33.80	8.87	42.67
1¼"	P2@.480	Ea	39.10	11.80	50.90
1½"	P2@.540	Ea	44.60	13.30	57.90
2"	P2@.600	Ea	59.60	14.80	74.40

Braided stainless steel pipe connector, threaded ends

Description	Craft@Hrs	Unit	Material $	Labor $	Total $
1/2"	P2@.210	Ea	24.30	5.17	29.47
3/4"	P2@.250	Ea	26.30	6.16	32.46
1"	P2@.300	Ea	28.40	7.39	35.79
1¼"	P2@.400	Ea	29.00	9.86	38.86
1½"	P2@.450	Ea	39.20	11.10	50.30
2"	P2@.500	Ea	50.60	12.30	62.90

Plumbing and Piping Specialties

Description	Craft@Hrs	Unit	Material $	Labor $	Total $

Braided stainless steel pipe connector, flanged ends

Description	Craft@Hrs	Unit	Material $	Labor $	Total $
2"	P2@.500	Ea	58.10	12.30	70.40
2½"	P2@.630	Ea	71.60	15.50	87.10
3"	P2@.750	Ea	92.50	18.50	111.00
4"	P2@1.50	Ea	122.00	37.00	159.00
6"	P2@2.50	Ea	220.00	61.60	281.60
8"	P2@3.00	Ea	491.00	73.90	564.90

Galvanized steel band pipe hanger

Description	Craft@Hrs	Unit	Material $	Labor $	Total $
1/2"	P2@.450	Ea	.56	11.10	11.66
3/4"	P2@.450	Ea	.56	11.10	11.66
1"	P2@.450	Ea	.56	11.10	11.66
1¼"	P2@.450	Ea	.56	11.10	11.66
1½"	P2@.450	Ea	.62	11.10	11.72
2"	P2@.450	Ea	.65	11.10	11.75
2½"	P2@.450	Ea	3.95	11.10	15.05
3"	P2@.450	Ea	4.32	11.10	15.42
4"	P2@.450	Ea	5.87	11.10	16.97
6"	P2@.500	Ea	11.70	12.30	24.00
8"	P2@.550	Ea	17.30	13.60	30.90
10"	P2@.650	Ea	21.00	16.00	37.00
12"	P2@.750	Ea	25.90	18.50	44.40

Trapeze pipe hanger

Description	Craft@Hrs	Unit	Material $	Labor $	Total $
24"	P2@1.00	Ea	20.40	24.60	45.00
36"	P2@1.25	Ea	26.10	30.80	56.90
48"	P2@1.50	Ea	39.60	37.00	76.60

Wall bracket support with anchors and bolts

Description	Craft@Hrs	Unit	Material $	Labor $	Total $
6 x 6"	P2@.600	Ea	9.14	14.80	23.94
8 x 8"	P2@.650	Ea	10.50	16.00	26.50
10 x 10"	P2@.700	Ea	11.70	17.20	28.90
12 x 12"	P2@.800	Ea	14.30	19.70	34.00
15 x 15"	P2@.950	Ea	16.40	23.40	39.80
18 x 18"	P2@1.10	Ea	23.10	27.10	50.20
24 x 24"	P2@1.60	Ea	33.60	39.40	73.00

Galvanized U-bolts with nuts

Description	Craft@Hrs	Unit	Material $	Labor $	Total $
1/2"	P2@.160	Ea	1.38	3.94	5.32
3/4"	P2@.165	Ea	1.38	4.07	5.45
1"	P2@.170	Ea	1.56	4.19	5.75
1¼"	P2@.175	Ea	1.24	4.31	5.55
1½"	P2@.180	Ea	1.98	4.44	6.42
2"	P2@.190	Ea	2.13	4.68	6.81
2½"	P2@.210	Ea	3.57	5.17	8.74
3"	P2@.250	Ea	3.75	6.16	9.91
4"	P2@.300	Ea	4.02	7.39	11.41
6"	P2@.400	Ea	7.47	9.86	17.33

Description	Craft@Hrs	Unit	Material $	Labor $	Total $
Nail-on wire pipe hooks					
1/2"	P2@.100	Ea	.05	2.46	2.51
3/4"	P2@.100	Ea	.06	2.46	2.52
1"	P2@.100	Ea	.07	2.46	2.53
Galvanized steel pipe sleeves					
1"	P2@.120	Ea	1.79	2.96	4.75
1¼"	P2@.125	Ea	2.00	3.08	5.08
1½"	P2@.125	Ea	2.25	3.08	5.33
2"	P2@.130	Ea	2.43	3.20	5.63
2½"	P2@.150	Ea	2.47	3.70	6.17
3"	P2@.180	Ea	2.53	4.44	6.97
4"	P2@.220	Ea	2.88	5.42	8.30
5"	P2@.250	Ea	3.57	6.16	9.73
6"	P2@.270	Ea	3.86	6.65	10.51
8"	P2@.270	Ea	4.44	6.65	11.09
10"	P2@.290	Ea	5.17	7.15	12.32
12"	P2@.310	Ea	7.01	7.64	14.65
14"	P2@.330	Ea	13.10	8.13	21.23
Dial-type pressure gauge					
2½"	P2@.200	Ea	15.60	4.93	20.53
3½"	P2@.200	Ea	27.10	4.93	32.03
Riser clamp					
1/2"	P2@.100	Ea	3.27	2.46	5.73
3/4"	P2@.100	Ea	3.27	2.46	5.73
1"	P2@.100	Ea	3.30	2.46	5.76
1¼"	P2@.105	Ea	4.01	2.59	6.60
1½"	P2@.110	Ea	4.23	2.71	6.94
2"	P2@.115	Ea	4.45	2.83	7.28
2½"	P2@.120	Ea	4.69	2.96	7.65
3"	P2@.120	Ea	5.13	2.96	8.09
4"	P2@.125	Ea	6.45	3.08	9.53
5"	P2@.180	Ea	8.58	4.44	13.02
6"	P2@.200	Ea	10.70	4.93	15.63
8"	P2@.200	Ea	17.40	4.93	22.33
10"	P2@.250	Ea	25.90	6.16	32.06
12"	P2@.250	Ea	30.70	6.16	36.86
15 PSIG float and thermostatic steam trap, threaded					
3/4"	P2@.250	Ea	127.00	6.16	133.16
1"	P2@.300	Ea	150.00	7.39	157.39
1¼"	P2@.400	Ea	184.00	9.86	193.86
1½"	P2@.450	Ea	254.00	11.10	265.10
2"	P2@.500	Ea	484.00	12.30	496.30

Description	Craft@Hrs	Unit	Material $	Labor $	Total $

15 PSIG inverted bucket steam trap, threaded

Description	Craft@Hrs	Unit	Material $	Labor $	Total $
1/2"	P2@.210	Ea	86.50	5.17	91.67
3/4"	P2@.250	Ea	86.50	6.16	92.66
1"	P2@.300	Ea	152.00	7.39	159.39
1¼"	P2@.400	Ea	185.00	9.86	194.86
1½"	P2@.450	Ea	269.00	11.10	280.10
2"	P2@.500	Ea	490.00	12.30	502.30

Thermometer with well

Description	Craft@Hrs	Unit	Material $	Labor $	Total $
7"	P2@.250	Ea	14.30	6.16	20.46
9"	P2@.250	Ea	14.80	6.16	20.96

Class 125 bronze body 2-piece ball valve, solder ends, brazed joints

Description	Craft@Hrs	Unit	Material $	Labor $	Total $
1/2"	P2@.240	Ea	6.10	5.91	12.01
3/4"	P2@.300	Ea	9.70	7.39	17.09
1"	P2@.360	Ea	12.40	8.87	21.27
1¼"	P2@.480	Ea	21.70	11.80	33.50
1½"	P2@.540	Ea	27.10	13.30	40.40
2"	P2@.600	Ea	34.90	14.80	49.70

Class 125 bronze body 2-piece ball valve, solder ends

Description	Craft@Hrs	Unit	Material $	Labor $	Total $
1/2"	P2@.240	Ea	6.10	5.91	12.01
3/4"	P2@.300	Ea	9.70	7.39	17.09
1"	P2@.360	Ea	12.40	8.87	21.27
1¼"	P2@.480	Ea	21.70	11.80	33.50
1½"	P2@.540	Ea	27.10	13.30	40.40
2"	P2@.600	Ea	34.90	14.80	49.70

Class 125 bronze body 2-piece ball valve, threaded ends

Description	Craft@Hrs	Unit	Material $	Labor $	Total $
1/2"	P2@.240	Ea	6.10	5.91	12.01
3/4"	P2@.300	Ea	9.70	7.39	17.09
1"	P2@.360	Ea	12.40	8.87	21.27
1¼"	P2@.480	Ea	21.70	11.80	33.50
1½"	P2@.540	Ea	27.10	13.30	40.40
2"	P2@.600	Ea	34.90	14.80	49.70

Class 150, 600 pound W.O.G. bronze body 2-piece ball valve

Description	Craft@Hrs	Unit	Material $	Labor $	Total $
1/2"	P2@.210	Ea	5.52	5.17	10.69
3/4"	P2@.250	Ea	9.04	6.16	15.20
1"	P2@.300	Ea	11.50	7.39	18.89
1¼"	P2@.400	Ea	19.20	9.86	29.06
1½"	P2@.450	Ea	24.60	11.10	35.70
2"	P2@.500	Ea	30.80	12.30	43.10

Description	Craft@Hrs	Unit	Material $	Labor $	Total $

200 PSIG iron body butterfly valve, lug type

Description	Craft@Hrs	Unit	Material $	Labor $	Total $
2"	P2@.200	Ea	77.60	4.93	82.53
2½"	P2@.300	Ea	81.00	7.39	88.39
3"	P2@.400	Ea	89.80	9.86	99.66
4"	P2@.500	Ea	120.00	12.30	132.30
6"	P2@.750	Ea	163.00	18.50	181.50
8"	P2@.950	Ea	213.00	23.40	236.40
10"	P2@1.40	Ea	328.00	34.50	362.50
12"	P2@1.80	Ea	477.00	44.40	521.40

200 PSIG iron body butterfly valve, wafer type

Description	Craft@Hrs	Unit	Material $	Labor $	Total $
2"	P2@.200	Ea	66.80	4.93	71.73
2½"	P2@.300	Ea	69.50	7.39	76.89
3"	P2@.400	Ea	76.30	9.86	86.16
4"	P2@.500	Ea	95.20	12.30	107.50
6"	P2@.750	Ea	148.00	18.50	166.50
8"	P2@.950	Ea	210.00	23.40	233.40
10"	P2@1.40	Ea	295.00	34.50	329.50
12"	P2@1.80	Ea	414.00	44.40	458.40

Class 125 bronze body swing check valve, solder ends with brazed joints

Description	Craft@Hrs	Unit	Material $	Labor $	Total $
1/2"	P2@.240	Ea	32.70	5.91	38.61
3/4"	P2@.300	Ea	41.00	7.39	48.39
1"	P2@.360	Ea	55.20	8.87	64.07
1¼"	P2@.480	Ea	71.90	11.80	83.70
1½"	P2@.540	Ea	95.30	13.30	108.60
2"	P2@.600	Ea	140.00	14.80	154.80
2½"	P2@1.00	Ea	222.00	24.60	246.60
3"	P2@1.50	Ea	329.00	37.00	366.00

Class 125 bronze body swing check valve, solder ends with soft-soldered joints

Description	Craft@Hrs	Unit	Material $	Labor $	Total $
1/2"	P2@.200	Ea	32.70	4.93	37.63
3/4"	P2@.249	Ea	41.00	6.14	47.14
1"	P2@.299	Ea	55.20	7.37	62.57
1¼"	P2@.398	Ea	71.90	9.81	81.71
1½"	P2@.448	Ea	95.30	11.00	106.30
2"	P2@.498	Ea	140.00	12.30	152.30
2½"	P2@.830	Ea	222.00	20.50	242.50
3"	P2@1.24	Ea	329.00	30.60	359.60

Class 125 bronze body swing check valve, threaded joints

Description	Craft@Hrs	Unit	Material $	Labor $	Total $
1/2"	P2@.210	Ea	20.80	5.17	25.97
3/4"	P2@.250	Ea	25.10	6.16	31.26
1"	P2@.300	Ea	34.60	7.39	41.99
1¼"	P2@.400	Ea	47.90	9.86	57.76
1½"	P2@.450	Ea	57.10	11.10	68.20
2"	P2@.500	Ea	83.30	12.30	95.60

Plumbing and Piping Specialties

Description	Craft@Hrs	Unit	Material $	Labor $	Total $

Class 300 bronze body swing check valve, threaded joints

Description	Craft@Hrs	Unit	Material $	Labor $	Total $
1/2"	P2@.210	Ea	32.80	5.17	37.97
3/4"	P2@.250	Ea	40.90	6.16	47.06
1"	P2@.300	Ea	55.10	7.39	62.49
1¼"	P2@.400	Ea	71.80	9.86	81.66
1½"	P2@.450	Ea	95.20	11.10	106.30
2"	P2@.500	Ea	141.00	12.30	153.30

Class 125 iron body swing check valve, flanged joints

Description	Craft@Hrs	Unit	Material $	Labor $	Total $
2"	P2@.500	Ea	174.00	12.30	186.30
2½"	P2@.600	Ea	221.00	14.80	235.80
3"	P2@.750	Ea	240.00	18.50	258.50
4"	P2@1.35	Ea	378.00	33.30	411.30
6"	P2@2.50	Ea	648.00	61.60	709.60
8"	P2@3.00	Ea	1,220.00	73.90	1,293.90
10"	P2@4.00	Ea	2,080.00	98.60	2,178.60
12"	P2@4.50	Ea	3,250.00	111.00	3,361.00

Class 250 iron body swing check valve, flanged joints

Description	Craft@Hrs	Unit	Material $	Labor $	Total $
2"	P2@.500	Ea	524.00	12.30	536.30
2½"	P2@.600	Ea	615.00	14.80	629.80
3"	P2@.750	Ea	768.00	18.50	786.50
4"	P2@1.35	Ea	985.00	33.30	1,018.30
6"	P2@2.50	Ea	1,780.00	61.60	1,841.60
8"	P2@3.00	Ea	2,990.00	73.90	3,063.90

Class 125 iron body silent check valve, wafer type

Description	Craft@Hrs	Unit	Material $	Labor $	Total $
2"	P2@.500	Ea	97.20	12.30	109.50
2½"	P2@.600	Ea	112.00	14.80	126.80
3"	P2@.750	Ea	124.00	18.50	142.50
4"	P2@1.35	Ea	167.00	33.30	200.30

Class 125 iron body silent check valve, flanged joints

Description	Craft@Hrs	Unit	Material $	Labor $	Total $
2"	P2@.500	Ea	104.00	12.30	116.30
2½"	P2@.600	Ea	120.00	14.80	134.80
3"	P2@.750	Ea	132.00	18.50	150.50
4"	P2@1.35	Ea	247.00	33.30	280.30
6"	P2@2.50	Ea	336.00	61.60	397.60
8"	P2@3.00	Ea	578.00	73.90	651.90
10"	P2@4.00	Ea	880.00	98.60	978.60

Description	Craft@Hrs	Unit	Material $	Labor $	Total $

Class 250 iron body silent check valve, flanged joints

Description	Craft@Hrs	Unit	Material $	Labor $	Total $
2"	P2@.500	Ea	104.00	12.30	116.30
2½"	P2@.600	Ea	120.00	14.80	134.80
3"	P2@.750	Ea	132.00	18.50	150.50
4"	P2@1.35	Ea	178.00	33.30	211.30
6"	P2@2.50	Ea	321.00	61.60	382.60
8"	P2@3.00	Ea	706.00	73.90	779.90
10"	P2@4.00	Ea	1,070.00	98.60	1,168.60

Class 125 bronze body gate valve, solder ends

Description	Craft@Hrs	Unit	Material $	Labor $	Total $
1/2"	P2@.240	Ea	15.00	5.91	20.91
3/4"	P2@.300	Ea	18.40	7.39	25.79
1"	P2@.360	Ea	23.00	8.87	31.87
1¼"	P2@.480	Ea	38.80	11.80	50.60
1½"	P2@.540	Ea	44.30	13.30	57.60
2"	P2@.600	Ea	61.30	14.80	76.10
2½"	P2@1.00	Ea	143.00	24.60	167.60
3"	P2@1.50	Ea	204.00	37.00	241.00

Class 125 bronze body gate valve, threaded joints

Description	Craft@Hrs	Unit	Material $	Labor $	Total $
1/2"	P2@.240	Ea	13.90	5.91	19.81
3/4"	P2@.300	Ea	17.10	7.39	24.49
1"	P2@.360	Ea	23.20	8.87	32.07
1¼"	P2@.480	Ea	33.00	11.80	44.80
1½"	P2@.540	Ea	40.10	13.30	53.40
2"	P2@.600	Ea	50.00	14.80	64.80
2½"	P2@1.00	Ea	133.00	24.60	157.60
3"	P2@1.50	Ea	184.00	37.00	221.00

Class 300 bronze body gate valve, threaded joints

Description	Craft@Hrs	Unit	Material $	Labor $	Total $
1/2"	P2@.210	Ea	44.90	5.17	50.07
3/4"	P2@.250	Ea	53.40	6.16	59.56
1"	P2@.300	Ea	76.10	7.39	83.49
1¼"	P2@.400	Ea	101.00	9.86	110.86
1½"	P2@.450	Ea	119.00	11.10	130.10
2"	P2@.500	Ea	172.00	12.30	184.30
2½"	P2@.750	Ea	541.00	18.50	559.50
3"	P2@.950	Ea	691.00	23.40	714.40

Plumbing and Piping Specialties

Description	Craft@Hrs	Unit	Material $	Labor $	Total $
Class 125 iron body gate valve, flanged ends					
2½"	P2@.600	Ea	210.00	14.80	224.80
3"	P2@.750	Ea	228.00	18.50	246.50
4"	P2@1.35	Ea	359.00	33.30	392.30
6"	P2@2.50	Ea	733.00	61.60	794.60
8"	P2@3.00	Ea	1,320.00	73.90	1,393.90
10"	P2@4.00	Ea	2,630.00	98.60	2,728.60
12"	P2@4.50	Ea	3,490.00	111.00	3,601.00
Class 250 iron body gate valve, flanged ends					
2½"	P2@.600	Ea	476.00	14.80	490.80
3"	P2@.750	Ea	530.00	18.50	548.50
4"	P2@1.35	Ea	782.00	33.30	815.30
6"	P2@2.50	Ea	1,300.00	61.60	1,361.60
8"	P2@3.00	Ea	2,280.00	73.90	2,353.90
10"	P2@4.00	Ea	3,440.00	98.60	3,538.60
12"	P2@4.50	Ea	5,360.00	111.00	5,471.00
Class 125 bronze body globe valve, solder ends					
1/2"	P2@.240	Ea	27.80	5.91	33.71
3/4"	P2@.300	Ea	36.60	7.39	43.99
1"	P2@.360	Ea	49.70	8.87	58.57
1¼"	P2@.480	Ea	64.90	11.80	76.70
1½"	P2@.540	Ea	88.00	13.30	101.30
2"	P2@.600	Ea	127.00	14.80	141.80
Class 125 bronze body globe valve, solder ends					
1/2"	P2@.200	Ea	23.20	4.93	28.13
3/4"	P2@.249	Ea	30.50	6.14	36.64
1"	P2@.299	Ea	41.40	7.37	48.77
1¼"	P2@.398	Ea	64.90	9.81	74.71
1½"	P2@.448	Ea	88.00	11.00	99.00
2"	P2@.498	Ea	127.00	12.30	139.30
Class 125 bronze body globe valve, threaded ends					
1/2"	P2@.210	Ea	19.40	5.17	24.57
3/4"	P2@.250	Ea	24.60	6.16	30.76
1"	P2@.300	Ea	32.90	7.39	40.29
1¼"	P2@.400	Ea	43.60	9.86	53.46
1½"	P2@.450	Ea	56.30	11.10	67.40
2"	P2@.500	Ea	85.70	12.30	98.00

Description	Craft@Hrs	Unit	Material $	Labor $	Total $

Class 300 bronze body globe valve, threaded ends

1/2"	P2@.210	Ea	41.90	5.17	47.07
3/4"	P2@.250	Ea	58.40	6.16	64.56
1"	P2@.300	Ea	70.70	7.39	78.09
1¼"	P2@.400	Ea	97.90	9.86	107.76
1½"	P2@.450	Ea	138.00	11.10	149.10
2"	P2@.500	Ea	215.00	12.30	227.30

Class 125 iron body globe valve, flanged ends

2½"	P2@.600	Ea	337.00	14.80	351.80
3"	P2@.750	Ea	390.00	18.50	408.50
4"	P2@1.35	Ea	558.00	33.30	591.30
6"	P2@2.50	Ea	1,020.00	61.60	1,081.60
8"	P2@3.00	Ea	1,990.00	73.90	2,063.90
10"	P2@4.00	Ea	3,110.00	98.60	3,208.60

Class 250 iron body globe valve, flanged ends

2½"	P2@.600	Ea	654.00	14.80	668.80
3"	P2@.750	Ea	678.00	18.50	696.50
4"	P2@1.35	Ea	990.00	33.30	1,023.30
6"	P2@2.50	Ea	1,780.00	61.60	1,841.60
8"	P2@3.00	Ea	3,020.00	73.90	3,093.90

Class 150 bronze body strainer, threaded ends

1/2"	P2@.230	Ea	15.10	5.67	20.77
3/4"	P2@.260	Ea	19.60	6.41	26.01
1"	P2@.330	Ea	24.80	8.13	32.93
1¼"	P2@.440	Ea	31.00	10.80	41.80
1½"	P2@.495	Ea	40.20	12.20	52.40
2"	P2@.550	Ea	65.20	13.60	78.80

Class 125 iron body strainer, flanged ends

2"	P2@.500	Ea	121.00	12.30	133.30
2½"	P2@.600	Ea	122.00	14.80	136.80
3"	P2@.750	Ea	145.00	18.50	163.50
4"	P2@1.35	Ea	258.00	33.30	291.30
6"	P2@2.50	Ea	530.00	61.60	591.60
8"	P2@3.00	Ea	846.00	73.90	919.90

Class 250 iron body strainer, flanged ends

2"	P2@.500	Ea	166.00	12.30	178.30
2½"	P2@.600	Ea	172.00	14.80	186.80
3"	P2@.750	Ea	207.00	18.50	225.50
4"	P2@1.35	Ea	376.00	33.30	409.30
6"	P2@2.50	Ea	663.00	61.60	724.60
8"	P2@3.00	Ea	1,090.00	73.90	1,163.90

Plumbing and Piping Specialties

Description	Craft@Hrs	Unit	Material $	Labor $	Total $
Bronze body circuit balancer valve, threaded ends					
1/2"	P2@.210	Ea	9.80	5.17	14.97
3/4"	P2@.250	Ea	11.50	6.16	17.66
1"	P2@.300	Ea	16.60	7.39	23.99
1¼"	P2@.400	Ea	21.00	9.86	30.86
Lever handled gas valve, threaded ends					
1/2"	P2@.210	Ea	21.90	5.17	27.07
3/4"	P2@.250	Ea	25.00	6.16	31.16
1"	P2@.300	Ea	31.00	7.39	38.39
1¼"	P2@.400	Ea	42.40	9.86	52.26
1½"	P2@.450	Ea	45.90	11.10	57.00
2"	P2@.500	Ea	67.30	12.30	79.60
Plug-type gas valve, flanged					
2"	P2@.500	Ea	83.70	12.30	96.00
2½"	P2@.600	Ea	110.00	14.80	124.80
3"	P2@.750	Ea	139.00	18.50	157.50
4"	P2@1.35	Ea	185.00	33.30	218.30
6"	P2@2.50	Ea	446.00	61.60	507.60
Hose bibb					
Low quality, 3/4"	P2@.350	Ea	13.70	8.62	22.32
Med. quality, 3/4"	P2@.350	Ea	20.00	8.62	28.62
High quality, 3/4"	P2@.350	Ea	22.40	8.62	31.02
Water pressure reducing valve, threaded ends					
1/2"	P2@.210	Ea	18.60	5.17	23.77
3/4"	P2@.250	Ea	21.40	6.16	27.56
1"	P2@.300	Ea	102.00	7.39	109.39
1¼"	P2@.400	Ea	115.00	9.86	124.86
1½"	P2@.450	Ea	200.00	11.10	211.10
2"	P2@.500	Ea	375.00	12.30	387.30
Reduced pressure backflow preventer, threaded ends					
3/4"	P2@1.00	Ea	94.30	24.60	118.90
1"	P2@1.25	Ea	107.00	30.80	137.80
1¼"	P2@1.50	Ea	258.00	37.00	295.00
1½"	P2@2.00	Ea	269.00	49.30	318.30
2"	P2@2.12	Ea	314.00	52.20	366.20
Water hammer arrester, threaded ends					
1/2"	P2@.160	Ea	20.10	3.94	24.04
3/4"	P2@.190	Ea	23.10	4.68	27.78
1"	P2@.230	Ea	59.10	5.67	64.77
1¼"	P2@.300	Ea	68.50	7.39	75.89
1½"	P2@.340	Ea	193.40	8.38	201.78
2"	P2@.380	Ea	149.00	9.36	158.36

Description	Craft@Hrs	Unit	Material $	Labor $	Total $

Carbon steel weldolet, Schedule 40

Description	Craft@Hrs	Unit	Material $	Labor $	Total $
1/2"	P2@.330	Ea	7.14	8.13	15.27
3/4"	P2@.500	Ea	7.31	12.30	19.61
1"	P2@.670	Ea	7.66	16.50	24.16
1¼"	P2@.830	Ea	9.31	20.50	29.81
1½"	P2@1.00	Ea	9.31	24.60	33.91
2"	P2@1.33	Ea	9.70	32.80	42.50
2½"	P2@1.67	Ea	22.00	41.10	63.10
3"	P2@2.00	Ea	25.60	49.30	74.90
4"	P2@2.67	Ea	32.70	65.80	98.50
6"	P2@4.00	Ea	90.80	98.60	189.40
8"	P2@4.80	Ea	164.00	118.00	282.00
10"	P2@6.00	Ea	374.00	148.00	522.00
12"	P2@7.20	Ea	691.00	177.00	868.00

Carbon steel weldolet, Schedule 80

Description	Craft@Hrs	Unit	Material $	Labor $	Total $
1/2"	P2@.440	Ea	7.80	10.80	18.60
3/4"	P2@.670	Ea	8.02	16.50	24.52
1"	P2@.890	Ea	8.72	21.90	30.62
1¼"	P2@1.11	Ea	11.10	27.40	38.50
1½"	P2@1.33	Ea	11.10	32.80	43.90
2"	P2@1.77	Ea	11.80	43.60	55.40
2½"	P2@2.22	Ea	26.60	54.70	81.30
3"	P2@2.66	Ea	27.30	65.50	92.80
4"	P2@3.55	Ea	34.10	87.50	121.60
6"	P2@5.32	Ea	142.00	131.00	273.00
8"	P2@6.38	Ea	257.00	157.00	414.00

Carbon steel threadolet, Schedule 40

Description	Craft@Hrs	Unit	Material $	Labor $	Total $
3/4"	P2@.330	Ea	3.73	8.13	11.86
1"	P2@.440	Ea	4.39	10.80	15.19
1¼"	P2@.560	Ea	4.92	13.80	18.72
1½"	P2@.670	Ea	6.97	16.50	23.47
2"	P2@.890	Ea	7.57	21.90	29.47
2½"	P2@1.11	Ea	8.72	27.40	36.12

Carbon steel threadolet, Schedule 80

Description	Craft@Hrs	Unit	Material $	Labor $	Total $
3/4"	P2@.440	Ea	6.59	10.80	17.39
1"	P2@.590	Ea	7.29	14.50	21.79
1¼"	P2@.740	Ea	8.96	18.20	27.16
1½"	P2@.890	Ea	40.40	21.90	62.30
2"	P2@1.18	Ea	40.40	29.10	69.50
2½"	P2@1.48	Ea	46.90	36.50	83.40

Plumbing Equipment, Fixtures and Accessories

Description	Craft@Hrs	Unit	Material $	Labor $	Total $

White enameled steel bathtub with trim

60"	P2@3.50	Ea	289.00	86.20	375.20

Fiberglass bathtub with trim

60"	P2@3.00	Ea	416.00	73.90	489.90

Fiberglass one-piece combination bathtub/shower with trim

60"	P2@5.50	Ea	395.00	136.00	531.00

Add for shower over tub

--	P2@.500	Ea	130.00	12.30	142.30

Wall cleanout plug, with chrome cover

2"	P2@.350	Ea	19.80	8.62	28.42
3"	P2@.350	Ea	18.40	8.62	27.02
4"	P2@.400	Ea	33.00	9.86	42.86

Bronze floor cleanout plug

2"	P2@.250	Ea	10.70	6.16	16.86
3"	P2@.250	Ea	18.40	6.16	24.56
4"	P2@.300	Ea	33.00	7.39	40.39

Cast iron area drain with flat grate and clamp

4"	P2@1.00	Ea	91.10	24.60	115.70
6"	P2@1.15	Ea	92.20	28.30	120.50
8"	P2@1.25	Ea	97.80	30.80	128.60

Cast iron floor drain with chrome grate

4"	P2@1.00	Ea	29.90	24.60	54.50
5"	P2@1.10	Ea	30.10	27.10	57.20
6"	P2@1.15	Ea	38.30	28.30	66.60
8"	P2@1.25	Ea	68.90	30.80	99.70

Cast iron planter drain

4"	P2@1.00	Ea	14.70	24.60	39.30
5"	P2@1.10	Ea	16.90	27.10	44.00
6"	P2@1.15	Ea	20.80	28.30	49.10
8"	P2@1.25	Ea	29.70	30.80	60.50

Description	Craft@Hrs	Unit	Material $	Labor $	Total $
Plastic planter drain					
4"	P2@.800	Ea	4.52	19.70	24.22
5"	P2@1.00	Ea	4.81	24.60	29.41
6"	P2@1.05	Ea	7.91	25.90	33.81
8"	P2@1.10	Ea	12.00	27.10	39.10
Cast iron roof and overflow drain with clamp					
2"	P2@1.20	Ea	53.50	29.60	83.10
3"	P2@1.20	Ea	53.50	29.60	83.10
4"	P2@1.25	Ea	53.50	30.80	84.30
5"	P2@1.40	Ea	60.60	34.50	95.10
6"	P2@1.60	Ea	60.60	39.40	100.00
Wall hung stainless steel refrigerated drinking fountain					
--	P2@2.00	Ea	748.00	49.30	797.30
Free-standing stainless steel recessed drinking fountain					
--	P2@2.50	Ea	776.00	61.60	837.60
Wall hung stainless steel un-refrigerated drinking fountain					
--	P2@1.50	Ea	445.00	37.00	482.00
Add-on cost for remote chiller					
--	P2@2.65	Ea	650.00	65.30	715.30
Galvanized roof flashing					
1"	P2@.200	Ea	3.10	4.93	8.03
1¼"	P2@.200	Ea	3.10	4.93	8.03
1½"	P2@.250	Ea	3.75	6.16	9.91
2"	P2@.250	Ea	4.51	6.16	10.67
2½"	P2@.250	Ea	4.86	6.16	11.02
3"	P2@.250	Ea	5.22	6.16	11.38
4"	P2@.250	Ea	5.75	6.16	11.91
5"	P2@.300	Ea	6.00	7.39	13.39
6"	P2@.350	Ea	6.25	8.62	14.87
8"	P2@.400	Ea	8.50	9.86	18.36
Type B double wall water heater flue					
2"	P2@.090	LF	2.00	2.22	4.22
3"	P2@.100	LF	2.47	2.46	4.93
4"	P2@.110	LF	3.29	2.71	6.00
6"	P2@.130	LF	3.76	3.20	6.96

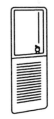

Plumbing Equipment, Fixtures and Accessories

Description	Craft@Hrs	Unit	Material $	Labor $	Total $
Grease and oil interceptor					
4 GPM	P2@4.00	Ea	315.00	98.60	413.60
10 GPM	P2@5.00	Ea	438.00	123.00	561.00
15 GPM	P2@7.00	Ea	514.00	172.00	686.00
20 GPM	P2@8.00	Ea	763.00	197.00	960.00
Hair and lint interceptor					
1½"	P2@.650	Ea	118.00	16.00	134.00
2"	P2@.750	Ea	179.00	18.50	197.50
Wall hung white vitreous china lavatory					
19 x 17"	P2@1.60	Ea	434.00	39.40	473.40
White vitreous china countertop lavatory					
19 x 17"	P2@1.75	Ea	285.00	43.10	328.10
White enameled steel countertop lavatory					
18" round	P2@1.70	Ea	140.00	41.90	181.90
White plastic countertop lavatory					
18" oval	P2@1.70	Ea	92.00	41.90	133.90
Stainless steel countertop bar sink with pantry faucet					
15 x 15"	P2@1.70	Ea	250.00	41.90	291.90
All bronze ¾" to 1½" in-line NPT pump					
1/12 HP	P2@1.50	Ea	195.00	37.00	232.00
1/6 HP	P2@1.50	Ea	379.00	37.00	416.00
1/4 HP	P2@1.50	Ea	543.00	37.00	580.00
Enameled steel floor sink					
9 x 9"	P2@1.00	Ea	17.20	24.60	41.80
12 x 12"	P2@1.00	Ea	22.50	24.60	47.10
15 x 15"	P2@1.15	Ea	25.40	28.30	53.70
18 x 18"	P2@1.25	Ea	37.40	30.80	68.20
24 x 24"	P2@1.50	Ea	50.10	37.00	87.10
Stainless steel 2-compartment kitchen sink					
21 x 24"	P2@1.50	Ea	185.00	37.00	222.00
21 x 32"	P2@1.60	Ea	210.00	39.40	249.40
Vitreous china 2-compartment sink with faucet and spray					
21 x 24"	P2@1.70	Ea	362.00	41.90	403.90
21 x 32"	P2@1.80	Ea	378.00	44.40	422.40

Description	Craft@Hrs	Unit	Material $	Labor $	Total $
Cast iron floor mounted service sink with wall mounted faucet					
22 x 18"	P2@3.00	Ea	250.00	73.90	323.90
Cast iron wall hung service sink with wall mounted faucet					
22 x 18"	P2@2.65	Ea	290.00	65.30	355.30
Composition wall hung laundry sink with wall mounted faucet					
24 x 20"	P2@2.80	Ea	180.00	69.00	249.00
Composition floor mounted laundry sink with faucet and stand					
24 x 20"	P2@2.50	Ea	165.00	61.60	226.60
Precast basin shower with valve and trim					
32 x 32"	P2@4.00	Ea	308.00	98.60	406.60
One-piece fiberglass shower with valve and trim					
32 x 32"	P2@3.25	Ea	206.00	80.10	286.10
White vitreous china wall hung urinal with flush valve					
--	P2@1.60	Ea	183.00	39.40	222.40
White vitreous china stall type urinal with flush valve					
--	P2@2.60	Ea	344.00	64.10	408.10
White vitreous china floor mounted water closet with flush tank					
--	P2@1.50	Ea	216.00	37.00	253.00
White vitreous china floor mounted water closet with flush valve					
--	P2@1.75	Ea	255.00	43.10	298.10
Vitreous china wall hung water closet with flush valve and carrier					
--	P2@2.00	Ea	350.00	49.30	399.30

Plumbing Equipment, Fixtures and Accessories

Description	Craft@Hrs	Unit	Material $	Labor $	Total $
Electric storage-type water heater					
6 gallon	P2@.750	Ea	305.00	18.50	323.50
10 gallon	P2@.800	Ea	327.00	19.70	346.70
15 gallon	P2@1.00	Ea	346.00	24.60	370.60
20 gallon	P2@1.50	Ea	364.00	37.00	401.00
30 gallon	P2@1.75	Ea	397.00	43.10	440.10
Gas-fired storage-type water heater					
30 gallon	P2@1.00	Ea	304.00	24.60	328.60
40 gallon	P2@1.50	Ea	355.00	37.00	392.00
50 gallon	P2@1.75	Ea	413.00	43.10	456.10
100 gallon	P2@2.00	Ea	1,000.00	49.30	1,049.30

(For equipment furnished and set by others)

Kitchen equipment booster heater

1,000 watt	P2@4.00	Ea	94.00	98.60	192.60

Commercial dishwasher

Built-in	P2@5.00	Ea	635.00	123.00	758.00

Garbage disposal

1/2 HP	P2@2.00	Ea	75.00	49.30	124.30
3/4 HP	P2@2.10	Ea	160.00	51.70	211.70

Kitchen appliance gas trim

1/2"	P2@1.15	Ea	18.10	28.30	46.40
3/4"	P2@1.30	Ea	22.90	32.00	54.90
1"	P2@1.60	Ea	26.50	39.40	65.90
1¼"	P2@2.10	Ea	43.40	51.70	95.10
1½"	P2@2.50	Ea	55.30	61.60	116.90
2"	P2@3.00	Ea	73.60	73.90	147.50

Hot and cold water supply

1/2"	P2@1.10	Ea	16.90	27.10	44.00
3/4"	P2@1.55	Ea	24.00	38.20	62.20
1"	P2@1.90	Ea	32.50	46.80	79.30
1¼"	P2@2.50	Ea	45.80	61.60	107.40
1½"	P2@3.00	Ea	57.80	73.90	131.70

Continuous waste

2-part	P2@.250	Ea	25.80	6.16	31.96
3-part	P2@.350	Ea	43.50	8.62	52.12
4-part	P2@.450	Ea	56.40	11.10	67.50

Indirect waste

1/2"	P2@1.05	Ea	3.50	25.90	29.40
3/4"	P2@1.50	Ea	5.84	37.00	42.84
1"	P2@1.90	Ea	9.35	46.80	56.15
1¼"	P2@2.15	Ea	14.00	53.00	67.00
1½"	P2@2.60	Ea	18.20	64.10	82.30
2"	P2@3.00	Ea	28.10	73.90	102.00

Kitchen fixture waste tailpiece

1½"	P2@.100	Ea	5.40	2.46	7.86

Kitchen fixture trap with solder bushing

1½"	P2@.250	Ea	16.40	6.16	22.56
2"	P2@.300	Ea	22.90	7.39	30.29

Water Coil Piping

Engineering drawings of coil piping details have a bad reputation among HVAC contractors and estimators. They're notorious for what they leave out. They'll rarely show more than one coil bank, no matter how big the system is. Furthermore, the drawings hardly ever call out sizes for either the piping or the control valves. Don't be taken in by the apparent simplicity of the system as shown in these drawings. It's likely to be only the tip of the iceberg. For example, unless the air handling capacity of the system is less than 16,000 CFM the single coil bank shown won't be adequate. Add one or two more coil banks and you're looking at a lot more piping -- and a more complex system that takes longer to install.

You probably haven't even decided which equipment supplier to use. This is hardly the time for you to start researching heating and cooling coils. Nevertheless, you need better, more complete and realistic data to come up with a competitive estimate.

We'll see how and where to track down the hard data that you have to take the time to find. I'll also pass along a few tips on estimating water coil piping. They'll help you avoid leaving something out of your estimates. That's a real pitfall for any beginner. It's those little (but essential) items, so easily overlooked, that are so deadly to a profit margin. Finally, at the end of this section there are three pages of diagrams with tables. The data given there, combined with the data you collected earlier, forms the basis for informed guesswork.

The Hard Data

There are two things you absolutely must know to estimate water coil piping. First, the size of the branch piping to the coils. Second, the CFM rating of the air handling units for the system.

To find the pipe sizes, look at either the floor plans or the details for the equipment room. They'll list the sizes of the branch run-out pipes. Once you know them, you can make a good guess at the right size for the control valves. (See the diagrams and tables on pages 203, 204 and 205.) The only information you need is the CFM ratings of the air handling units and the branch piping sizes to the coils. If the system's capacity is over 16,000 CFM, you need two or three coil banks.

A Few Tips on Estimating Water Coil Piping

1) Coil connection sizes seldom match branch pipe sizes. Be sure to include the reducing fittings you'll need in any estimate.

2) Two-way and three-way control valves are both usually one pipe size smaller than the pipe where they're installed. That means you'll need either two or three reducing fittings per control valve. Be sure you include their cost in your estimates.

Using the Diagrams

In the following diagrams, for clarity some items are not included. These items are: balance valves, shut-off valves, reducers, strainers, gauges and gauge taps. Any details you need about these items for your estimate are in the engineer's coil piping details.

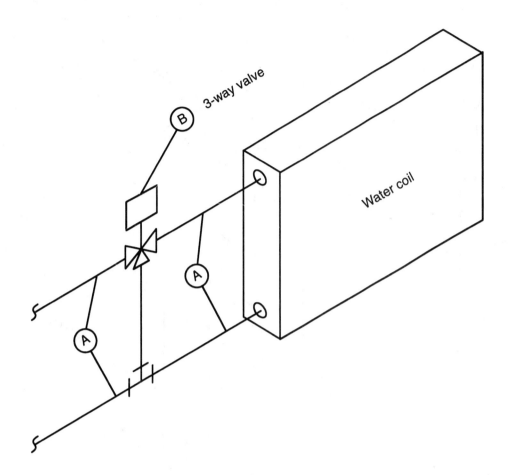

If Ⓐ is:	Then Ⓑ is:
1"	¾"
1¼"	1"
1½"	1¼"
2"	1½"
2½"	2"
3"	2½"

Typical water coil piping for A.H. units up to 16,000 CFM

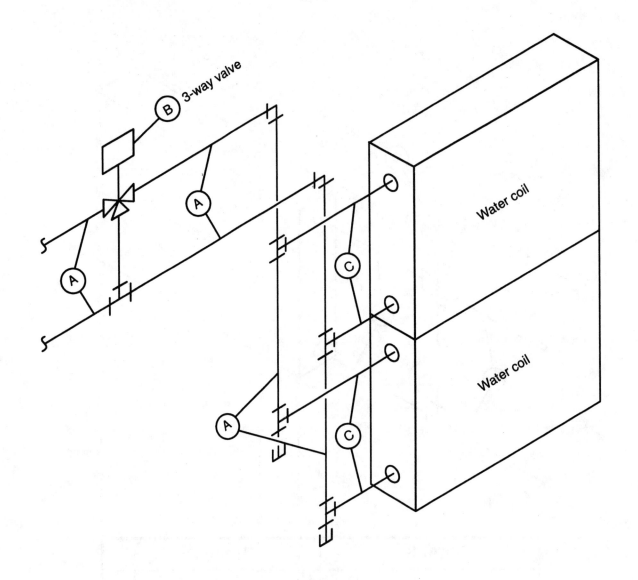

If Ⓐ is:	Then Ⓑ is:	And Ⓒ is:
2½"	2"	2"
3"	2½"	2½"
4"	3"	3"

Typical water coil piping for A.H. units from 16,000 to 26,000 CFM

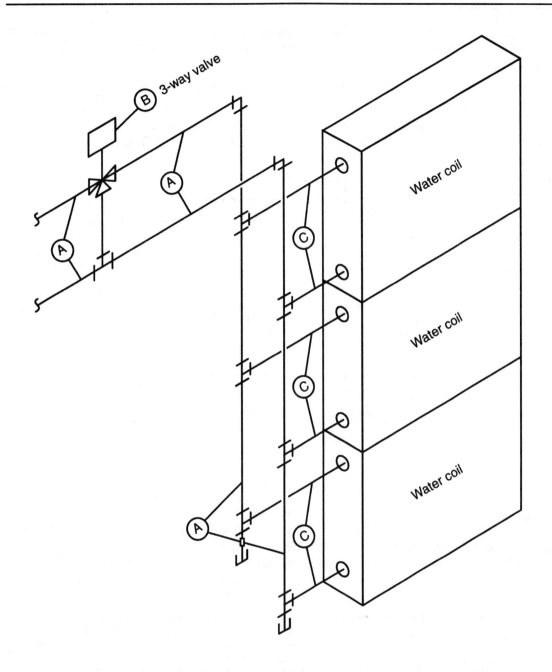

If Ⓐ is:	Then Ⓑ is:	And Ⓒ is:
4"	3"	2½"
6"	4"	3"
8"	6"	4"
10"	8"	6"

Typical water coil piping for A.H. units over 26,000 CFM

Galvanized Steel Ductwork

The costs for fabricating and installing galvanized steel ductwork are usually based on the total weights of the duct and fittings. Because of cost differences, fitting weights should be kept separate from straight duct weights.

Round spiral duct: Normally-used duct gauge/size relationships for low-pressure systems, up to 2 inches static pressure, are as follows:

Duct Diameter (inches)	U.S. Standard Gauge
Up to 12	26
13 through 24	24
26 through 36	22

Rectangular duct: Normally-used duct gauge/size relationships for low-pressure systems, up to 2 inches static pressure, are as follows:

Duct Size (inches)	U.S. Standard Gauge
Up to 12	26
13 through 30	24
31 through 54	22
55 through 72	20

The fitting weights in the following tables are in accordance with the above gauges.

Other material costs: All weights in the tables are net weights; they do not include bracing, cleats, scrap, hangers, end closures, sealants and miscellaneous hardware. Add 15% to calculated duct/fitting weights to cover these items if ductwork is purchased. Add 20% if ducting is manufactured in contractor's shop. The additional 5% covers the costs for scrap which is already included in an outside vendor's price.

Ductwork costs: Typical costs for purchased ductwork are as follows:

Straight duct, less than 1,000 pounds per order	$1.10/lb.
Straight duct, over 1,000 pounds per order	$ .93/lb.
Fittings, less than 1,000 pounds per order	$2.21/lb
Fittings, over 1,000 pounds per order	$1.93/lb.
Lined ducts and fittings, add	$ .50/lb.*
Delivery costs	$ .05/lb.

*Duct sizes shown on drawings are always inside, or net, dimensions and must be increased in size when estimating lined ductwork.

If ductwork is manufactured in the contractor's fabrication shop, the above costs can be reduced by about 20%, depending on the shop's rate of productivity.

Installation costs: An average crew can install approximately 25 pounds of duct and fittings per manhour under normal conditions. See "Applying Correction Factors" on page 6 for situations that do not conform to the definition of a standard labor unit.

EXAMPLE:

What is the cost to furnish and install 2,585 pounds of unlined duct and 560 pounds of unlined fittings? The ductwork will be purchased from an outside vendor. A 15% allowance for miscellaneous material is included in the weights.

Material:

2,585 pounds of straight duct x $.93/lb.	=	$2,404.05
560 pounds of fittings x $2.21/lb.	=	1,237.60
Delivery cost: 2,585 + 560 x $.05/lb.	=	157.25
Total material cost	=	$3,798.90*

Labor:

$$\frac{2,585 + 560}{25} = 125.8 \text{ MH}; \ 125.8 \text{ MH} \times \$25.39 \qquad = \$3,194.06$$

Total installed cost: $3,798.90 + $3,194.06 = $6,992.96*

Sales tax not included.

Galvanized Steel Spiral Ductwork

Weights of Galvanized Steel Spiral Duct (pounds per LF)					
Diameter (inches)	U.S. Standard Gauge				
	26	24	22	20	18
3	.76	1.01	1.23	1.46	—
4	1.02	1.35	1.64	1.94	—
5	1.28	1.69	2.06	2.42	—
6	1.54	2.03	2.47	2.91	3.88
7	1.79	2.37	2.88	3.40	4.52
8	2.05	2.71	3.29	3.86	5.17
9	2.31	3.05	3.71	4.37	5.82
10	2.57	3.39	4.12	4.86	6.47
12	3.08	4.07	4.95	5.83	7.75
14	3.59	4.74	5.77	6.81	9.05
16	4.11	5.42	6.60	7.78	10.34
18	4.63	6.10	7.43	8.76	11.64
20	5.15	6.78	8.25	9.73	12.93
22	5.65	7.46	9.08	10.71	14.93
24	6.16	8.14	9.91	11.68	15.52
26	6.67	8.82	10.73	12.66	16.82
28	7.18	9.50	11.56	13.63	18.11
30	7.71	10.18	12.38	14.60	19.41
32	8.22	10.84	13.21	15.58	20.71
34	—	11.52	14.05	16.55	22.00
36	—	12.20	14.90	17.53	23.29

Weights of Galvanized Steel Round Spiral Fittings (pounds per piece)				
Diameter (inches)	90° elbow	45° elbow	Coupling	Reducer
3	1.3	1.0	0.5	1.0
4	2.2	1.3	0.6	1.2
5	3.3	1.9	0.7	1.4
6	4.3	2.5	0.9	1.8
7	5.8	3.3	1.0	2.0
8	7.3	4.3	1.2	2.4
9	8.8	5.3	2.6	3.3
10	11.8	7.5	2.9	4.4
12	16.3	10.0	3.5	5.3
14	22.0	13.0	4.1	6.2
16	28.3	15.8	4.6	6.9
18	34.5	19.0	5.2	7.8
20	41.5	23.5	5.8	8.7
22	48.3	27.5	6.4	9.6
24	57.5	32.0	6.9	10.3
26	68.8	37.5	7.5	11.2
28	76.8	42.8	8.1	12.1
30	87.0	48.0	8.7	13.0
32	99.5	55.0	9.2	13.8
34	112.0	61.0	9.8	14.7
36	161.0	89.5	10.4	15.6

Largest run diam. (inches)	Weights of Galvanized Steel Round Spiral Fittings (pounds per piece) Tee with reducing run and branch Branch diameter (inches)													
	3	4	5	6	7	8	9	10	12	14	16	18	20	22
4	2.3													
5	2.8	3.0												
6	3.2	3.5	3.7											
7	3.7	3.9	4.1	4.5										
8	4.1	4.4	4.7	5.0	5.3									
9	4.5	4.8	5.1	5.5	5.8	6.2								
10	6.4	6.7	7.1	7.5	7.8	8.2	8.6							
12	7.7	8.1	8.5	8.9	9.4	9.9	10.3	10.7						
14	8.9	9.4	10.0	10.5	11.0	11.5	12.1	12.6	13.6					
16	10.2	10.8	11.4	12.0	12.5	13.1	13.7	14.3	15.5	16.7				
18	11.5	12.1	12.7	13.5	14.1	14.8	15.5	16.1	17.5	18.8	19.2			
20	12.7	13.4	14.2	14.9	17.7	16.4	17.2	17.9	19.4	20.9	22.4	23.9		
22	14.0	14.8	15.6	16.4	17.2	18.1	18.9	19.7	21.3	23.0	24.6	26.2	27.9	
24	15.3	16.1	17.0	17.8	18.7	19.6	20.5	21.4	23.2	25.0	26.8	28.5	30.3	32.1
26	16.5	17.5	18.5	19.5	20.4	21.3	22.3	23.3	25.2	26.2	29.1	31.1	33.0	34.9
28	17.7	18.7	19.8	20.8	21.9	22.9	23.9	25.0	27.0	29.2	31.2	33.2	35.4	37.4
30	18.9	20.0	21.1	22.2	23.3	24.5	25.5	26.7	28.9	31.1	33.3	35.5	37.7	40.0
32	20.4	21.6	22.6	24.0	25.2	26.4	27.6	28.8	31.2	33.6	36.0	38.4	40.8	42.2
34	21.6	22.9	24.2	25.4	26.7	28.0	29.3	30.5	33.1	35.6	38.1	40.7	42.2	45.6
36	29.6	31.4	33.1	34.8	36.6	38.6	40.1	41.8	45.3	48.8	52.2	55.7	59.2	62.7

Galvanized Steel Round Spiral Fittings

Run diam. (inches)	Weights of Galvanized Steel Round Spiral Fittings (pounds per piece) Cross with reducing run and branch Branch diameter (inches)													
	3	4	5	6	7	8	9	10	12	14	16	18	20	22
4	2.6													
5	3.2	3.6												
6	3.8	4.2	4.6											
7	4.4	4.9	5.4	5.8										
8	4.9	5.4	6.0	6.5	7.0									
9	5.4	6.0	6.6	7.2	7.7	8.3								
10	7.6	8.2	8.9	9.5	10.2	10.8	11.4							
12	8.8	9.6	10.3	11.0	11.7	12.5	13.2	13.8						
14	10.1	10.9	11.7	12.6	13.4	14.2	15.1	15.9	17.5					
16	11.3	12.2	13.1	14.0	15.0	15.9	16.8	17.7	19.5	21.4				
18	12.5	13.5	14.5	15.5	16.5	17.5	18.5	19.5	21.5	23.5	25.5			
20	13.8	15.0	16.1	17.2	18.2	19.4	20.6	21.7	23.9	26.2	28.4	30.6		
22	15.2	16.4	17.7	18.9	20.1	21.4	22.6	23.8	26.2	28.7	31.2	33.6	36.1	
24	16.6	17.9	19.3	20.6	22.0	23.3	24.6	26.0	28.7	31.5	34.1	36.7	39.4	42.1
26	18.0	19.4	20.9	22.3	23.8	25.2	26.7	28.2	31.1	33.9	36.9	39.8	42.7	45.6
28	19.3	20.8	22.4	24.0	25.5	27.1	28.6	30.2	32.3	36.4	39.6	42.7	45.8	48.9
30	20.6	22.3	23.9	25.6	27.3	28.9	30.6	32.3	35.6	38.9	42.3	45.6	49.9	52.2
32	22.2	24.0	25.8	27.6	29.4	31.2	33.0	34.8	38.4	42.0	45.6	49.1	52.8	56.5
34	23.5	25.4	27.3	29.2	31.1	33.1	34.9	36.9	40.7	44.5	48.3	52.1	55.9	59.8
36	32.2	34.8	37.4	40.0	42.8	45.3	48.0	50.5	55.7	61.0	66.2	71.5	76.7	82.0

Size (inches)	Weights of 26 Gauge Galvanized Steel Rectangular Duct (pounds per LF)								
	4	5	6	7	8	9	10	11	12
4	1.21								
5	1.36	1.51							
6	1.51	1.66	1.81						
7	1.66	1.81	1.96	2.11					
8	1.81	1.96	2.11	2.27	2.42				
9	1.96	2.11	2.27	2.42	2.57	2.72			
10	2.11	2.27	2.42	2.57	2.72	2.87	3.02		
11	2.27	2.42	2.57	2.72	2.87	3.02	3.17	3.32	
12	2.42	2.57	2.72	2.87	3.02	3.17	3.32	3.47	3.62

Weights of 24 Gauge Galvanized Steel Rectangular Duct (pounds per LF)									
Size (inches)	14	16	18	20	22	24	26	28	30
6	3.85	4.24	4.62	5.01	5.40	5.78	6.17	6.55	6.94
8	4.24	4.62	5.01	5.40	5.78	6.17	6.55	6.94	7.32
10	4.62	5.01	5.40	5.78	6.17	6.55	6.94	7.32	7.71
12	5.01	5.40	5.78	6.17	6.55	6.94	7.32	7.71	8.09
14	5.40	5.78	6.17	6.55	6.94	7.32	7.71	8.09	8.48
16	5.78	6.17	6.55	6.94	7.32	7.71	8.09	8.48	8.86
18	6.17	6.55	6.94	7.32	7.71	8.09	8.48	8.86	9.25
20	6.55	6.94	7.32	7.71	8.09	8.48	8.86	9.25	9.64
22	6.94	7.32	7.71	8.09	8.48	8.86	9.25	9.64	10.00
24	7.32	7.71	8.09	8.48	8.86	9.25	9.64	10.00	10.40
26	7.71	8.09	8.48	8.86	9.25	9.64	10.00	10.40	10.80
28	8.09	8.48	8.86	9.25	9.64	10.00	10.40	10.80	11.20
30	8.48	8.86	9.25	9.64	10.00	10.40	10.80	11.20	11.60

Weights of 22 Gauge Galvanized Steel Rectangular Duct (pounds per LF)											
Size (inches)	32	34	36	38	40	42	44	46	48	50	52	54
8	9.37	9.84	10.3	10.8	11.3	11.7	12.2	12.7	13.1	13.6	14.1	14.5
10	9.84	10.3	10.8	11.3	11.7	12.2	12.7	13.1	13.6	14.1	14.5	15.0
12	10.3	10.8	11.3	11.7	12.2	12.7	13.1	13.6	14.1	14.5	15.0	15.5
14	10.8	11.3	11.7	12.2	12.7	13.1	13.6	14.1	14.5	15.0	15.5	15.9
16	11.3	11.7	12.2	12.7	13.1	13.6	14.1	14.5	15.0	15.5	15.9	16.4
18	11.7	12.2	12.7	13.1	13.6	14.1	14.5	15.0	15.5	15.9	16.4	16.9
20	12.2	12.7	13.1	13.6	14.1	14.5	15.0	15.5	15.9	16.4	16.9	17.3
22	12.7	13.1	13.6	14.1	14.5	15.0	15.5	15.9	16.4	16.9	17.3	17.8
24	13.1	13.6	14.1	14.5	15.0	15.5	15.9	16.4	16.9	17.3	17.8	18.3
26	13.6	14.1	14.5	15.0	15.5	15.9	16.4	16.9	17.3	17.8	18.3	18.7
28	14.1	14.5	15.0	15.5	15.9	16.4	16.9	17.3	17.8	18.3	18.7	19.2
30	14.5	15.0	15.5	15.9	16.4	16.9	17.3	17.8	18.3	18.7	19.2	19.7
32	15.0	15.5	15.9	16.4	16.9	17.3	17.8	18.3	18.7	19.2	19.7	20.2
34	15.5	15.9	16.4	16.9	17.3	17.8	18.3	18.7	19.2	19.7	20.2	20.6
36	15.9	16.4	16.9	17.3	17.8	18.3	18.7	19.2	19.7	20.2	20.6	21.1
38	16.4	16.9	17.3	17.8	18.3	18.7	19.2	19.7	20.2	20.6	21.1	21.6
40	16.9	17.3	17.8	18.3	18.7	19.2	19.7	20.2	20.6	21.1	21.6	22.0
42	17.3	17.8	18.3	18.7	19.2	19.7	20.2	20.6	21.1	21.6	22.0	22.5
44	17.8	18.3	18.7	19.2	19.7	20.2	20.6	21.1	21.6	22.0	22.5	23.0
46	18.3	18.7	19.2	19.7	20.2	20.6	21.1	21.6	22.0	22.5	23.0	23.4
48	18.7	19.2	19.7	20.2	20.6	21.1	21.6	22.0	22.5	23.0	23.4	23.9
50	19.2	19.7	20.2	20.6	21.1	21.6	22.0	22.5	23.0	23.4	23.9	24.4
52	19.7	20.2	20.6	21.1	21.6	22.0	22.5	23.0	23.4	23.9	24.4	24.8
54	20.2	20.6	21.1	21.6	22.0	22.5	23.0	23.4	23.9	24.4	24.8	25.3

Galvanized Steel Rectangular Duct

Weights of 20 Gauge Galvanized Steel Rectangular Duct (pounds per LF)								
Size (inches)	56	58	60	62	64	66	68	70	72
18	20.4	21.0	21.5	22.1	22.6	23.2	23.7	24.3	24.8
20	21.0	21.5	22.1	22.6	23.2	23.7	24.3	24.8	25.4
22	21.5	22.1	22.6	23.2	23.7	24.3	24.8	25.4	25.9
24	22.1	22.6	23.2	23.7	24.3	24.8	25.4	25.9	26.5
26	22.6	23.2	23.7	24.3	24.8	25.4	25.9	26.5	27.0
28	23.2	23.7	24.3	24.8	25.4	25.9	26.5	27.0	27.6
30	23.7	24.3	24.8	25.4	25.9	26.5	27.0	27.6	28.2
32	24.3	24.8	25.4	25.9	26.5	27.0	27.6	28.2	28.7
34	24.8	25.4	25.9	26.5	27.0	27.6	28.2	28.7	29.3
36	25.4	25.9	26.5	27.0	27.6	28.2	28.7	29.3	29.8
38	25.9	26.5	27.0	27.6	28.2	28.7	29.3	29.8	30.4
40	26.5	27.0	27.6	28.2	28.7	29.3	29.8	30.4	30.9
42	27.0	27.6	28.2	28.7	29.3	29.8	30.4	30.9	31.5
44	27.6	28.2	28.7	29.3	29.8	30.4	30.9	31.5	32.0
46	28.2	28.7	29.3	29.8	30.4	30.9	31.5	32.0	32.6
48	28.7	29.3	29.8	30.4	30.9	31.5	32.0	32.6	33.1
50	29.3	29.8	30.4	30.9	31.5	32.0	32.6	33.1	33.7
52	29.8	30.4	30.9	31.5	32.0	32.6	33.1	33.7	34.2
54	30.4	30.9	31.5	32.0	32.6	33.1	33.7	34.2	34.8
56	30.9	31.5	32.0	32.6	33.1	33.7	34.2	34.8	35.3
58	31.5	32.0	32.6	33.1	33.7	34.2	34.8	35.3	35.9
60	32.0	32.6	33.1	33.7	34.2	34.8	35.3	35.9	36.4
62	32.6	33.1	33.7	34.2	34.8	35.3	35.9	36.4	37.0
64	33.1	33.7	34.2	34.8	35.3	35.9	36.4	37.0	37.5
66	33.7	34.2	34.8	35.3	35.9	36.4	37.0	37.5	38.1
68	34.2	34.8	35.3	35.9	36.4	37.0	37.5	38.1	38.6
70	34.8	35.3	35.9	36.4	37.0	37.5	38.1	38.6	39.2
72	35.3	35.9	36.4	37.0	37.5	38.1	38.6	39.2	39.7

	\multicolumn{9}{c}{**Weights of Galvanized Steel Rectangular 90 Degree Elbow** (pounds per piece)}
Size (inches)	**4**	**6**	**8**	**10**	**12**	**14**	**16**	**18**	**20**
4	1.05	1.45	1.95	2.60	3.30	5.30	6.50	7.80	9.25
6	1.45	1.95	2.60	3.30	3.70	5.80	7.75	8.75	9.75
8	1.95	2.60	3.30	3.70	4.25	6.50	8.50	9.75	10.4
10	2.60	3.30	3.70	4.25	4.85	7.30	9.50	10.7	11.2
12	3.30	3.70	4.25	4.85	5.60	8.20	10.5	11.7	12.2
14	5.30	5.80	6.50	7.30	8.20	9.25	11.5	12.6	13.2
16	6.50	7.50	8.50	9.50	10.5	11.5	12.5	13.4	14.4
18	7.80	8.75	9.75	10.7	11.7	12.6	13.4	14.5	15.7
20	9.25	9.75	10.4	11.2	12.2	13.8	14.4	15.7	17.2
22	10.8	11.8	12.8	13.8	14.8	25.8	16.8	17.8	18.8
24	12.6	13.6	14.6	15.6	16.6	17.6	18.6	19.6	20.5
26	14.4	15.4	16.4	17.4	18.4	19.4	20.4	21.4	22.3
28	16.4	17.4	18.4	19.4	20.4	21.4	22.4	23.4	24.3
30	18.5	19.5	20.5	21.5	22.5	23.5	24.5	25.5	26.4
32	24.3	25.7	27.0	28.4	29.7	31.1	32.4	33.7	35.0
34	28.2	29.4	30.6	31.8	33.0	34.3	35.5	36.7	37.9
36	31.3	32.5	33.7	34.9	36.1	37.3	38.5	39.7	40.9
38	34.5	35.7	36.9	38.1	39.3	40.5	41.7	42.9	44.1
40	37.9	39.1	40.3	41.5	42.7	43.9	45.1	46.3	47.5
42	41.4	42.6	43.8	45.0	46.2	47.4	48.6	49.8	51.0
44	45.1	46.4	47.6	48.8	50.0	51.2	52.4	53.6	54.8
46	49.0	50.2	51.4	52.6	53.8	55.0	56.2	57.4	58.6
48	53.0	54.2	55.4	56.6	57.8	59.0	60.2	61.4	62.6
50	57.2	58.5	59.7	60.9	62.1	63.3	64.5	65.7	66.9

Note: Elbow weights do not include turning vanes.

Galvanized Steel Rectangular 90 Degree Elbows

Weights of Galvanized Steel Rectangular 90 Degree Elbows (pounds per piece)									
Size (inches)	22	24	26	28	30	32	34	36	38
22	19.6	21.9	24.1	26.3	28.5	37.0	40.8	44.5	48.3
24	21.9	24.2	26.4	28.6	30.9	39.5	43.6	47.2	50.8
26	24.1	26.4	28.7	30.9	33.2	42.0	46.4	49.8	53.3
28	26.3	28.6	30.9	33.3	35.6	44.5	49.2	52.5	55.8
30	28.5	30.9	33.2	35.6	37.9	47.0	52.0	55.1	58.3
32	37.0	39.5	42.0	44.5	47.0	49.5	52.0	57.8	60.8
34	40.8	43.6	46.4	49.2	52.0	54.8	57.6	60.5	63.3
36	44.5	47.2	49.8	52.5	55.1	57.8	60.5	63.1	65.8
38	48.3	50.8	53.3	55.8	58.3	60.8	63.3	65.8	68.3
40	52.1	54.6	57.1	59.6	62.1	64.6	67.1	69.6	72.1
42	55.9	58.4	60.9	63.4	65.9	68.4	70.9	73.4	75.9
44	59.6	62.1	64.6	67.1	69.6	72.1	74.6	77.1	79.6
46	63.4	65.9	68.4	70.9	73.4	75.9	78.4	80.9	83.4
48	67.2	69.7	72.2	74.7	77.2	79.7	82.2	84.7	87.2
50	70.9	73.4	75.9	78.4	80.9	83.4	85.9	88.4	90.9
52	74.8	77.3	79.8	82.3	84.8	87.3	89.8	92.3	94.7
54	78.5	81.0	83.5	86.0	88.5	91.0	93.5	96.0	98.5
56	97.2	100	103	106	109	112	115	118	121
58	103	106	109	112	115	118	121	124	127
60	110	113	116	119	122	125	128	131	134
62	116	119	122	125	128	131	134	137	140
64	122	125	129	132	135	138	141	144	147
66	128	131	135	138	141	144	148	151	154
68	135	138	142	145	148	151	154	157	160
70	141	144	148	151	154	157	160	163	166
72	147	150	154	157	160	163	166	169	172

Note: Elbow weights do not include turning vanes.

Weights of Galvanized Steel Rectangular 90 Degree Elbows (pounds per piece)									
Size (inches)	40	42	44	46	48	50	52	54	56
40	74.6	78.4	82.1	85.9	89.7	93.4	97.2	102	125
42	78.4	82.3	86.2	90.1	94.0	97.9	102	106	130
44	82.1	86.2	90.3	94.4	98.5	102	106	110	135
46	85.9	90.1	94.4	98.7	102	106	110	113	140
48	89.7	94.0	98.5	102	106	110	113	117	144
50	93.4	97.9	102	106	110	113	117	122	149
52	97.2	102	106	110	113	117	122	126	154
54	102	106	110	113	117	122	126	131	159
56	126	130	134	139	145	150	155	160	165
58	132	137	142	147	151	156	161	166	170
60	137	142	147	152	157	162	167	172	176
62	143	148	153	158	163	168	173	178	182
64	150	155	160	165	169	174	179	184	188
66	156	161	166	171	176	181	186	191	195
68	163	168	173	178	182	187	192	197	201
70	170	175	180	185	189	194	199	204	208
72	176	181	185	190	194	199	204	210	215

Note: Elbow weights do not include turning vanes.

Weights of Galvanized Steel Rectangular 90 Degree Elbows (pounds per piece)

Size (inches)	58	60	62	64	66	68	70	72
58	176	182	187	194	200	207	214	221
60	182	187	194	200	207	214	221	227
62	187	194	200	207	214	221	227	233
64	194	200	207	214	221	227	233	239
66	200	207	214	221	227	233	239	246
68	207	214	221	227	233	239	246	252
70	214	221	227	233	239	246	252	259
72	221	227	233	239	246	252	259	266

Note: Elbow weights do not include turning vanes.

Weights of Galvanized Steel Rectangular Drops and Tap-In Tees (pounds per piece)

Size (inches)	6	8	10	12	16	20	24	28	32
6	1.81	2.11	2.42	2.72	4.24	5.01	5.78	6.55	8.90
8	2.11	2.42	2.72	3.02	4.62	5.40	6.17	6.94	9.37
10	2.42	2.72	3.02	3.32	5.01	5.78	6.55	7.32	9.84
12	2.72	3.02	3.32	3.62	5.40	6.17	6.94	7.71	10.3
16	4.24	4.62	5.01	5.40	6.17	6.94	7.71	8.48	11.3
20	5.01	5.40	5.78	6.17	6.94	7.71	8.48	9.25	12.2
24	5.78	6.17	6.55	6.94	7.71	8.48	9.25	10.0	13.1
28	6.55	6.94	7.32	7.71	8.48	9.25	10.0	10.8	14.1
32	8.90	9.37	9.84	10.3	11.3	12.2	13.1	14.1	15.0
36	9.84	10.3	10.8	11.3	12.2	13.1	14.1	15.0	15.9
40	10.8	11.3	11.7	12.2	13.1	14.1	15.0	15.9	16.9
44	11.7	12.2	12.7	13.1	14.1	15.0	15.9	16.9	17.8
48	12.7	13.1	13.6	14.1	15.0	15.9	16.9	17.8	18.7
52	13.6	14.1	14.5	15.0	15.9	16.9	17.8	18.7	19.7
56	16.0	17.6	18.2	18.8	19.6	21.0	22.1	23.2	24.3
60	18.7	19.0	19.3	19.9	21.0	22.1	23.2	24.3	25.4

Note: Weights of drops and tap-in tees do not include splitter dampers.

Galvanized Steel Spiral Duct

Costs for Steel Spiral Duct. Costs per linear foot of duct, based on quantities less than 1,000 pounds. Add 50% to the material cost for lined duct. These costs include delivery, typical bracing, cleats, scrap, hangers, end closures, sealants and miscellaneous hardware.

Description	Craft@Hrs	Unit	Material $	Labor $	Total $

Galvanized steel spiral duct 26 gauge

Description	Craft@Hrs	Unit	Material $	Labor $	Total $
3"	S2@.030	LF	.89	.76	1.65
4"	S2@.041	LF	1.19	1.04	2.23
5"	S2@.051	LF	1.49	1.29	2.78
6"	S2@.061	LF	1.78	1.55	3.33
7"	S2@.072	LF	2.07	1.83	3.90
8"	S2@.082	LF	2.38	2.08	4.46
9"	S2@.092	LF	2.68	2.34	5.02
10"	S2@.103	LF	2.98	2.62	5.60
12"	S2@.123	LF	3.57	3.12	6.69
14" - 16"	S2@.174	LF	4.46	4.42	8.88
18" - 20"	S2@.195	LF	5.67	4.95	10.62
22" - 24"	S2@.236	LF	6.83	5.99	12.82
26" - 28"	S2@.277	LF	8.01	7.03	15.04
30" - 32"	S2@.318	LF	9.21	8.07	17.28

Galvanized steel spiral duct 24 gauge

Description	Craft@Hrs	Unit	Material $	Labor $	Total $
3"	S2@.041	LF	1.18	1.04	2.22
4"	S2@.054	LF	1.56	1.37	2.93
5"	S2@.067	LF	1.96	1.70	3.66
6"	S2@.081	LF	2.35	2.06	4.41
7"	S2@.095	LF	2.74	2.41	5.15
8"	S2@.108	LF	3.14	2.74	5.88
9"	S2@.122	LF	3.53	3.10	6.63
10"	S2@.135	LF	3.92	3.43	7.35
12"	S2@.163	LF	4.72	4.14	8.86
14" - 16"	S2@.203	LF	5.88	5.15	11.03
18" - 20"	S2@.257	LF	7.46	6.53	13.99
22" - 24"	S2@.312	LF	9.02	7.92	16.94
26" - 28"	S2@.366	LF	10.60	9.29	19.89
30" - 32"	S2@.419	LF	12.10	10.60	22.70
34" - 36"	S2@.474	LF	13.80	12.00	25.80

Galvanized steel spiral duct 22 gauge

3"	S2@.049	LF	1.43	1.24	2.67
4"	S2@.065	LF	1.91	1.65	3.56
5"	S2@.082	LF	2.39	2.08	4.47
6"	S2@.099	LF	2.86	2.51	5.37
7"	S2@.115	LF	3.33	2.92	6.25
8"	S2@.132	LF	3.81	3.35	7.16
9"	S2@.148	LF	4.29	3.76	8.05
10"	S2@.165	LF	4.77	4.19	8.96
12"	S2@.198	LF	5.73	5.03	10.76
14" - 16"	S2@.247	LF	7.15	6.27	13.42
18" - 20"	S2@.313	LF	9.07	7.95	17.02
22" - 24"	S2@.380	LF	11.00	9.65	20.65
26" - 28"	S2@.444	LF	12.90	11.30	24.20
30" - 32"	S2@.491	LF	14.80	12.50	27.30
34" - 36"	S2@.576	LF	16.80	14.60	31.40

Galvanized steel spiral duct 20 gauge

3"	S2@.058	LF	1.70	1.47	3.17
4"	S2@.078	LF	2.25	1.98	4.23
5"	S2@.097	LF	2.81	2.46	5.27
6"	S2@.116	LF	3.37	2.95	6.32
7"	S2@.136	LF	3.94	3.45	7.39
8"	S2@.154	LF	4.47	3.91	8.38
9"	S2@.175	LF	5.05	4.44	9.49
10"	S2@.194	LF	5.62	4.93	10.55
12"	S2@.233	LF	6.75	5.92	12.67
14" - 16"	S2@.291	LF	8.44	7.39	15.83
18" - 20"	S2@.369	LF	10.70	9.37	20.07
22" - 24"	S2@.446	LF	12.90	11.30	24.20
26" - 28"	S2@.525	LF	15.20	13.30	28.50
30" - 32"	S2@.604	LF	17.40	15.30	32.70
34" - 36"	S2@.681	LF	19.70	17.30	37.00

Galvanized steel spiral duct 18 gauge

6"	S2@.155	LF	4.49	3.94	8.43
7"	S2@.181	LF	5.23	4.60	9.83
8"	S2@.207	LF	5.98	5.26	11.24
9"	S2@.233	LF	6.74	5.92	12.66
10"	S2@.259	LF	7.49	6.58	14.07
12"	S2@.310	LF	8.97	7.87	16.84
14" - 16"	S2@.387	LF	11.20	9.83	21.03
18" - 20"	S2@.491	LF	14.20	12.50	26.70
22" - 24"	S2@.608	LF	17.60	15.40	33.00
26" - 28"	S2@.696	LF	20.20	17.70	37.90
30" - 32"	S2@.803	LF	23.20	20.40	43.60
34" - 36"	S2@.905	LF	26.20	23.00	49.20

Galvanized Steel Spiral Duct Fittings

Costs for Steel Spiral Duct Fittings. Costs per fitting, based on quantities less than 1,000 pounds. For quantities over 1,000 pounds, deduct 15% from material costs. For lined duct add 50%. These costs include delivery, typical bracing, cleats, scrap, hangers, end closures, sealants and miscellaneous hardware.

Description	Craft@Hrs	Unit	Material $	Labor $	Total $
Galvanized steel spiral duct 90 degree elbows					
3"	S2@.052	Ea	3.00	1.32	4.32
4"	S2@.088	Ea	5.05	2.23	7.28
5"	S2@.132	Ea	7.59	3.35	10.94
6"	S2@.172	Ea	9.90	4.37	14.27
7"	S2@.232	Ea	13.30	5.89	19.19
8"	S2@.292	Ea	16.80	7.41	24.21
9"	S2@.352	Ea	20.30	8.94	29.24
10"	S2@.470	Ea	27.10	11.90	39.00
12"	S2@.653	Ea	37.50	16.60	54.10
14" - 16"	S2@1.01	Ea	57.70	25.60	83.30
18" - 20"	S2@1.52	Ea	87.40	38.60	126.00
22" - 24"	S2@2.11	Ea	121.00	53.60	174.60
26" - 28"	S2@2.91	Ea	168.00	73.90	241.90
30" - 32"	S2@3.73	Ea	215.00	94.70	309.70
34" - 36"	S2@5.47	Ea	314.00	139.00	453.00
Galvanized steel spiral duct 45 degree elbow					
3"	S2@.040	Ea	2.30	1.02	3.32
4"	S2@.052	Ea	3.00	1.32	4.32
5"	S2@.076	Ea	4.36	1.93	6.29
6"	S2@.100	Ea	5.75	2.54	8.29
7"	S2@.132	Ea	7.59	3.35	10.94
8"	S2@.172	Ea	9.90	4.37	14.27
9"	S2@.212	Ea	12.10	5.38	17.48
10"	S2@.300	Ea	17.20	7.62	24.82
12"	S2@.400	Ea	23.00	10.20	33.20
14" - 16"	S2@.576	Ea	33.00	14.60	47.60
18" - 20"	S2@.850	Ea	49.00	21.60	70.60
22" - 24"	S2@1.19	Ea	68.40	30.20	98.60
26" - 28"	S2@1.61	Ea	92.30	40.90	133.20
30" - 32"	S2@2.06	Ea	118.00	52.30	170.30
34" - 36"	S2@3.01	Ea	173.00	76.40	249.40

Description	Craft@Hrs	Unit	Material $	Labor $	Total $

Galvanized steel spiral duct coupling

Description	Craft@Hrs	Unit	Material $	Labor $	Total $
3"	S2@.020	Ea	1.15	.51	1.66
4"	S2@.024	Ea	1.39	.61	2.00
5"	S2@.028	Ea	1.60	.71	2.31
6"	S2@.036	Ea	2.06	.91	2.97
7"	S2@.040	Ea	2.30	1.02	3.32
8"	S2@.048	Ea	2.75	1.22	3.97
9"	S2@.104	Ea	5.98	2.64	8.62
10"	S2@.116	Ea	6.67	2.95	9.62
12"	S2@.140	Ea	8.05	3.55	11.60
14" - 16"	S2@.174	Ea	10.00	4.42	14.42
18" - 20"	S2@.220	Ea	12.60	5.59	18.19
22" - 24"	S2@.266	Ea	15.30	6.75	22.05
26" - 28"	S2@.312	Ea	17.90	7.92	25.82
30" - 32"	S2@.357	Ea	20.60	9.06	29.66
34" - 36"	S2@.403	Ea	23.20	10.20	33.40

Galvanized steel spiral duct with reducer

Description	Craft@Hrs	Unit	Material $	Labor $	Total $
3"	S2@.040	Ea	2.30	1.02	3.32
4"	S2@.048	Ea	2.75	1.22	3.97
5"	S2@.056	Ea	3.22	1.42	4.64
6"	S2@.072	Ea	4.15	1.83	5.98
7"	S2@.080	Ea	4.59	2.03	6.62
8"	S2@.096	Ea	5.52	2.44	7.96
9"	S2@.132	Ea	7.59	3.35	10.94
10"	S2@.176	Ea	10.10	4.47	14.57
12"	S2@.212	Ea	12.10	5.38	17.48
14" - 16"	S2@.262	Ea	15.00	6.65	21.65
18" - 20"	S2@.330	Ea	19.00	8.38	27.38
22" - 24"	S2@.398	Ea	22.90	10.10	33.00
26" - 28"	S2@.465	Ea	26.80	11.80	38.60
30" - 32"	S2@.536	Ea	30.70	13.60	44.30
34" - 36"	S2@.606	Ea	34.80	15.40	50.20

Galvanized Steel Spiral Tees

Costs for Steel Spiral Tees with Reducing Branch. Costs per tee, based on quantities less than 1,000 pounds. Deduct 15% from the material cost for quantities over 1,000 pounds. Add 50% to the material cost for lined duct. These costs include delivery, typical bracing, cleats, scrap, hangers, end closures, sealants and miscellaneous hardware.

Description	Craft@Hrs	Unit	Material $	Labor $	Total $

Galvanized steel spiral duct tees with 3" reducing branch

Description	Craft@Hrs	Unit	Material $	Labor $	Total $
4"	S2@.109	Ea	3.91	2.77	6.68
5"	S2@.112	Ea	4.59	2.84	7.43
6"	S2@.128	Ea	5.29	3.25	8.54
7"	S2@.148	Ea	5.98	3.76	9.74
8"	S2@.164	Ea	6.67	4.16	10.83
9"	S2@.180	Ea	7.12	4.57	11.69
10"	S2@.256	Ea	11.20	6.50	17.70
12"	S2@.308	Ea	13.30	7.82	21.12
14" - 16"	S2@.382	Ea	16.80	9.70	26.50
18" - 20"	S2@.485	Ea	21.20	12.30	33.50
22" - 24"	S2@.585	Ea	25.70	14.90	40.60
26" - 28"	S2@.683	Ea	30.00	17.30	47.30
30" - 32"	S2@.786	Ea	34.50	20.00	54.50
34" - 36"	S2@1.02	Ea	44.90	25.90	70.80

Galvanized steel spiral duct tees with 4" reducing branch

Description	Craft@Hrs	Unit	Material $	Labor $	Total $
5"	S2@.120	Ea	5.05	3.05	8.10
6"	S2@.140	Ea	5.98	3.55	9.53
7"	S2@.156	Ea	6.45	3.96	10.41
8"	S2@.176	Ea	7.35	4.47	11.82
9"	S2@.192	Ea	7.81	4.87	12.68
10"	S2@.268	Ea	11.90	6.80	18.70
12"	S2@.324	Ea	14.50	8.23	22.73
14" - 16"	S2@.403	Ea	18.00	10.20	28.20
18" - 20"	S2@.508	Ea	22.80	12.90	35.70
22" - 24"	S2@.619	Ea	27.50	15.70	43.20
26" - 28"	S2@.724	Ea	32.30	18.40	50.70
30" - 32"	S2@.830	Ea	37.10	21.10	58.20
34" - 36"	S2@1.08	Ea	48.40	27.40	75.80

Galvanized steel spiral duct tees with 5" reducing branch

Description	Craft@Hrs	Unit	Material $	Labor $	Total $
6"	S2@.148	Ea	6.45	3.76	10.21
7"	S2@.164	Ea	7.12	4.16	11.28
8"	S2@.188	Ea	8.05	4.77	12.82
9"	S2@.204	Ea	8.74	5.18	13.92
10"	S2@.284	Ea	12.90	7.21	20.11
12"	S2@.340	Ea	15.40	8.63	24.03
14" - 16"	S2@.428	Ea	19.40	10.90	30.30
18" - 20"	S2@.538	Ea	24.60	13.70	38.30
22" - 24"	S2@.651	Ea	29.50	16.50	46.00
26" - 28"	S2@.764	Ea	34.60	19.40	54.00
30" - 32"	S2@.873	Ea	39.80	22.20	62.00
34" - 36"	S2@1.15	Ea	52.00	29.20	81.20

Description	Craft@Hrs	Unit	Material $	Labor $	Total $

Galvanized steel spiral duct tees with 6" reducing branch

Description	Craft@Hrs	Unit	Material $	Labor $	Total $
7"	S2@.180	Ea	8.05	4.57	12.62
8"	S2@.200	Ea	8.74	5.08	13.82
9"	S2@.220	Ea	9.43	5.59	15.02
10"	S2@.300	Ea	13.80	7.62	21.42
12"	S2@.356	Ea	16.60	9.04	25.64
14" - 16"	S2@.449	Ea	20.70	11.40	32.10
18" - 20"	S2@.566	Ea	26.10	14.40	40.50
22" - 24"	S2@.683	Ea	31.40	17.30	48.70
26" - 28"	S2@.807	Ea	36.90	20.50	57.40
30" - 32"	S2@.924	Ea	42.40	23.50	65.90
34" - 36"	S2@1.20	Ea	55.20	30.50	85.70

Galvanized steel spiral duct tees with 7" reducing branch

Description	Craft@Hrs	Unit	Material $	Labor $	Total $
8"	S2@.164	Ea	9.43	4.16	13.59
9"	S2@.180	Ea	10.30	4.57	14.87
10"	S2@.252	Ea	14.50	6.40	20.90
12"	S2@.304	Ea	17.50	7.72	25.22
14" - 16"	S2@.381	Ea	21.90	9.67	31.57
18" - 20"	S2@.482	Ea	27.60	12.20	39.80
22" - 24"	S2@.579	Ea	33.20	14.70	47.90
26" - 28"	S2@.683	Ea	39.40	17.30	56.70
30" - 32"	S2@.784	Ea	45.00	19.90	64.90
34" - 36"	S2@1.02	Ea	58.80	25.90	84.70

Galvanized steel spiral duct tees with 8" reducing branch

Description	Craft@Hrs	Unit	Material $	Labor $	Total $
9"	S2@.228	Ea	13.10	5.79	18.89
10"	S2@.268	Ea	15.40	6.80	22.20
12"	S2@.324	Ea	18.60	8.23	26.83
14" - 16"	S2@.401	Ea	23.10	10.20	33.30
18" - 20"	S2@.508	Ea	29.30	12.90	42.20
22" - 24"	S2@.617	Ea	35.30	15.70	51.00
26" - 28"	S2@.722	Ea	41.60	18.30	59.90
30" - 32"	S2@.830	Ea	47.80	21.10	68.90
34" - 36"	S2@1.08	Ea	62.20	27.40	89.60

Galvanized steel spiral duct tees with 9" reducing branch

Description	Craft@Hrs	Unit	Material $	Labor $	Total $
10"	S2@.284	Ea	16.40	7.21	23.61
12"	S2@.340	Ea	19.50	8.63	28.13
14" - 16"	S2@.426	Ea	24.60	10.80	35.40
18" - 20"	S2@.540	Ea	30.90	13.70	44.60
22" - 24"	S2@.649	Ea	37.30	16.50	53.80
26" - 28"	S2@.760	Ea	43.70	19.30	63.00
30" - 32"	S2@.875	Ea	50.30	22.20	72.50
34" - 36"	S2@1.15	Ea	65.80	29.20	95.00

Galvanized Steel Spiral Tees

Description	Craft@Hrs	Unit	Material $	Labor $	Total $
Galvanized steel spiral duct tees with 10" reducing branch					
12"	S2@.360	Ea	20.70	9.14	29.84
14" - 16"	S2@.447	Ea	25.70	11.30	37.00
18" - 20"	S2@.564	Ea	32.50	14.30	46.80
22" - 24"	S2@.683	Ea	39.40	17.30	56.70
26" - 28"	S2@.805	Ea	46.10	20.40	66.50
30" - 32"	S2@.924	Ea	53.10	23.50	76.60
34" - 36"	S2@1.20	Ea	69.20	30.50	99.70
Galvanized steel spiral duct tees with 12" reducing branch					
14" - 16"	S2@.572	Ea	28.30	14.50	42.80
18" - 20"	S2@.681	Ea	35.80	17.30	53.10
22" - 24"	S2@.820	Ea	43.30	20.80	64.10
26" - 28"	S2@.965	Ea	50.60	24.50	75.10
30" - 32"	S2@1.11	Ea	58.30	28.20	86.50
34" - 36"	S2@1.44	Ea	76.10	36.60	112.70
Galvanized steel spiral duct tees with 14" reducing branch					
14" - 16"	S2@.572	Ea	32.80	14.50	47.30
18" - 20"	S2@.681	Ea	39.10	17.30	56.40
22" - 24"	S2@.820	Ea	47.20	20.80	68.00
26" - 28"	S2@.965	Ea	55.50	24.50	80.00
30" - 32"	S2@1.11	Ea	63.60	28.20	91.80
34" - 36"	S2@1.44	Ea	83.00	36.60	119.60
Galvanized steel spiral duct tees with 16" reducing branch					
18" - 20"	S2@.739	Ea	42.40	18.80	61.20
22" - 24"	S2@.888	Ea	51.10	22.50	73.60
26" - 28"	S2@1.04	Ea	60.00	26.40	86.40
30" - 32"	S2@1.20	Ea	69.00	30.50	99.50
34" - 36"	S2@1.56	Ea	89.90	39.60	129.50
Galvanized steel spiral duct tees with 18" reducing branch					
18" - 20"	S2@.837	Ea	48.00	21.30	69.30
22" - 24"	S2@.954	Ea	55.00	24.20	79.20
26" - 28"	S2@1.12	Ea	64.60	28.40	93.00
30" - 32"	S2@1.29	Ea	74.30	32.80	107.10
34" - 36"	S2@1.68	Ea	96.80	42.70	139.50
Galvanized steel spiral duct tees with 20" reducing branch					
22" - 24"	S2@1.02	Ea	59.00	25.90	84.90
26" - 28"	S2@1.20	Ea	69.30	30.50	99.80
30" - 32"	S2@1.38	Ea	79.50	35.00	114.50
34" - 36"	S2@1.81	Ea	104.00	46.00	150.00

Description	Craft@Hrs	Unit	Material $	Labor $	Total $

Galvanized steel spiral duct tees with 22" reducing branch

Description	Craft@Hrs	Unit	Material $	Labor $	Total $
22" - 24"	S2@1.57	Ea	65.40	39.90	105.30
26" - 28"	S2@1.32	Ea	76.10	33.50	109.60
30" - 32"	S2@1.48	Ea	84.90	37.60	122.50
34" - 36"	S2@1.92	Ea	111.00	48.70	159.70

Galvanized Steel Spiral Tees

Costs for Steel Spiral Tees with Reducing Run and Branch. Costs per tee, based on quantities less than 1,000 pounds. Deduct 15% from the material cost for quantities over 1,000 pounds. Add 50% to the material cost for lined duct. These costs include delivery, typical bracing, cleats, scrap, hangers, end closures, sealants and miscellaneous hardware.

Description	Craft@Hrs	Unit	Material $	Labor $	Total $

Galvanized steel spiral duct tees with 3" run and branch

Description	Craft@Hrs	Unit	Material $	Labor $	Total $
4"	S2@.109	Ea	6.31	2.77	9.08
5"	S2@.112	Ea	6.45	2.84	9.29
6"	S2@.128	Ea	7.35	3.25	10.60
7"	S2@.148	Ea	8.50	3.76	12.26
8"	S2@.164	Ea	9.43	4.16	13.59
9"	S2@.180	Ea	10.30	4.57	14.87
10"	S2@.256	Ea	14.70	6.50	21.20
12"	S2@.308	Ea	17.70	7.82	25.52
14" - 16"	S2@.382	Ea	21.90	9.70	31.60
18" - 20"	S2@.485	Ea	27.90	12.30	40.20
22" - 24"	S2@.585	Ea	33.70	14.90	48.60
26" - 28"	S2@.683	Ea	39.40	17.30	56.70
30" - 32"	S2@.786	Ea	45.20	20.00	65.20
34" - 36"	S2@1.02	Ea	58.80	25.90	84.70

Galvanized steel spiral duct tees with 4" run and branch

Description	Craft@Hrs	Unit	Material $	Labor $	Total $
5"	S2@.120	Ea	6.90	3.05	9.95
6"	S2@.140	Ea	8.05	3.55	11.60
7"	S2@.156	Ea	8.97	3.96	12.93
8"	S2@.176	Ea	10.10	4.47	14.57
9"	S2@.192	Ea	11.00	4.87	15.87
10"	S2@.268	Ea	15.40	6.80	22.20
12"	S2@.324	Ea	18.60	8.23	26.83
14" - 16"	S2@.403	Ea	23.20	10.20	33.40
18" - 20"	S2@.508	Ea	29.30	12.90	42.20
22" - 24"	S2@.619	Ea	35.40	15.70	51.10
26" - 28"	S2@.724	Ea	41.70	18.40	60.10
30" - 32"	S2@.830	Ea	47.80	21.10	68.90
34" - 36"	S2@1.08	Ea	62.40	27.40	89.80

Galvanized steel spiral duct tees with 5" run and branch

Description	Craft@Hrs	Unit	Material $	Labor $	Total $
6"	S2@.148	Ea	8.50	3.76	12.26
7"	S2@.164	Ea	9.43	4.16	13.59
8"	S2@.188	Ea	10.80	4.77	15.57
9"	S2@.204	Ea	11.70	5.18	16.88
10"	S2@.284	Ea	16.40	7.21	23.61
12"	S2@.340	Ea	19.50	8.63	28.13
14" - 16"	S2@.428	Ea	24.60	10.90	35.50
18" - 20"	S2@.538	Ea	30.80	13.70	44.50
22" - 24"	S2@.651	Ea	37.50	16.50	54.00
26" - 28"	S2@.764	Ea	44.10	19.40	63.50
30" - 32"	S2@.873	Ea	50.20	22.20	72.40
34" - 36"	S2@1.15	Ea	65.80	29.20	95.00

Description	Craft@Hrs	Unit	Material $	Labor $	Total $

Galvanized steel spiral duct tees with 6" run and branch

Description	Craft@Hrs	Unit	Material $	Labor $	Total $
7"	S2@.180	Ea	10.30	4.57	14.87
8"	S2@.200	Ea	11.50	5.08	16.58
9"	S2@.220	Ea	12.60	5.59	18.19
10"	S2@.300	Ea	17.20	7.62	24.82
12"	S2@.356	Ea	20.50	9.04	29.54
14" - 16"	S2@.449	Ea	25.80	11.40	37.20
18" - 20"	S2@.566	Ea	32.60	14.40	47.00
22" - 24"	S2@.683	Ea	39.40	17.30	56.70
26" - 28"	S2@.807	Ea	46.20	20.50	66.70
30" - 32"	S2@.924	Ea	53.10	23.50	76.60
34" - 36"	S2@1.20	Ea	69.20	30.50	99.70

Galvanized steel spiral duct tees with 7" run and branch

Description	Craft@Hrs	Unit	Material $	Labor $	Total $
8"	S2@.212	Ea	12.10	5.38	17.48
9"	S2@.232	Ea	13.30	5.89	19.19
10"	S2@.312	Ea	17.90	7.92	25.82
12"	S2@.376	Ea	21.60	9.55	31.15
14" - 16"	S2@.470	Ea	27.00	11.90	38.90
18" - 20"	S2@.636	Ea	36.60	16.10	52.70
22" - 24"	S2@.717	Ea	41.40	18.20	59.60
26" - 28"	S2@.845	Ea	48.50	21.50	70.00
30" - 32"	S2@.969	Ea	55.70	24.60	80.30
34" - 36"	S2@1.27	Ea	72.70	32.20	104.90

Galvanized steel spiral duct tees with 8" run and branch

Description	Craft@Hrs	Unit	Material $	Labor $	Total $
9"	S2@.248	Ea	14.20	6.30	20.50
10"	S2@.328	Ea	18.90	8.33	27.23
12"	S2@.396	Ea	22.70	10.10	32.80
14" - 16"	S2@.493	Ea	28.20	12.50	40.70
18" - 20"	S2@.623	Ea	35.80	15.80	51.60
22" - 24"	S2@.751	Ea	43.30	19.10	62.40
26" - 28"	S2@.882	Ea	50.70	22.40	73.10
30" - 32"	S2@1.02	Ea	58.30	25.90	84.20
34" - 36"	S2@1.33	Ea	76.50	33.80	110.30

Galvanized steel spiral duct tees with 9" run and branch

Description	Craft@Hrs	Unit	Material $	Labor $	Total $
10"	S2@.344	Ea	19.70	8.73	28.43
12"	S2@.412	Ea	23.60	10.50	34.10
14" - 16"	S2@.515	Ea	29.60	13.10	42.70
18" - 20"	S2@.653	Ea	37.60	16.60	54.20
22" - 24"	S2@.779	Ea	45.30	19.80	65.10
26" - 28"	S2@.924	Ea	53.00	23.50	76.50
30" - 32"	S2@1.06	Ea	60.90	26.90	87.80
34" - 36"	S2@1.39	Ea	79.70	35.30	115.00

Galvanized Steel Spiral Tees

Description	Craft@Hrs	Unit	Material $	Labor $	Total $
Galvanized steel spiral duct tees with 10" run and branch					
12"	S2@.427	Ea	12.30	10.80	23.10
14" - 16"	S2@.538	Ea	15.50	13.70	29.20
18" - 20"	S2@.681	Ea	39.00	17.30	56.30
22" - 24"	S2@.820	Ea	47.20	20.80	68.00
26" - 28"	S2@.965	Ea	55.40	24.50	79.90
30" - 32"	S2@1.11	Ea	63.60	28.20	91.80
34" - 36"	S2@1.44	Ea	82.90	36.60	119.50
Galvanized steel spiral duct tees with 12" run and branch					
14" - 16"	S2@.581	Ea	33.30	14.80	48.10
18" - 20"	S2@.739	Ea	42.40	18.80	61.20
22" - 24"	S2@.888	Ea	51.10	22.50	73.60
26" - 28"	S2@1.04	Ea	59.90	26.40	86.30
30" - 32"	S2@1.20	Ea	69.00	30.50	99.50
34" - 36"	S2@1.57	Ea	90.00	39.90	129.90
Galvanized steel spiral duct tees with 14" run and branch					
14" - 16"	S2@.666	Ea	38.30	16.90	55.20
18" - 20"	S2@.794	Ea	45.60	20.20	65.80
22" - 24"	S2@.959	Ea	55.10	24.30	79.40
26" - 28"	S2@1.11	Ea	63.50	28.20	91.70
30" - 32"	S2@1.29	Ea	74.30	32.80	107.10
34" - 36"	S2@1.69	Ea	96.80	42.90	139.70
Galvanized steel spiral duct tees with 16" run and branch					
18" - 20"	S2@.833	Ea	47.80	21.10	68.90
22" - 24"	S2@1.03	Ea	59.20	26.20	85.40
26" - 28"	S2@1.20	Ea	69.30	30.50	99.80
30" - 32"	S2@1.39	Ea	79.70	35.30	115.00
34" - 36"	S2@1.81	Ea	104.00	46.00	150.00
Galvanized steel spiral duct tees with 18" run and branch					
18" - 20"	S2@.956	Ea	55.00	24.30	79.30
22" - 24"	S2@1.09	Ea	62.90	27.70	90.60
26" - 28"	S2@1.29	Ea	74.00	32.80	106.80
30" - 32"	S2@1.48	Ea	84.90	37.60	122.50
34" - 36"	S2@1.93	Ea	111.00	49.00	160.00
Galvanized steel spiral duct tees with 20" run and branch					
22" - 24"	S2@1.16	Ea	66.90	29.50	96.40
26" - 28"	S2@1.37	Ea	78.50	34.80	113.30
30" - 32"	S2@1.57	Ea	90.20	39.90	130.10
34" - 36"	S2@2.03	Ea	117.00	51.50	168.50

Description	Craft@Hrs	Unit	Material $	Labor $	Total $

Galvanized steel spiral duct tees with 22" run and branch

Description	Craft@Hrs	Unit	Material $	Labor $	Total $
22" - 24"	S2@1.28	Ea	73.70	32.50	106.20
26" - 28"	S2@1.45	Ea	83.00	36.80	119.80
30" - 32"	S2@1.64	Ea	94.50	41.60	136.10
34" - 36"	S2@2.16	Ea	124.00	54.80	178.80

Galvanized Steel Spiral Tees

Costs for Steel Spiral Crosses with Reducing Branches. Costs per cross, based on quantities less than 1,000 pounds. Deduct 15% from the material cost for quantities over 1,000 pounds. Add 50% to the material cost for lined duct. These costs include delivery, typical bracing, cleats, scrap, hangers, end closures, sealants and miscellaneous hardware.

Description	Craft@Hrs	Unit	Material $	Labor $	Total $

Galvanized steel spiral duct crosses with 3" branch

Description	Craft@Hrs	Unit	Material $	Labor $	Total $
4"	S2@.104	Ea	5.98	2.64	8.62
5"	S2@.128	Ea	7.35	3.25	10.60
6"	S2@.152	Ea	8.74	3.86	12.60
7"	S2@.176	Ea	10.10	4.47	14.57
8"	S2@.196	Ea	11.20	4.98	16.18
9"	S2@.216	Ea	12.30	5.48	17.78
10"	S2@.304	Ea	17.50	7.72	25.22
12"	S2@.352	Ea	20.30	8.94	29.24
14" - 16"	S2@.428	Ea	24.70	10.90	35.60
18" - 20"	S2@.525	Ea	30.20	13.30	43.50
22" - 24"	S2@.634	Ea	36.60	16.10	52.70
26" - 28"	S2@.745	Ea	42.90	18.90	61.80
30" - 32"	S2@.856	Ea	49.20	21.70	70.90
34" - 36"	S2@1.11	Ea	64.00	28.20	92.20

Galvanized steel spiral duct crosses with 4" branch

Description	Craft@Hrs	Unit	Material $	Labor $	Total $
5"	S2@.144	Ea	8.28	3.66	11.94
6"	S2@.168	Ea	9.65	4.27	13.92
7"	S2@.196	Ea	11.20	4.98	16.18
8"	S2@.216	Ea	12.30	5.48	17.78
9"	S2@.240	Ea	13.80	6.09	19.89
10"	S2@.328	Ea	18.90	8.33	27.23
12"	S2@.384	Ea	22.20	9.75	31.95
14" - 16"	S2@.461	Ea	26.60	11.70	38.30
18" - 20"	S2@.568	Ea	32.70	14.40	47.10
22" - 24"	S2@.685	Ea	39.50	17.40	56.90
26" - 28"	S2@.805	Ea	46.10	20.40	66.50
30" - 32"	S2@.927	Ea	53.20	23.50	76.70
34" - 36"	S2@1.20	Ea	69.20	30.50	99.70

Galvanized steel spiral duct crosses with 5" branch

Description	Craft@Hrs	Unit	Material $	Labor $	Total $
6"	S2@.184	Ea	10.60	4.67	15.27
7"	S2@.216	Ea	12.30	5.48	17.78
8"	S2@.240	Ea	13.80	6.09	19.89
9"	S2@.264	Ea	15.20	6.70	21.90
10"	S2@.356	Ea	20.50	9.04	29.54
12"	S2@.412	Ea	23.60	10.50	34.10
14" - 16"	S2@.497	Ea	28.40	12.60	41.00
18" - 20"	S2@.613	Ea	35.10	15.60	50.70
22" - 24"	S2@.741	Ea	42.50	18.80	61.30
26" - 28"	S2@.867	Ea	49.70	22.00	71.70
30" - 32"	S2@.993	Ea	57.20	25.20	82.40
34" - 36"	S2@1.29	Ea	74.40	32.80	107.20

Description	Craft@Hrs	Unit	Material $	Labor $	Total $

Galvanized steel spiral duct crosses with 6" branch

Description	Craft@Hrs	Unit	Material $	Labor $	Total $
7"	S2@.232	Ea	13.30	5.89	19.19
8"	S2@.260	Ea	14.90	6.60	21.50
9"	S2@.288	Ea	16.60	7.31	23.91
10"	S2@.380	Ea	21.80	9.65	31.45
12"	S2@.440	Ea	25.30	11.20	36.50
14" - 16"	S2@.532	Ea	30.50	13.50	44.00
18" - 20"	S2@.653	Ea	37.60	16.60	54.20
22" - 24"	S2@.790	Ea	45.40	20.10	65.50
26" - 28"	S2@.927	Ea	53.20	23.50	76.70
30" - 32"	S2@1.06	Ea	61.10	26.90	88.00
34" - 36"	S2@1.38	Ea	79.50	35.00	114.50

Galvanized steel spiral duct crosses with 7" branch

Description	Craft@Hrs	Unit	Material $	Labor $	Total $
8"	S2@.280	Ea	16.00	7.11	23.11
9"	S2@.308	Ea	17.70	7.82	25.52
10"	S2@.408	Ea	23.40	10.40	33.80
12"	S2@.470	Ea	26.90	11.90	38.80
14" - 16"	S2@.566	Ea	32.60	14.40	47.00
18" - 20"	S2@.692	Ea	39.90	17.60	57.50
22" - 24"	S2@.841	Ea	48.30	21.40	69.70
26" - 28"	S2@.986	Ea	56.70	25.00	81.70
30" - 32"	S2@1.13	Ea	65.20	28.70	93.90
34" - 36"	S2@1.48	Ea	84.90	37.60	122.50

Galvanized steel spiral duct crosses with 8" branch

Description	Craft@Hrs	Unit	Material $	Labor $	Total $
9"	S2@.310	Ea	19.10	7.87	26.97
10"	S2@.403	Ea	24.90	10.20	35.10
12"	S2@.470	Ea	28.60	11.90	40.50
14" - 16"	S2@.561	Ea	34.60	14.20	48.80
18" - 20"	S2@.690	Ea	42.40	17.50	59.90
22" - 24"	S2@.835	Ea	51.40	21.20	72.60
26" - 28"	S2@.980	Ea	60.10	24.90	85.00
30" - 32"	S2@1.12	Ea	69.10	28.40	97.50
34" - 36"	S2@1.47	Ea	90.10	37.30	127.40

Galvanized steel spiral duct crosses with 9" branch

Description	Craft@Hrs	Unit	Material $	Labor $	Total $
10"	S2@.427	Ea	26.10	10.80	36.90
12"	S2@.495	Ea	30.30	12.60	42.90
14" - 16"	S2@.596	Ea	36.70	15.10	51.80
18" - 20"	S2@.732	Ea	44.90	18.60	63.50
22" - 24"	S2@.882	Ea	54.20	22.40	76.60
26" - 28"	S2@1.04	Ea	63.50	26.40	89.90
30" - 32"	S2@1.19	Ea	73.10	30.20	103.30
34" - 36"	S2@1.55	Ea	95.30	39.40	134.70

Galvanized Steel Spiral Crosses

Description	Craft@Hrs	Unit	Material $	Labor $	Total $
Galvanized steel spiral duct crosses with 12" branch					
14" - 16"	S2@.692	Ea	42.50	17.60	60.10
18" - 20"	S2@.850	Ea	52.20	21.60	73.80
22" - 24"	S2@1.03	Ea	63.10	26.20	89.30
26" - 28"	S2@1.18	Ea	72.80	30.00	102.80
30" - 32"	S2@1.38	Ea	85.00	35.00	120.00
34" - 36"	S2@1.80	Ea	111.00	45.70	156.70
Galvanized steel spiral duct crosses with 14" branch					
14" - 16"	S2@1.81	Ea	21.60	46.00	67.60
18" - 20"	S2@2.11	Ea	25.10	53.60	78.70
22" - 24"	S2@2.55	Ea	30.50	64.70	95.20
26" - 28"	S2@2.98	Ea	35.60	75.70	111.30
30" - 32"	S2@3.43	Ea	40.90	87.10	128.00
34" - 36"	S2@4.47	Ea	53.40	113.00	166.40
Galvanized steel spiral duct crosses with 16" branch					
18" - 20"	S2@1.08	Ea	61.90	27.40	89.30
22" - 24"	S2@1.30	Ea	75.00	33.00	108.00
26" - 28"	S2@1.53	Ea	87.90	38.80	126.70
30" - 32"	S2@1.76	Ea	101.00	44.70	145.70
34" - 36"	S2@2.31	Ea	132.00	58.70	190.70
Galvanized steel spiral duct crosses with 18" branch					
18" - 20"	S2@1.22	Ea	70.30	31.00	101.30
22" - 24"	S2@1.40	Ea	80.70	35.50	116.20
26" - 28"	S2@1.65	Ea	94.80	41.90	136.70
30" - 32"	S2@1.89	Ea	109.00	48.00	157.00
34" - 36"	S2@2.47	Ea	142.00	62.70	204.70
Galvanized steel spiral duct crosses with 20" branch					
22" - 24"	S2@1.51	Ea	86.70	38.30	125.00
26" - 28"	S2@1.77	Ea	102.00	44.90	146.90
30" - 32"	S2@2.05	Ea	118.00	52.00	170.00
34" - 36"	S2@2.65	Ea	152.00	67.30	219.30
Galvanized steel spiral duct crosses with 22" branch					
22" - 24"	S2@1.68	Ea	96.80	42.70	139.50
26" - 28"	S2@1.89	Ea	109.00	48.00	157.00
30" - 32"	S2@2.17	Ea	124.00	55.10	179.10
34" - 36"	S2@2.84	Ea	164.00	72.10	236.10

Costs for Steel 26 Gauge Rectangular Duct. Costs per linear foot of duct, based on quantities less than 1,000 pounds of prefabricated duct. Deduct 15% from the material cost for quantities over 1,000 pounds. Add 50% to the material cost for lined duct. These costs include delivery, typical bracing, cleats, scrap, hangers, end closures, sealants and miscellaneous hardware.

Description	Craft@Hrs	Unit	Material $	Labor $	Total $

Galvanized steel 4" rectangular duct 26 gauge

Description	Craft@Hrs	Unit	Material $	Labor $	Total $
4"	S2@.048	Ea	1.41	1.22	2.63
5"	S2@.054	Ea	1.57	1.37	2.94
6"	S2@.060	Ea	1.75	1.52	3.27
7"	S2@.066	Ea	1.93	1.68	3.61
8"	S2@.072	Ea	2.10	1.83	3.93
9"	S2@.078	Ea	2.27	1.98	4.25
10"	S2@.084	Ea	2.45	2.13	4.58
11"	S2@.091	Ea	2.64	2.31	4.95
12"	S2@.097	Ea	2.81	2.46	5.27

Galvanized steel 5" rectangular duct 26 gauge

Description	Craft@Hrs	Unit	Material $	Labor $	Total $
5"	S2@.060	Ea	1.75	1.52	3.27
6"	S2@.066	Ea	1.93	1.68	3.61
7"	S2@.072	Ea	2.10	1.83	3.93
8"	S2@.078	Ea	2.27	1.98	4.25
9"	S2@.084	Ea	2.45	2.13	4.58
10"	S2@.091	Ea	2.64	2.31	4.95
11"	S2@.097	Ea	2.81	2.46	5.27
12"	S2@.103	Ea	2.98	2.62	5.60

Galvanized steel 6" rectangular duct 26 gauge

Description	Craft@Hrs	Unit	Material $	Labor $	Total $
6"	S2@.072	Ea	2.10	1.83	3.93
7"	S2@.078	Ea	2.27	1.98	4.25
8"	S2@.084	Ea	2.45	2.13	4.58
9"	S2@.091	Ea	2.64	2.31	4.95
10"	S2@.097	Ea	2.81	2.46	5.27
11"	S2@.103	Ea	2.98	2.62	5.60
12"	S2@.109	Ea	3.15	2.77	5.92

Galvanized steel 7" rectangular duct 26 gauge

Description	Craft@Hrs	Unit	Material $	Labor $	Total $
7"	S2@.084	Ea	2.45	2.13	4.58
8"	S2@.091	Ea	2.64	2.31	4.95
9"	S2@.097	Ea	2.81	2.46	5.27
10"	S2@.103	Ea	2.98	2.62	5.60
11"	S2@.109	Ea	3.15	2.77	5.92
12"	S2@.115	Ea	3.32	2.92	6.24

Galvanized steel 8" rectangular duct 26 gauge

Description	Craft@Hrs	Unit	Material $	Labor $	Total $
8"	S2@.097	Ea	2.81	2.46	5.27
9"	S2@.103	Ea	2.98	2.62	5.60
10"	S2@.109	Ea	3.15	2.77	5.92
11"	S2@.115	Ea	3.32	2.92	6.24
12"	S2@.121	Ea	3.50	3.07	6.57

Galvanized Steel Rectangular Duct

Description	Craft@Hrs	Unit	Material $	Labor $	Total $
Galvanized steel 9" rectangular duct 26 gauge					
9"	S2@.109	Ea	3.15	2.77	5.92
10"	S2@.115	Ea	3.32	2.92	6.24
11"	S2@.121	Ea	3.50	3.07	6.57
12"	S2@.127	Ea	3.67	3.22	6.89
Galvanized steel 10" rectangular duct 26 gauge					
10"	S2@.129	Ea	3.50	3.28	6.78
11"	S2@.127	Ea	3.67	3.22	6.89
12"	S2@.133	Ea	3.84	3.38	7.22
Galvanized steel 11" rectangular duct 26 gauge					
11"	S2@.133	Ea	3.84	3.38	7.22
12"	S2@.139	Ea	4.02	3.53	7.55
Galvanized steel 12" rectangular duct 26 gauge					
12"	S2@.145	Ea	4.20	3.68	7.88

Costs for Galvanized 24 Gauge Rectangular Duct. Costs per linear foot of duct, based on quantities less than 1,000 pounds of prefabricated duct. Deduct 15% from the material cost for over 1,000 pounds. Add 50% to the material cost for lined duct. These costs include delivery, typical bracing, cleats, scrap, hangers, end closures.

Description	Craft@Hrs	Unit	Material $	Labor $	Total $

14" wide galvanized steel rectangular duct 24 gauge

Description	Craft@Hrs	Unit	Material $	Labor $	Total $
6"	S2@.154	LF	4.50	3.91	8.41
8"	S2@.170	LF	4.96	4.32	9.28
10"	S2@.184	LF	5.40	4.67	10.07
12"	S2@.200	LF	5.86	5.08	10.94
14" - 16"	S2@.223	LF	6.53	5.66	12.19
18" - 20"	S2@.254	LF	7.42	6.45	13.87
22" - 24"	S2@.285	LF	8.33	7.24	15.57
26" - 28"	S2@.316	LF	9.23	8.02	17.25
30"	S2@.339	LF	9.90	8.61	18.51

16" wide galvanized steel rectangular duct 24 gauge

Description	Craft@Hrs	Unit	Material $	Labor $	Total $
6"	S2@.170	LF	4.96	4.32	9.28
8"	S2@.184	LF	5.40	4.67	10.07
10"	S2@.200	LF	5.86	5.08	10.94
12"	S2@.216	LF	6.30	5.48	11.78
14" - 16"	S2@.239	LF	6.98	6.07	13.05
18" - 20"	S2@.269	LF	7.87	6.83	14.70
22" - 24"	S2@.300	LF	8.78	7.62	16.40
26" - 28"	S2@.331	LF	9.67	8.40	18.07
30"	S2@.354	LF	10.40	8.99	19.39

18" wide galvanized steel rectangular duct 24 gauge

Description	Craft@Hrs	Unit	Material $	Labor $	Total $
6"	S2@.184	LF	5.40	4.67	10.07
8"	S2@.200	LF	5.86	5.08	10.94
10"	S2@.216	LF	6.30	5.48	11.78
12"	S2@.231	LF	6.75	5.87	12.62
14" - 16"	S2@.254	LF	7.42	6.45	13.87
18" - 20"	S2@.285	LF	8.33	7.24	15.57
22" - 24"	S2@.316	LF	9.23	8.02	17.25
26" - 28"	S2@.346	LF	10.10	8.78	18.88
30"	S2@.370	LF	10.80	9.39	20.19

Galvanized Steel Rectangular Ductwork

Description	Craft@Hrs	Unit	Material $	Labor $	Total $

20" wide galvanized steel rectangular duct 24 gauge

Description	Craft@Hrs	Unit	Material $	Labor $	Total $
6"	S2@.200	LF	5.86	5.08	10.94
8"	S2@.216	LF	6.30	5.48	11.78
10"	S2@.231	LF	6.75	5.87	12.62
12"	S2@.247	LF	7.21	6.27	13.48
14" - 16"	S2@.269	LF	7.87	6.83	14.70
18" - 20"	S2@.300	LF	8.78	7.62	16.40
22" - 24"	S2@.331	LF	9.67	8.40	18.07
26" - 28"	S2@.362	LF	10.60	9.19	19.79
30"	S2@.385	LF	11.20	9.78	20.98

22" wide galvanized steel rectangular duct 24 gauge

Description	Craft@Hrs	Unit	Material $	Labor $	Total $
6"	S2@.216	LF	6.30	5.48	11.78
8"	S2@.231	LF	6.75	5.87	12.62
10"	S2@.247	LF	7.21	6.27	13.48
12"	S2@.262	LF	7.65	6.65	14.30
14" - 16"	S2@.285	LF	8.33	7.24	15.57
18" - 20"	S2@.316	LF	9.23	8.02	17.25
22" - 24"	S2@.346	LF	10.10	8.78	18.88
26" - 28"	S2@.377	LF	11.10	9.57	20.67
30"	S2@.400	LF	11.70	10.20	21.90

24" wide galvanized steel rectangular duct 24 gauge

Description	Craft@Hrs	Unit	Material $	Labor $	Total $
6"	S2@.231	LF	6.75	5.87	12.62
8"	S2@.247	LF	7.21	6.27	13.48
10"	S2@.262	LF	7.65	6.65	14.30
12"	S2@.278	LF	8.10	7.06	15.16
14" - 16"	S2@.300	LF	8.78	7.62	16.40
18" - 20"	S2@.331	LF	9.67	8.40	18.07
22" - 24"	S2@.362	LF	10.60	9.19	19.79
26" - 28"	S2@.392	LF	11.50	9.95	21.45
30"	S2@.415	LF	12.10	10.50	22.60

26" wide galvanized steel rectangular duct 24 gauge

Description	Craft@Hrs	Unit	Material $	Labor $	Total $
6"	S2@.247	LF	7.21	6.27	13.48
8"	S2@.262	LF	7.65	6.65	14.30
10"	S2@.278	LF	8.10	7.06	15.16
12"	S2@.292	LF	8.55	7.41	15.96
14" - 16"	S2@.316	LF	9.23	8.02	17.25
18" - 20"	S2@.346	LF	10.10	8.78	18.88
22" - 24"	S2@.377	LF	11.10	9.57	20.67
26" - 28"	S2@.407	LF	11.90	10.30	22.20
30"	S2@.431	LF	12.60	10.90	23.50

Description	Craft@Hrs	Unit	Material $	Labor $	Total $

28" wide galvanized steel rectangular duct 24 gauge

Description	Craft@Hrs	Unit	Material $	Labor $	Total $
6"	S2@.262	LF	7.65	6.65	14.30
8"	S2@.278	LF	8.10	7.06	15.16
10"	S2@.292	LF	8.55	7.41	15.96
12"	S2@.308	LF	9.01	7.82	16.83
14" - 16"	S2@.331	LF	9.67	8.40	18.07
18" - 20"	S2@.362	LF	10.60	9.19	19.79
22" - 24"	S2@.392	LF	11.50	9.95	21.45
26" - 28"	S2@.423	LF	12.40	10.70	23.10
30"	S2@.448	LF	13.10	11.40	24.50

30" wide galvanized steel rectangular duct 24 gauge

Description	Craft@Hrs	Unit	Material $	Labor $	Total $
6"	S2@.278	LF	8.10	7.06	15.16
8"	S2@.292	LF	8.55	7.41	15.96
10"	S2@.308	LF	9.01	7.82	16.83
12"	S2@.323	LF	9.45	8.20	17.65
14" - 16"	S2@.346	LF	10.10	8.78	18.88
18" - 20"	S2@.377	LF	11.10	9.57	20.67
22" - 24"	S2@.407	LF	11.90	10.30	22.20
26" - 28"	S2@.440	LF	12.80	11.20	24.00
30"	S2@.465	LF	13.50	11.80	25.30

Costs for Galvanized 22 Gauge Rectangular Duct. Costs per linear foot of duct, based on quantities less than 1,000 pounds of prefabricated duct. Deduct 15% from the material cost for over 1,000 pounds. Add 50% to the material cost for lined duct. These costs include delivery, typical bracing, cleats, scrap, hangers, end closures.

32" wide galvanized steel rectangular duct 22 gauge

Description	Craft@Hrs	Unit	Material $	Labor $	Total $
8"	S2@.374	LF	11.00	9.50	20.50
10"	S2@.393	LF	11.50	9.98	21.48
12"	S2@.412	LF	12.00	10.50	22.50
14" - 16"	S2@.442	LF	12.90	11.20	24.10
18" - 20"	S2@.478	LF	14.00	12.10	26.10
22" - 24"	S2@.517	LF	15.00	13.10	28.10
26" - 28"	S2@.553	LF	16.10	14.00	30.10
30" - 32"	S2@.589	LF	17.30	15.00	32.30
34" - 36"	S2@.628	LF	18.30	15.90	34.20
38" - 40"	S2@.664	LF	19.50	16.90	36.40
42" - 44"	S2@.702	LF	20.40	17.80	38.20
46" - 48"	S2@.739	LF	21.60	18.80	40.40
50" - 52"	S2@.777	LF	22.70	19.70	42.40
54"	S2@.807	LF	23.50	20.50	44.00

Galvanized Steel Rectangular Ductwork

Description	Craft@Hrs	Unit	Material $	Labor $	Total $

34" wide galvanized steel rectangular duct 22 gauge

Description	Craft@Hrs	Unit	Material $	Labor $	Total $
8"	S2@.393	LF	11.50	9.98	21.48
10"	S2@.412	LF	12.00	10.50	22.50
12"	S2@.431	LF	12.60	10.90	23.50
14" - 16"	S2@.461	LF	13.40	11.70	25.10
18" - 20"	S2@.497	LF	14.60	12.60	27.20
22" - 24"	S2@.534	LF	15.50	13.60	29.10
26" - 28"	S2@.572	LF	16.70	14.50	31.20
30" - 32"	S2@.608	LF	17.80	15.40	33.20
34" - 36"	S2@.645	LF	18.90	16.40	35.30
38" - 40"	S2@.683	LF	20.00	17.30	37.30
42" - 44"	S2@.722	LF	21.10	18.30	39.40
46" - 48"	S2@.758	LF	22.20	19.20	41.40
50" - 52"	S2@.796	LF	23.20	20.20	43.40
54"	S2@.824	LF	24.10	20.90	45.00

36" wide galvanized steel rectangular duct 22 gauge

Description	Craft@Hrs	Unit	Material $	Labor $	Total $
8"	S2@.412	LF	12.00	10.50	22.50
10"	S2@.431	LF	12.60	10.90	23.50
12"	S2@.453	LF	13.20	11.50	24.70
14" - 16"	S2@.478	LF	14.00	12.10	26.10
18" - 20"	S2@.517	LF	15.00	13.10	28.10
22" - 24"	S2@.553	LF	16.10	14.00	30.10
26" - 28"	S2@.589	LF	17.30	15.00	32.30
30" - 32"	S2@.628	LF	18.30	15.90	34.20
34" - 36"	S2@.664	LF	19.50	16.90	36.40
38" - 40"	S2@.702	LF	20.40	17.80	38.20
42" - 44"	S2@.739	LF	21.60	18.80	40.40
46" - 48"	S2@.777	LF	22.70	19.70	42.40
50" - 52"	S2@.816	LF	23.70	20.70	44.40
54"	S2@.841	LF	24.70	21.40	46.10

38" wide galvanized steel rectangular duct 22 gauge

Description	Craft@Hrs	Unit	Material $	Labor $	Total $
8"	S2@.431	LF	12.60	10.90	23.50
10"	S2@.453	LF	13.20	11.50	24.70
12"	S2@.470	LF	13.60	11.90	25.50
14" - 16"	S2@.497	LF	14.60	12.60	27.20
18" - 20"	S2@.534	LF	15.50	13.60	29.10
22" - 24"	S2@.572	LF	16.70	14.50	31.20
26" - 28"	S2@.608	LF	17.80	15.40	33.20
30" - 32"	S2@.645	LF	18.90	16.40	35.30
34" - 36"	S2@.683	LF	20.00	17.30	37.30
38" - 40"	S2@.722	LF	21.10	18.30	39.40
42" - 44"	S2@.758	LF	22.20	19.20	41.40
46" - 48"	S2@.796	LF	23.20	20.20	43.40
50" - 52"	S2@.833	LF	24.40	21.10	45.50
54"	S2@.863	LF	25.20	21.90	47.10

Description	Craft@Hrs	Unit	Material $	Labor $	Total $

40" wide galvanized steel rectangular duct 22 gauge

Description	Craft@Hrs	Unit	Material $	Labor $	Total $
8"	S2@.453	LF	13.20	11.50	24.70
10"	S2@.470	LF	13.60	11.90	25.50
12"	S2@.487	LF	14.30	12.40	26.70
14" - 16"	S2@.517	LF	15.00	13.10	28.10
18" - 20"	S2@.553	LF	16.10	14.00	30.10
22" - 24"	S2@.589	LF	17.30	15.00	32.30
26" - 28"	S2@.628	LF	18.30	15.90	34.20
30" - 32"	S2@.664	LF	19.50	16.90	36.40
34" - 36"	S2@.702	LF	20.40	17.80	38.20
38" - 40"	S2@.739	LF	21.60	18.80	40.40
42" - 44"	S2@.777	LF	22.70	19.70	42.40
46" - 48"	S2@.816	LF	23.70	20.70	44.40
50" - 52"	S2@.852	LF	24.90	21.60	46.50
54"	S2@.880	LF	25.70	22.30	48.00

42" wide galvanized steel rectangular duct 22 gauge

Description	Craft@Hrs	Unit	Material $	Labor $	Total $
8"	S2@.470	LF	13.60	11.90	25.50
10"	S2@.487	LF	14.30	12.40	26.70
12"	S2@.508	LF	14.80	12.90	27.70
14" - 16"	S2@.534	LF	15.50	13.60	29.10
18" - 20"	S2@.572	LF	16.70	14.50	31.20
22" - 24"	S2@.611	LF	17.80	15.50	33.30
26" - 28"	S2@.645	LF	18.90	16.40	35.30
30" - 32"	S2@.683	LF	20.00	17.30	37.30
34" - 36"	S2@.722	LF	21.10	18.30	39.40
38" - 40"	S2@.758	LF	22.20	19.20	41.40
42" - 44"	S2@.798	LF	23.20	20.30	43.50
46" - 48"	S2@.833	LF	24.40	21.10	45.50
50" - 52"	S2@.871	LF	25.50	22.10	47.60
54"	S2@.901	LF	26.20	22.90	49.10

44" wide galvanized steel rectangular duct 22 gauge

Description	Craft@Hrs	Unit	Material $	Labor $	Total $
8"	S2@.487	LF	14.30	12.40	26.70
10"	S2@.508	LF	14.80	12.90	27.70
12"	S2@.525	LF	15.30	13.30	28.60
14" - 16"	S2@.555	LF	16.10	14.10	30.20
18" - 20"	S2@.589	LF	17.30	15.00	32.30
22" - 24"	S2@.628	LF	18.30	15.90	34.20
26" - 28"	S2@.666	LF	19.50	16.90	36.40
30" - 32"	S2@.705	LF	20.40	17.90	38.30
34" - 36"	S2@.739	LF	21.60	18.80	40.40
38" - 40"	S2@.777	LF	22.70	19.70	42.40
42" - 44"	S2@.816	LF	23.70	20.70	44.40
46" - 48"	S2@.852	LF	24.90	21.60	46.50
50" - 52"	S2@.890	LF	26.00	22.60	48.60
54"	S2@.918	LF	26.80	23.30	50.10

Galvanized Steel Rectangular Ductwork

Description	Craft@Hrs	Unit	Material $	Labor $	Total $

46" wide galvanized steel rectangular duct 22 gauge

Description	Craft@Hrs	Unit	Material $	Labor $	Total $
8"	S2@.508	LF	14.80	12.90	27.70
10"	S2@.525	LF	15.30	13.30	28.60
12"	S2@.542	LF	15.90	13.80	29.70
14" - 16"	S2@.572	LF	16.70	14.50	31.20
18" - 20"	S2@.611	LF	17.80	15.50	33.30
22" - 24"	S2@.649	LF	18.90	16.50	35.40
26" - 28"	S2@.683	LF	20.00	17.30	37.30
30" - 32"	S2@.722	LF	21.10	18.30	39.40
34" - 36"	S2@.758	LF	22.20	19.20	41.40
38" - 40"	S2@.796	LF	23.20	20.20	43.40
42" - 44"	S2@.833	LF	24.40	21.10	45.50
46" - 48"	S2@.871	LF	25.50	22.10	47.60
50" - 52"	S2@.909	LF	26.60	23.10	49.70
54"	S2@.935	LF	27.40	23.70	51.10

48" wide galvanized steel rectangular duct 22 gauge

Description	Craft@Hrs	Unit	Material $	Labor $	Total $
8"	S2@.525	LF	15.30	13.30	28.60
10"	S2@.542	LF	15.90	13.80	29.70
12"	S2@.564	LF	16.50	14.30	30.80
14" - 16"	S2@.589	LF	17.30	15.00	32.30
18" - 20"	S2@.628	LF	18.30	15.90	34.20
22" - 24"	S2@.664	LF	19.50	16.90	36.40
26" - 28"	S2@.702	LF	20.40	17.80	38.20
30" - 32"	S2@.739	LF	21.60	18.80	40.40
34" - 36"	S2@.777	LF	22.70	19.70	42.40
38" - 40"	S2@.816	LF	23.70	20.70	44.40
42" - 44"	S2@.852	LF	24.90	21.60	46.50
46" - 48"	S2@.890	LF	26.00	22.60	48.60
50" - 52"	S2@.927	LF	27.20	23.50	50.70
54"	S2@.956	LF	27.90	24.30	52.20

50" wide galvanized steel rectangular duct 22 gauge

Description	Craft@Hrs	Unit	Material $	Labor $	Total $
8"	S2@.542	LF	15.90	13.80	29.70
10"	S2@.564	LF	16.50	14.30	30.80
12"	S2@.581	LF	16.90	14.80	31.70
14" - 16"	S2@.611	LF	17.80	15.50	33.30
18" - 20"	S2@.645	LF	18.90	16.40	35.30
22" - 24"	S2@.683	LF	20.00	17.30	37.30
26" - 28"	S2@.722	LF	21.10	18.30	39.40
30" - 32"	S2@.758	LF	22.20	19.20	41.40
34" - 36"	S2@.798	LF	23.20	20.30	43.50
38" - 40"	S2@.833	LF	24.40	21.10	45.50
42" - 44"	S2@.871	LF	25.50	22.10	47.60
46" - 48"	S2@.909	LF	26.60	23.10	49.70
50" - 52"	S2@.946	LF	27.60	24.00	51.60
54"	S2@.974	LF	28.40	24.70	53.10

Description	Craft@Hrs	Unit	Material $	Labor $	Total $

52" wide galvanized steel rectangular duct 22 gauge

Description	Craft@Hrs	Unit	Material $	Labor $	Total $
8"	S2@.564	LF	16.50	14.30	30.80
10"	S2@.581	LF	16.90	14.80	31.70
12"	S2@.598	LF	17.50	15.20	32.70
14" - 16"	S2@.628	LF	18.30	15.90	34.20
18" - 20"	S2@.662	LF	19.50	16.80	36.30
22" - 24"	S2@.700	LF	20.40	17.80	38.20
26" - 28"	S2@.739	LF	21.60	18.80	40.40
30" - 32"	S2@.777	LF	22.70	19.70	42.40
34" - 36"	S2@.816	LF	23.70	20.70	44.40
38" - 40"	S2@.852	LF	24.90	21.60	46.50
42" - 44"	S2@.890	LF	26.00	22.60	48.60
46" - 48"	S2@.927	LF	27.20	23.50	50.70
50" - 52"	S2@.976	LF	28.20	24.80	53.00
54"	S2@.991	LF	28.90	25.20	54.10

54" wide galvanized steel rectangular duct 22 gauge

Description	Craft@Hrs	Unit	Material $	Labor $	Total $
8"	S2@.581	LF	16.90	14.80	31.70
10"	S2@.598	LF	17.50	15.20	32.70
12"	S2@.619	LF	18.10	15.70	33.80
14" - 16"	S2@.645	LF	18.90	16.40	35.30
18" - 20"	S2@.683	LF	20.00	17.30	37.30
22" - 24"	S2@.722	LF	21.10	18.30	39.40
26" - 28"	S2@.758	LF	22.20	19.20	41.40
30" - 32"	S2@.796	LF	23.20	20.20	43.40
34" - 36"	S2@.833	LF	24.40	21.10	45.50
38" - 40"	S2@.871	LF	25.50	22.10	47.60
42" - 44"	S2@.909	LF	26.60	23.10	49.70
46" - 48"	S2@.946	LF	27.60	24.00	51.60
50" - 52"	S2@.982	LF	28.60	24.90	53.50
54"	S2@1.01	LF	29.60	25.60	55.20

Galvanized Steel Rectangular Ductwork

Costs for Galvanized 20 Gauge Rectangular Duct. Costs per linear foot of duct, based on quantities less than 1,000 pounds of prefabricated duct. Deduct 15% from the material cost for over 1,000 pounds. Add 50% to the material cost for lined duct. These costs include delivery, typical bracing, cleats, scrap, hangers, end closures, sealants and miscellaneous hardware. See installation costs that follow.

Description	Craft@Hrs	Unit	Material $	Labor $	Total $

56" wide galvanized steel rectangular duct 20 gauge

Description	Craft@Hrs	Unit	Material $	Labor $	Total $
18" - 20"	S2@.828	LF	24.40	21.00	45.40
22" - 24"	S2@.871	LF	25.70	22.10	47.80
26" - 28"	S2@.916	LF	26.90	23.30	50.20
30" - 32"	S2@.959	LF	28.30	24.30	52.60
34" - 36"	S2@1.00	LF	29.60	25.40	55.00
38" - 40"	S2@1.05	LF	30.90	26.70	57.60
42" - 44"	S2@1.09	LF	32.20	27.70	59.90
46" - 48"	S2@1.14	LF	33.50	28.90	62.40
50" - 52"	S2@1.18	LF	34.80	30.00	64.80
54" - 56"	S2@1.23	LF	36.10	31.20	67.30
58" - 60"	S2@1.27	LF	37.40	32.20	69.60
62" - 64"	S2@1.31	LF	38.80	33.30	72.10
66" - 68"	S2@1.36	LF	39.90	34.50	74.40
70" - 72"	S2@1.40	LF	41.30	35.50	76.80

58" wide galvanized steel rectangular duct 20 gauge

Description	Craft@Hrs	Unit	Material $	Labor $	Total $
18" - 20"	S2@.850	LF	25.00	21.60	46.60
22" - 24"	S2@.895	LF	26.40	22.70	49.10
26" - 28"	S2@.937	LF	27.60	23.80	51.40
30" - 32"	S2@.980	LF	29.00	24.90	53.90
34" - 36"	S2@1.02	LF	30.20	25.90	56.10
38" - 40"	S2@1.07	LF	31.50	27.20	58.70
42" - 44"	S2@1.11	LF	32.90	28.20	61.10
46" - 48"	S2@1.16	LF	34.10	29.50	63.60
50" - 52"	S2@1.02	LF	35.50	25.90	61.40
54" - 56"	S2@1.25	LF	36.80	31.70	68.50
58" - 60"	S2@1.29	LF	38.00	32.80	70.80
62" - 64"	S2@1.33	LF	39.40	33.80	73.20
66" - 68"	S2@1.38	LF	40.70	35.00	75.70
70" - 72"	S2@1.42	LF	42.00	36.10	78.10

Description	Craft@Hrs	Unit	Material $	Labor $	Total $

60" wide galvanized steel rectangular duct 20 gauge

Description	Craft@Hrs	Unit	Material $	Labor $	Total $
18" - 20"	S2@.871	LF	25.70	22.10	47.80
22" - 24"	S2@.916	LF	26.90	23.30	50.20
26" - 28"	S2@.959	LF	28.30	24.30	52.60
30" - 32"	S2@1.00	LF	29.60	25.40	55.00
34" - 36"	S2@1.05	LF	30.90	26.70	57.60
38" - 40"	S2@1.09	LF	32.20	27.70	59.90
42" - 44"	S2@1.14	LF	33.50	28.90	62.40
46" - 48"	S2@1.18	LF	34.80	30.00	64.80
50" - 52"	S2@1.23	LF	36.10	31.20	67.30
54" - 56"	S2@1.27	LF	37.40	32.20	69.60
58" - 60"	S2@1.31	LF	38.80	33.30	72.10
62" - 64"	S2@1.36	LF	39.90	34.50	74.40
66" - 68"	S2@1.40	LF	41.30	35.50	76.80
70" - 72"	S2@1.45	LF	42.70	36.80	79.50

62" wide galvanized steel rectangular duct 20 gauge

Description	Craft@Hrs	Unit	Material $	Labor $	Total $
18" - 20"	S2@.895	LF	26.40	22.70	49.10
22" - 24"	S2@.937	LF	27.60	23.80	51.40
26" - 28"	S2@.980	LF	29.00	24.90	53.90
30" - 32"	S2@1.02	LF	30.20	25.90	56.10
34" - 36"	S2@1.07	LF	31.50	27.20	58.70
38" - 40"	S2@1.11	LF	32.90	28.20	61.10
42" - 44"	S2@1.16	LF	34.10	29.50	63.60
46" - 48"	S2@1.20	LF	35.50	30.50	66.00
50" - 52"	S2@1.25	LF	36.80	31.70	68.50
54" - 56"	S2@1.29	LF	38.00	32.80	70.80
58" - 60"	S2@1.33	LF	39.40	33.80	73.20
62" - 64"	S2@1.38	LF	40.70	35.00	75.70
66" - 68"	S2@1.42	LF	42.00	36.10	78.10
70" - 72"	S2@1.47	LF	43.20	37.30	80.50

64" wide galvanized steel rectangular duct 20 gauge

Description	Craft@Hrs	Unit	Material $	Labor $	Total $
18" - 20"	S2@.916	LF	26.90	23.30	50.20
22" - 24"	S2@.959	LF	28.30	24.30	52.60
26" - 28"	S2@1.00	LF	29.60	25.40	55.00
30" - 32"	S2@1.05	LF	30.90	26.70	57.60
34" - 36"	S2@1.09	LF	32.20	27.70	59.90
38" - 40"	S2@1.14	LF	33.50	28.90	62.40
42" - 44"	S2@1.18	LF	34.80	30.00	64.80
46" - 48"	S2@1.23	LF	36.10	31.20	67.30
50" - 52"	S2@1.27	LF	37.40	32.20	69.60
54" - 56"	S2@1.31	LF	38.80	33.30	72.10
58" - 60"	S2@1.36	LF	39.90	34.50	74.40
62" - 64"	S2@1.40	LF	41.30	35.50	76.80
66" - 68"	S2@1.45	LF	42.70	36.80	79.50
70" - 72"	S2@1.49	LF	43.90	37.80	81.70

Galvanized Steel Rectangular Ductwork

Description	Craft@Hrs	Unit	Material $	Labor $	Total $
66" wide galvanized steel rectangular duct 20 gauge					
18" - 20"	S2@.937	LF	27.60	23.80	51.40
22" - 24"	S2@.980	LF	29.00	24.90	53.90
26" - 28"	S2@1.02	LF	30.20	25.90	56.10
30" - 32"	S2@1.07	LF	31.50	27.20	58.70
34" - 36"	S2@1.11	LF	32.90	28.20	61.10
38" - 40"	S2@1.16	LF	34.10	29.50	63.60
42" - 44"	S2@1.20	LF	35.50	30.50	66.00
.46" - 48"	S2@1.25	LF	36.80	31.70	68.50
50" - 52"	S2@1.29	LF	38.00	32.80	70.80
54" - 56"	S2@1.33	LF	39.40	33.80	73.20
58" - 60"	S2@1.38	LF	40.70	35.00	75.70
62" - 64"	S2@1.42	LF	42.00	36.10	78.10
66" - 68"	S2@1.47	LF	43.20	37.30	80.50
70" - 72"	S2@1.51	LF	44.50	38.30	82.80
68" wide galvanized steel rectangular duct 20 gauge					
18" - 20"	S2@.959	LF	28.30	24.30	52.60
22" - 24"	S2@1.00	LF	29.60	25.40	55.00
26" - 28"	S2@1.05	LF	30.90	26.70	57.60
30" - 32"	S2@1.09	LF	32.20	27.70	59.90
34" - 36"	S2@1.14	LF	33.50	28.90	62.40
38" - 40"	S2@1.18	LF	34.80	30.00	64.80
42" - 44"	S2@1.23	LF	36.10	31.20	67.30
46" - 48"	S2@1.27	LF	37.40	32.20	69.60
50" - 52"	S2@1.31	LF	38.80	33.30	72.10
54" - 56"	S2@1.36	LF	39.90	34.50	74.40
58" - 60"	S2@1.40	LF	41.30	35.50	76.80
62" - 64"	S2@1.45	LF	42.70	36.80	79.50
66" - 68"	S2@1.49	LF	43.90	37.80	81.70
70" - 72"	S2@1.53	LF	45.30	38.80	84.10
70" wide galvanized steel rectangular duct 20 gauge					
18" - 20"	S2@.980	LF	29.00	24.90	53.90
22" - 24"	S2@1.02	LF	30.20	25.90	56.10
26" - 28"	S2@1.07	LF	31.50	27.20	58.70
30" - 32"	S2@1.11	LF	32.90	28.20	61.10
34" - 36"	S2@1.16	LF	34.10	29.50	63.60
38" - 40"	S2@1.20	LF	35.50	30.50	66.00
42" - 44"	S2@1.25	LF	36.80	31.70	68.50
46" - 48"	S2@1.29	LF	38.00	32.80	70.80
50" - 52"	S2@1.33	LF	39.40	33.80	73.20
54" - 56"	S2@1.38	LF	40.70	35.00	75.70
58" - 60"	S2@1.42	LF	42.00	36.10	78.10
62" - 64"	S2@1.47	LF	43.20	37.30	80.50
66" - 68"	S2@1.51	LF	44.50	38.30	82.80
70" - 72"	S2@1.55	LF	45.80	39.40	85.20

Description	Craft@Hrs	Unit	Material $	Labor $	Total $

72" wide galvanized steel rectangular duct 20 gauge

Description	Craft@Hrs	Unit	Material $	Labor $	Total $
18" - 20"	S2@1.00	LF	29.60	25.40	55.00
22" - 24"	S2@1.05	LF	30.90	26.70	57.60
26" - 28"	S2@1.09	LF	32.20	27.70	59.90
30" - 32"	S2@1.14	LF	33.50	28.90	62.40
34" - 36"	S2@1.18	LF	34.80	30.00	64.80
38" - 40"	S2@1.23	LF	36.10	31.20	67.30
42" - 44"	S2@1.27	LF	37.40	32.20	69.60
46" - 48"	S2@1.31	LF	38.80	33.30	72.10
50" - 52"	S2@1.36	LF	39.90	34.50	74.40
54" - 56"	S2@1.40	LF	41.30	35.50	76.80
58" - 60"	S2@1.45	LF	42.70	36.80	79.50
62" - 64"	S2@1.49	LF	43.90	37.80	81.70
66" - 68"	S2@1.53	LF	45.30	38.80	84.10
70" - 72"	S2@1.58	LF	46.40	40.10	86.50

Galvanized Steel Rectangular Elbows

Costs for Galvanized Rectangular 90 Degree Elbows. Costs per elbow, based on quantities less than 1,000 pounds. Deduct 15% from the material cost for over 1,000 pounds. Add 50% to the material cost for lined duct. These costs include delivery, typical bracing, cleats, scrap, hangers, end closures, sealants and miscellaneous hardware. For turning vanes, please see page 267.

Description	Craft@Hrs	Unit	Material $	Labor $	Total $

4" galvanized steel rectangular duct 90 degree elbow

Description	Craft@Hrs	Unit	Material $	Labor $	Total $
4"	S2@.042	Ea	2.47	1.07	3.54
6"	S2@.058	Ea	3.40	1.47	4.87
8"	S2@.078	Ea	4.57	1.98	6.55
10"	S2@.104	Ea	6.09	2.64	8.73
12"	S2@.132	Ea	7.73	3.35	11.08
14" - 16"	S2@.236	Ea	13.80	5.99	19.79
18" - 20"	S2@.339	Ea	20.00	8.61	28.61
22" - 24"	S2@.468	Ea	27.40	11.90	39.30
26" - 28"	S2@.615	Ea	36.00	15.60	51.60
30" - 32"	S2@.854	Ea	50.20	21.70	71.90
34" - 36"	S2@1.19	Ea	69.80	30.20	100.00
38" - 40"	S2@1.45	Ea	84.80	36.80	121.60
42" - 44"	S2@1.73	Ea	101.00	43.90	144.90
46" - 48"	S2@2.04	Ea	119.00	51.80	170.80
50"	S2@2.28	Ea	134.00	57.90	191.90

6" galvanized steel rectangular duct 90 degree elbow

Description	Craft@Hrs	Unit	Material $	Labor $	Total $
4"	S2@.058	Ea	3.40	1.47	4.87
6"	S2@.078	Ea	4.57	1.98	6.55
8"	S2@.104	Ea	6.09	2.64	8.73
10"	S2@.132	Ea	7.73	3.35	11.08
12"	S2@.148	Ea	8.67	3.76	12.43
14" - 16"	S2@.266	Ea	15.70	6.75	22.45
18" - 20"	S2@.370	Ea	21.60	9.39	30.99
22" - 24"	S2@.506	Ea	29.70	12.80	42.50
26" - 28"	S2@.655	Ea	38.30	16.60	54.90
30" - 32"	S2@.905	Ea	52.80	23.00	75.80
34" - 36"	S2@1.24	Ea	72.50	31.50	104.00
38" - 40"	S2@1.49	Ea	87.50	37.80	125.30
42" - 44"	S2@1.78	Ea	104.00	45.20	149.20
46" - 48"	S2@2.09	Ea	122.00	53.10	175.10
50"	S2@2.34	Ea	137.00	59.40	196.40

Description	Craft@Hrs	Unit	Material $	Labor $	Total $

8" galvanized steel rectangular duct 90 degree elbow

Description	Craft@Hrs	Unit	Material $	Labor $	Total $
4"	S2@.078	Ea	4.57	1.98	6.55
6"	S2@.104	Ea	6.09	2.64	8.73
8"	S2@.132	Ea	7.73	3.35	11.08
10"	S2@.148	Ea	8.67	3.76	12.43
12"	S2@.170	Ea	9.97	4.32	14.29
14" - 16"	S2@.300	Ea	17.50	7.62	25.12
18" - 20"	S2@.403	Ea	23.60	10.20	33.80
22" - 24"	S2@.549	Ea	32.10	13.90	46.00
26" - 28"	S2@.694	Ea	40.80	17.60	58.40
30" - 32"	S2@.950	Ea	55.70	24.10	79.80
34" - 36"	S2@1.28	Ea	75.30	32.50	107.80
38" - 40"	S2@1.54	Ea	90.30	39.10	129.40
42" - 44"	S2@1.83	Ea	107.00	46.50	153.50
46" - 48"	S2@2.13	Ea	125.00	54.10	179.10
50"	S2@2.39	Ea	141.00	60.70	201.70

10" galvanized steel rectangular duct 90 degree elbow

Description	Craft@Hrs	Unit	Material $	Labor $	Total $
4"	S2@.104	Ea	6.09	2.64	8.73
6"	S2@.132	Ea	7.73	3.35	11.08
8"	S2@.148	Ea	8.67	3.76	12.43
10"	S2@.170	Ea	9.97	4.32	14.29
12"	S2@.194	Ea	11.30	4.93	16.23
14" - 16"	S2@.336	Ea	19.60	8.53	28.13
18" - 20"	S2@.438	Ea	25.70	11.10	36.80
22" - 24"	S2@.587	Ea	34.50	14.90	49.40
26" - 28"	S2@.737	Ea	43.10	18.70	61.80
30" - 32"	S2@.997	Ea	58.40	25.30	83.70
34" - 36"	S2@1.33	Ea	78.20	33.80	112.00
38" - 40"	S2@1.59	Ea	93.30	40.40	133.70
42" - 44"	S2@1.87	Ea	110.00	47.50	157.50
46" - 48"	S2@2.18	Ea	128.00	55.40	183.40
50"	S2@2.43	Ea	143.00	61.70	204.70

12" galvanized steel rectangular duct 90 degree elbow

Description	Craft@Hrs	Unit	Material $	Labor $	Total $
4"	S2@.132	Ea	7.73	3.35	11.08
6"	S2@.148	Ea	8.67	3.76	12.43
8"	S2@.170	Ea	9.97	4.32	14.29
10"	S2@.194	Ea	11.30	4.93	16.23
12"	S2@.224	Ea	13.10	5.69	18.79
18" - 20"	S2@.478	Ea	27.90	12.10	40.00
22" - 24"	S2@.628	Ea	36.80	15.90	52.70
26" - 28"	S2@.775	Ea	45.50	19.70	65.20
30" - 32"	S2@1.04	Ea	61.00	26.40	87.40
34" - 36"	S2@1.38	Ea	80.90	35.00	115.90
38" - 40"	S2@1.64	Ea	96.10	41.60	137.70
42" - 44"	S2@1.92	Ea	112.00	48.70	160.70
46" - 48"	S2@2.23	Ea	131.00	56.60	187.60
50"	S2@2.48	Ea	145.00	63.00	208.00

Galvanized Steel Rectangular Elbows

Description	Craft@Hrs	Unit	Material $	Labor $	Total $

14" galvanized steel rectangular duct 90 degree elbow

Description	Craft@Hrs	Unit	Material $	Labor $	Total $
4"	S2@.212	Ea	12.40	5.38	17.78
6"	S2@.232	Ea	13.60	5.89	19.49
8"	S2@.260	Ea	15.10	6.60	21.70
10"	S2@.292	Ea	17.00	7.41	24.41
12"	S2@.328	Ea	19.20	8.33	27.53
14" - 16"	S2@.415	Ea	24.40	10.50	34.90
18" - 20"	S2@.527	Ea	30.90	13.40	44.30
22" - 24"	S2@.867	Ea	39.10	22.00	61.10
26" - 28"	S2@.816	Ea	47.70	20.70	68.40
30" - 32"	S2@1.09	Ea	64.00	27.70	91.70
34" - 36"	S2@1.43	Ea	83.80	36.30	120.10
38" - 40"	S2@1.69	Ea	98.80	42.90	141.70
42" - 44"	S2@1.97	Ea	116.00	50.00	166.00
46" - 48"	S2@2.28	Ea	133.00	57.90	190.90
50"	S2@2.53	Ea	148.00	64.20	212.20

16" galvanized steel rectangular duct 90 degree elbow

Description	Craft@Hrs	Unit	Material $	Labor $	Total $
4"	S2@.260	Ea	15.10	6.60	21.70
6"	S2@.310	Ea	18.20	7.87	26.07
8"	S2@.340	Ea	19.90	8.63	28.53
10"	S2@.380	Ea	22.30	9.65	31.95
12"	S2@.420	Ea	24.60	10.70	35.30
14" - 16"	S2@.480	Ea	28.10	12.20	40.30
18" - 20"	S2@.555	Ea	32.60	14.10	46.70
22" - 24"	S2@.707	Ea	41.50	18.00	59.50
26" - 28"	S2@.856	Ea	50.10	21.70	71.80
30" - 32"	S2@1.14	Ea	66.70	28.90	95.60
34" - 36"	S2@1.48	Ea	86.70	37.60	124.30
38" - 40"	S2@1.73	Ea	102.00	43.90	145.90
42" - 44"	S2@2.02	Ea	118.00	51.30	169.30
46" - 48"	S2@2.32	Ea	137.00	58.90	195.90
50"	S2@2.58	Ea	150.00	65.50	215.50

Description	Craft@Hrs	Unit	Material $	Labor $	Total $

18" galvanized steel rectangular duct 90 degree elbow

Description	Craft@Hrs	Unit	Material $	Labor $	Total $
4"	S2@.312	Ea	18.30	7.92	26.22
6"	S2@.350	Ea	20.60	8.89	29.49
8"	S2@.390	Ea	22.90	9.90	32.80
10"	S2@.427	Ea	25.10	10.80	35.90
12"	S2@.470	Ea	27.40	11.90	39.30
14" - 16"	S2@.519	Ea	30.50	13.20	43.70
18" - 20"	S2@.604	Ea	35.40	15.30	50.70
22" - 24"	S2@.747	Ea	43.90	19.00	62.90
26" - 28"	S2@.895	Ea	52.40	22.70	75.10
30" - 32"	S2@1.18	Ea	69.30	30.00	99.30
34" - 36"	S2@1.53	Ea	89.50	38.80	128.30
38" - 40"	S2@1.78	Ea	104.00	45.20	149.20
42" - 44"	S2@2.07	Ea	121.00	52.60	173.60
46" - 48"	S2@2.37	Ea	140.00	60.20	200.20
50"	S2@2.63	Ea	154.00	66.80	220.80

20" galvanized steel rectangular duct 90 degree elbow

Description	Craft@Hrs	Unit	Material $	Labor $	Total $
4"	S2@.370	Ea	21.60	9.39	30.99
6"	S2@.390	Ea	22.90	9.90	32.80
8"	S2@.415	Ea	24.40	10.50	34.90
10"	S2@.448	Ea	26.30	11.40	37.70
12"	S2@.487	Ea	28.50	12.40	40.90
14" - 16"	S2@.553	Ea	32.40	14.00	46.40
18" - 20"	S2@.658	Ea	38.60	16.70	55.30
22" - 24"	S2@.786	Ea	46.00	20.00	66.00
26" - 28"	S2@.931	Ea	54.50	23.60	78.10
30" - 32"	S2@1.23	Ea	72.00	31.20	103.20
34" - 36"	S2@1.58	Ea	92.20	40.10	132.30
38" - 40"	S2@1.83	Ea	108.00	46.50	154.50
42" - 44"	S2@2.11	Ea	124.00	53.60	177.60
46" - 48"	S2@2.42	Ea	142.00	61.40	203.40
50"	S2@2.67	Ea	158.00	67.80	225.80

22" galvanized steel rectangular duct 90 degree elbow

Description	Craft@Hrs	Unit	Material $	Labor $	Total $
22" - 24"	S2@.828	Ea	48.60	21.00	69.60
26" - 28"	S2@1.01	Ea	59.00	25.60	84.60
30" - 32"	S2@1.31	Ea	76.70	33.30	110.00
34" - 36"	S2@1.71	Ea	99.90	43.40	143.30
38" - 40"	S2@2.01	Ea	118.00	51.00	169.00
42" - 44"	S2@2.31	Ea	134.00	58.70	192.70
46" - 48"	S2@2.61	Ea	152.00	66.30	218.30
50" - 52"	S2@2.91	Ea	170.00	73.90	243.90
54" - 56"	S2@3.51	Ea	206.00	89.10	295.10
58" - 60"	S2@4.26	Ea	250.00	108.00	358.00
62" - 64"	S2@4.76	Ea	278.00	121.00	399.00
66" - 68"	S2@5.25	Ea	308.00	133.00	441.00
70" - 72"	S2@5.76	Ea	337.00	146.00	483.00

Galvanized Steel Rectangular Elbows

Description	Craft@Hrs	Unit	Material $	Labor $	Total $

24" galvanized steel rectangular duct 90 degree elbow

Description	Craft@Hrs	Unit	Material $	Labor $	Total $
22" - 24"	S2@.922	Ea	54.00	23.40	77.40
26" - 28"	S2@1.10	Ea	64.40	27.90	92.30
30" - 32"	S2@1.41	Ea	82.40	35.80	118.20
34" - 36"	S2@1.81	Ea	106.00	46.00	152.00
38" - 40"	S2@2.10	Ea	124.00	53.30	177.30
42" - 44"	S2@2.41	Ea	141.00	61.20	202.20
46" - 48"	S2@2.71	Ea	159.00	68.80	227.80
50" - 52"	S2@3.01	Ea	176.00	76.40	252.40
54" - 56"	S2@3.62	Ea	212.00	91.90	303.90
58" - 60"	S2@4.38	Ea	257.00	111.00	368.00
62" - 64"	S2@4.87	Ea	285.00	124.00	409.00
66" - 68"	S2@5.38	Ea	314.00	137.00	451.00
70" - 72"	S2@5.87	Ea	345.00	149.00	494.00

26" galvanized steel rectangular duct 90 degree elbow

Description	Craft@Hrs	Unit	Material $	Labor $	Total $
22" - 24"	S2@1.01	Ea	59.10	25.60	84.70
26" - 28"	S2@1.19	Ea	69.90	30.20	100.10
30" - 32"	S2@1.50	Ea	88.10	38.10	126.20
34" - 36"	S2@1.92	Ea	112.00	48.70	160.70
38" - 40"	S2@2.21	Ea	129.00	56.10	185.10
42" - 44"	S2@2.51	Ea	147.00	63.70	210.70
46" - 48"	S2@2.81	Ea	164.00	71.30	235.30
50" - 52"	S2@3.11	Ea	183.00	79.00	262.00
54" - 56"	S2@3.73	Ea	219.00	94.70	313.70
58" - 60"	S2@4.50	Ea	264.00	114.00	378.00
62" - 64"	S2@5.02	Ea	293.00	127.00	420.00
66" - 68"	S2@5.53	Ea	325.00	140.00	465.00
70" - 72"	S2@6.04	Ea	354.00	153.00	507.00

28" galvanized steel rectangular duct 90 degree elbow

Description	Craft@Hrs	Unit	Material $	Labor $	Total $
22" - 24"	S2@1.10	Ea	64.30	27.90	92.20
26" - 28"	S2@1.28	Ea	75.20	32.50	107.70
30" - 32"	S2@1.60	Ea	93.70	40.60	134.30
34" - 36"	S2@2.03	Ea	119.00	51.50	170.50
38" - 40"	S2@2.31	Ea	134.00	58.70	192.70
42" - 44"	S2@2.61	Ea	152.00	66.30	218.30
46" - 48"	S2@2.91	Ea	170.00	73.90	243.90
50" - 52"	S2@3.21	Ea	189.00	81.50	270.50
54" - 56"	S2@3.84	Ea	226.00	97.50	323.50
58" - 60"	S2@4.61	Ea	271.00	117.00	388.00
62" - 64"	S2@5.15	Ea	300.00	131.00	431.00
66" - 68"	S2@5.66	Ea	331.00	144.00	475.00
70" - 72"	S2@6.15	Ea	360.00	156.00	516.00

Description	Craft@Hrs	Unit	Material $	Labor $	Total $

30" galvanized steel rectangular duct 90 degree elbow

Description	Craft@Hrs	Unit	Material $	Labor $	Total $
22" - 24"	S2@1.19	Ea	69.40	30.20	99.60
26" - 28"	S2@1.37	Ea	80.60	34.80	115.40
30" - 32"	S2@1.70	Ea	99.50	43.20	142.70
34" - 36"	S2@2.14	Ea	125.00	54.30	179.30
38" - 40"	S2@2.41	Ea	141.00	61.20	202.20
42" - 44"	S2@2.71	Ea	159.00	68.80	227.80
46" - 48"	S2@3.01	Ea	176.00	76.40	252.40
50" - 52"	S2@3.31	Ea	194.00	84.00	278.00
54" - 56"	S2@3.95	Ea	231.00	100.00	331.00
58" - 60"	S2@4.74	Ea	277.00	120.00	397.00
62" - 64"	S2@5.25	Ea	308.00	133.00	441.00
66" - 68"	S2@5.79	Ea	338.00	147.00	485.00
70" - 72"	S2@6.28	Ea	368.00	159.00	527.00

32" galvanized steel rectangular duct 90 degree elbow

Description	Craft@Hrs	Unit	Material $	Labor $	Total $
22" - 24"	S2@1.53	Ea	89.60	38.80	128.40
26" - 28"	S2@1.73	Ea	101.00	43.90	144.90
30" - 32"	S2@1.93	Ea	113.00	49.00	162.00
34" - 36"	S2@2.25	Ea	132.00	57.10	189.10
38" - 40"	S2@2.51	Ea	147.00	63.70	210.70
42" - 44"	S2@2.81	Ea	164.00	71.30	235.30
46" - 48"	S2@3.11	Ea	182.00	79.00	261.00
50" - 52"	S2@3.41	Ea	200.00	86.60	286.60
54" - 56"	S2@4.06	Ea	237.00	103.00	340.00
58" - 60"	S2@4.85	Ea	284.00	123.00	407.00
62" - 64"	S2@5.38	Ea	314.00	137.00	451.00
66" - 68"	S2@5.89	Ea	346.00	150.00	496.00
70" - 72"	S2@6.40	Ea	375.00	162.00	537.00

34" galvanized steel rectangular duct 90 degree elbow

Description	Craft@Hrs	Unit	Material $	Labor $	Total $
22" - 24"	S2@1.67	Ea	98.00	42.40	140.40
26" - 28"	S2@1.91	Ea	112.00	48.50	160.50
30" - 32"	S2@2.12	Ea	124.00	53.80	177.80
34" - 36"	S2@2.36	Ea	138.00	59.90	197.90
38" - 40"	S2@2.60	Ea	152.00	66.00	218.00
42" - 44"	S2@2.91	Ea	170.00	73.90	243.90
46" - 48"	S2@3.21	Ea	187.00	81.50	268.50
50" - 52"	S2@3.51	Ea	206.00	89.10	295.10
54" - 56"	S2@4.17	Ea	245.00	106.00	351.00
58" - 60"	S2@4.97	Ea	292.00	126.00	418.00
62" - 64"	S2@5.49	Ea	322.00	139.00	461.00
66" - 68"	S2@6.04	Ea	354.00	153.00	507.00
70" - 72"	S2@6.51	Ea	381.00	165.00	546.00

Galvanized Steel Rectangular Elbows

Description	Craft@Hrs	Unit	Material $	Labor $	Total $

36" galvanized steel rectangular duct 90 degree elbow

Description	Craft@Hrs	Unit	Material $	Labor $	Total $
22" - 24"	S2@1.83	Ea	108.00	46.50	154.50
26" - 28"	S2@2.04	Ea	120.00	51.80	171.80
30" - 32"	S2@2.26	Ea	132.00	57.40	189.40
34" - 36"	S2@2.47	Ea	145.00	62.70	207.70
38" - 40"	S2@2.70	Ea	159.00	68.60	227.60
42" - 44"	S2@3.01	Ea	176.00	76.40	252.40
46" - 48"	S2@3.31	Ea	194.00	84.00	278.00
50" - 52"	S2@3.61	Ea	211.00	91.70	302.70
54" - 56"	S2@4.27	Ea	250.00	108.00	358.00
58" - 60"	S2@5.10	Ea	298.00	129.00	427.00
62" - 64"	S2@5.61	Ea	329.00	142.00	471.00
66" - 68"	S2@6.15	Ea	360.00	156.00	516.00
70" - 72"	S2@6.64	Ea	389.00	169.00	558.00

38" galvanized steel rectangular duct 90 degree elbow

Description	Craft@Hrs	Unit	Material $	Labor $	Total $
22" - 24"	S2@1.98	Ea	116.00	50.30	166.30
26" - 28"	S2@2.18	Ea	128.00	55.40	183.40
30" - 32"	S2@2.38	Ea	140.00	60.40	200.40
34" - 36"	S2@2.58	Ea	151.00	65.50	216.50
38" - 40"	S2@2.81	Ea	164.00	71.30	235.30
42" - 44"	S2@3.11	Ea	182.00	79.00	261.00
46" - 48"	S2@3.41	Ea	200.00	86.60	286.60
50" - 52"	S2@3.71	Ea	217.00	94.20	311.20
54" - 56"	S2@4.38	Ea	257.00	111.00	368.00
58" - 60"	S2@5.21	Ea	306.00	132.00	438.00
62" - 64"	S2@5.74	Ea	335.00	146.00	481.00
66" - 68"	S2@6.28	Ea	368.00	159.00	527.00
70" - 72"	S2@6.75	Ea	396.00	171.00	567.00

40" galvanized steel rectangular duct 90 degree elbow

Description	Craft@Hrs	Unit	Material $	Labor $	Total $
40" - 42"	S2@3.06	Ea	180.00	77.70	257.70
44" - 46"	S2@3.36	Ea	196.00	85.30	281.30
48" - 50"	S2@3.66	Ea	214.00	92.90	306.90
52" - 54"	S2@3.98	Ea	233.00	101.00	334.00
56" - 58"	S2@5.17	Ea	302.00	131.00	433.00
60" - 62"	S2@5.59	Ea	328.00	142.00	470.00
64" - 66"	S2@6.11	Ea	358.00	155.00	513.00
68" - 70"	S2@6.66	Ea	391.00	169.00	560.00
72"	S2@7.05	Ea	412.00	179.00	591.00

Description	Craft@Hrs	Unit	Material $	Labor $	Total $

42" galvanized steel rectangular duct 90 degree elbow

Description	Craft@Hrs	Unit	Material $	Labor $	Total $
40" - 42"	S2@3.21	Ea	189.00	81.50	270.50
44" - 46"	S2@3.52	Ea	207.00	89.40	296.40
48" - 50"	S2@3.83	Ea	226.00	97.20	323.20
52" - 54"	S2@4.16	Ea	244.00	106.00	350.00
56" - 58"	S2@5.34	Ea	312.00	136.00	448.00
60" - 62"	S2@5.81	Ea	339.00	148.00	487.00
64" - 66"	S2@6.32	Ea	371.00	160.00	531.00
68" - 70"	S2@6.85	Ea	402.00	174.00	576.00
72"	S2@7.22	Ea	424.00	183.00	607.00

44" galvanized steel rectangular duct 90 degree elbow

Description	Craft@Hrs	Unit	Material $	Labor $	Total $
40" - 42"	S2@3.36	Ea	196.00	85.30	281.30
44" - 46"	S2@3.69	Ea	216.00	93.70	309.70
48" - 50"	S2@4.01	Ea	234.00	102.00	336.00
52" - 54"	S2@4.32	Ea	253.00	110.00	363.00
56" - 58"	S2@5.51	Ea	324.00	140.00	464.00
60" - 62"	S2@6.00	Ea	351.00	152.00	503.00
64" - 66"	S2@6.51	Ea	381.00	165.00	546.00
68" - 70"	S2@7.05	Ea	414.00	179.00	593.00
72"	S2@7.39	Ea	433.00	188.00	621.00

46" galvanized steel rectangular duct 90 degree elbow

Description	Craft@Hrs	Unit	Material $	Labor $	Total $
40" - 42"	S2@3.52	Ea	207.00	89.40	296.40
44" - 46"	S2@3.86	Ea	227.00	98.00	325.00
48" - 50"	S2@4.16	Ea	244.00	106.00	350.00
52" - 54"	S2@4.32	Ea	262.00	110.00	372.00
56" - 58"	S2@5.72	Ea	334.00	145.00	479.00
60" - 62"	S2@6.19	Ea	362.00	157.00	519.00
64" - 66"	S2@6.70	Ea	394.00	170.00	564.00
68" - 70"	S2@7.26	Ea	425.00	184.00	609.00
72"	S2@7.60	Ea	445.00	193.00	638.00

48" galvanized steel rectangular duct 90 degree elbow

Description	Craft@Hrs	Unit	Material $	Labor $	Total $
40" - 42"	S2@3.67	Ea	215.00	93.20	308.20
44" - 46"	S2@4.01	Ea	234.00	102.00	336.00
48" - 50"	S2@4.32	Ea	253.00	110.00	363.00
52" - 54"	S2@4.61	Ea	269.00	117.00	386.00
56" - 58"	S2@5.91	Ea	347.00	150.00	497.00
60" - 62"	S2@6.40	Ea	375.00	162.00	537.00
64" - 66"	S2@6.90	Ea	404.00	175.00	579.00
68" - 70"	S2@7.41	Ea	435.00	188.00	623.00
72"	S2@7.77	Ea	455.00	197.00	652.00

Galvanized Steel Rectangular Elbows

Description	Craft@Hrs	Unit	Material $	Labor $	Total $

50" galvanized steel rectangular duct 90 degree elbow

Description	Craft@Hrs	Unit	Material $	Labor $	Total $
40" - 42"	S2@3.82	Ea	225.00	97.00	322.00
44" - 46"	S2@4.16	Ea	244.00	106.00	350.00
48" - 50"	S2@4.46	Ea	262.00	113.00	375.00
52" - 54"	S2@4.78	Ea	279.00	121.00	400.00
56" - 58"	S2@6.11	Ea	358.00	155.00	513.00
60" - 62"	S2@6.60	Ea	387.00	168.00	555.00
64" - 66"	S2@7.09	Ea	416.00	180.00	596.00
68" - 70"	S2@7.62	Ea	446.00	193.00	639.00
72"	S2@7.94	Ea	465.00	202.00	667.00

52" galvanized steel rectangular duct 90 degree elbow

Description	Craft@Hrs	Unit	Material $	Labor $	Total $
40" - 42"	S2@3.98	Ea	233.00	101.00	334.00
44" - 46"	S2@4.32	Ea	253.00	110.00	363.00
48" - 50"	S2@4.61	Ea	269.00	117.00	386.00
52" - 54"	S2@4.95	Ea	290.00	126.00	416.00
56" - 58"	S2@6.32	Ea	371.00	160.00	531.00
60" - 62"	S2@6.79	Ea	398.00	172.00	570.00
64" - 66"	S2@7.30	Ea	428.00	185.00	613.00
68" - 70"	S2@7.81	Ea	458.00	198.00	656.00
72"	S2@8.16	Ea	477.00	207.00	684.00

54" galvanized steel rectangular duct 90 degree elbow

Description	Craft@Hrs	Unit	Material $	Labor $	Total $
40" - 42"	S2@4.16	Ea	244.00	106.00	350.00
44" - 46"	S2@4.46	Ea	262.00	113.00	375.00
48" - 50"	S2@4.78	Ea	279.00	121.00	400.00
52" - 54"	S2@5.15	Ea	300.00	131.00	431.00
56" - 58"	S2@6.51	Ea	381.00	165.00	546.00
60" - 62"	S2@7.00	Ea	410.00	178.00	588.00
64" - 66"	S2@7.49	Ea	440.00	190.00	630.00
68" - 70"	S2@8.01	Ea	470.00	203.00	673.00
72"	S2@8.41	Ea	492.00	214.00	706.00

56" galvanized steel rectangular duct 90 degree elbow

Description	Craft@Hrs	Unit	Material $	Labor $	Total $
40" - 42"	S2@5.10	Ea	298.00	129.00	427.00
44" - 46"	S2@5.49	Ea	322.00	139.00	461.00
48" - 50"	S2@5.85	Ea	344.00	149.00	493.00
52" - 54"	S2@6.26	Ea	367.00	159.00	526.00
56" - 58"	S2@6.68	Ea	393.00	170.00	563.00
60" - 62"	S2@7.15	Ea	420.00	182.00	602.00
64" - 66"	S2@7.66	Ea	449.00	194.00	643.00
68" - 70"	S2@8.18	Ea	478.00	208.00	686.00
72"	S2@8.58	Ea	503.00	218.00	721.00

Description	Craft@Hrs	Unit	Material $	Labor $	Total $

58" galvanized steel rectangular duct 90 degree elbow

58" - 60"	S2@7.15	Ea	420.00	182.00	602.00
62" - 64"	S2@7.62	Ea	446.00	193.00	639.00
66" - 68"	S2@8.13	Ea	476.00	206.00	682.00
70" - 72"	S2@8.69	Ea	510.00	221.00	731.00

60" galvanized steel rectangular duct 90 degree elbow

58" - 60"	S2@7.37	Ea	432.00	187.00	619.00
62" - 64"	S2@7.88	Ea	461.00	200.00	661.00
66" - 68"	S2@8.41	Ea	493.00	214.00	707.00
70" - 72"	S2@8.95	Ea	524.00	227.00	751.00

62" galvanized steel rectangular duct 90 degree elbow

58" - 60"	S2@7.62	Ea	446.00	193.00	639.00
62" - 64"	S2@8.13	Ea	476.00	206.00	682.00
66" - 68"	S2@8.69	Ea	510.00	221.00	731.00
70" - 72"	S2@9.18	Ea	539.00	233.00	772.00

64" galvanized steel rectangular duct 90 degree elbow

58" - 60"	S2@7.88	Ea	461.00	200.00	661.00
62" - 64"	S2@8.41	Ea	493.00	214.00	707.00
66" - 68"	S2@8.95	Ea	524.00	227.00	751.00
70" - 72"	S2@9.44	Ea	553.00	240.00	793.00

66" galvanized steel rectangular duct 90 degree elbow

58" - 60"	S2@8.13	Ea	476.00	206.00	682.00
62" - 64"	S2@8.69	Ea	510.00	221.00	731.00
66" - 68"	S2@9.18	Ea	539.00	233.00	772.00
70" - 72"	S2@9.69	Ea	568.00	246.00	814.00

68" galvanized steel rectangular duct 90 degree elbow

58" - 60"	S2@8.41	Ea	493.00	214.00	707.00
62" - 64"	S2@8.95	Ea	524.00	227.00	751.00
66" - 68"	S2@9.44	Ea	553.00	240.00	793.00
70" - 72"	S2@9.95	Ea	584.00	253.00	837.00

70" galvanized steel rectangular duct 90 degree elbow

58" - 60"	S2@8.69	Ea	510.00	221.00	731.00
62" - 64"	S2@9.18	Ea	539.00	233.00	772.00
66" - 68"	S2@9.69	Ea	568.00	246.00	814.00
70" - 72"	S2@10.2	Ea	599.00	259.00	858.00

Galvanized Steel Rectangular Elbows

Description	Craft@Hrs	Unit	Material $	Labor $	Total $
72" galvanized steel rectangular duct 90 degree elbow					
58" - 60"	S2@8.95	Ea	524.00	227.00	751.00
62" - 64"	S2@9.44	Ea	553.00	240.00	793.00
66" - 68"	S2@9.95	Ea	584.00	253.00	837.00
70" - 72"	S2@10.5	Ea	616.00	267.00	883.00

Costs for Galvanized Rectangular Drops and Tap-in Tees. Costs per drop or tap-in tee, based on quantities less than 1,000 pounds. Deduct 15% from the material cost for over 1,000 pounds. Add 50% to the material cost for the lined duct. These costs include delivery, typical bracing, cleats, scrap, hangers, end closures, sealants and miscellaneous hardware. For turning vanes, please see page 267.

Description	Craft@Hrs	Unit	Material $	Labor $	Total $

6" wide galvanized steel rectangular duct drops and tap-in tees

Description	Craft@Hrs	Unit	Material $	Labor $	Total $
6"	S2@.072	Ea	4.61	1.83	6.44
8"	S2@.084	Ea	5.38	2.13	7.51
10"	S2@.097	Ea	6.18	2.46	8.64
12" - 16"	S2@.139	Ea	8.87	3.53	12.40
20" - 24"	S2@.216	Ea	13.70	5.48	19.18
28" - 32"	S2@.309	Ea	19.60	7.85	27.45
36" - 40"	S2@.412	Ea	26.30	10.50	36.80
44" - 48"	S2@.489	Ea	31.10	12.40	43.50
52" - 56"	S2@.591	Ea	37.70	15.00	52.70
60"	S2@.747	Ea	47.60	19.00	66.60

8" wide galvanized steel rectangular duct drops and tap-in tees

Description	Craft@Hrs	Unit	Material $	Labor $	Total $
6"	S2@.084	Ea	5.38	2.13	7.51
8"	S2@.097	Ea	6.18	2.46	8.64
10"	S2@.109	Ea	6.93	2.77	9.70
12" - 16"	S2@.152	Ea	9.73	3.86	13.59
20" - 24"	S2@.231	Ea	14.70	5.87	20.57
28" - 32"	S2@.326	Ea	20.80	8.28	29.08
36" - 40"	S2@.432	Ea	27.50	11.00	38.50
44" - 48"	S2@.506	Ea	32.20	12.80	45.00
52" - 56"	S2@.634	Ea	40.40	16.10	56.50
60"	S2@.760	Ea	48.40	19.30	67.70

10" wide galvanized steel rectangular duct drops and tap-in tees

Description	Craft@Hrs	Unit	Material $	Labor $	Total $
6"	S2@.097	Ea	6.18	2.46	8.64
8"	S2@.109	Ea	6.93	2.77	9.70
10"	S2@.121	Ea	7.69	3.07	10.76
12" - 16"	S2@.167	Ea	10.60	4.24	14.84
20" - 24"	S2@.246	Ea	15.80	6.25	22.05
28" - 32"	S2@.343	Ea	21.90	8.71	30.61
36" - 40"	S2@.450	Ea	28.60	11.40	40.00
44" - 48"	S2@.525	Ea	33.40	13.30	46.70
52" - 56"	S2@.653	Ea	41.60	16.60	58.20
60"	S2@.773	Ea	49.20	19.60	68.80

Galvanized Steel Rectangular Drops and Tees

Description	Craft@Hrs	Unit	Material $	Labor $	Total $

12" wide galvanized steel rectangular duct drops and tap-in tees

Description	Craft@Hrs	Unit	Material $	Labor $	Total $
6"	S2@.109	Ea	6.93	2.77	9.70
8"	S2@.121	Ea	7.69	3.07	10.76
10"	S2@.133	Ea	8.46	3.38	11.84
12" - 16"	S2@.180	Ea	11.50	4.57	16.07
20" - 24"	S2@.262	Ea	16.70	6.65	23.35
28" - 32"	S2@.360	Ea	22.90	9.14	32.04
36" - 40"	S2@.470	Ea	29.80	11.90	41.70
44" - 48"	S2@.544	Ea	34.60	13.80	48.40
52" - 56"	S2@.675	Ea	43.00	17.10	60.10
60"	S2@.794	Ea	50.60	20.20	70.80

16" wide galvanized steel rectangular duct drops and tap-in tees

Description	Craft@Hrs	Unit	Material $	Labor $	Total $
6"	S2@.170	Ea	10.80	4.32	15.12
8"	S2@.184	Ea	11.80	4.67	16.47
10"	S2@.200	Ea	12.80	5.08	17.88
12" - 16"	S2@.231	Ea	14.70	5.87	20.57
20" - 24"	S2@.293	Ea	18.60	7.44	26.04
28" - 32"	S2@.396	Ea	25.10	10.10	35.20
36" - 40"	S2@.506	Ea	32.20	12.80	45.00
44" - 48"	S2@.581	Ea	37.10	14.80	51.90
52" - 56"	S2@.709	Ea	45.30	18.00	63.30
60"	S2@.841	Ea	53.50	21.40	74.90

20" wide galvanized steel rectangular duct drops and tap-in tees

Description	Craft@Hrs	Unit	Material $	Labor $	Total $
6"	S2@.200	Ea	12.80	5.08	17.88
8"	S2@.216	Ea	13.70	5.48	19.18
10"	S2@.231	Ea	14.70	5.87	20.57
12" - 16"	S2@.262	Ea	16.70	6.65	23.35
20" - 24"	S2@.324	Ea	20.70	8.23	28.93
28" - 32"	S2@.428	Ea	27.30	10.90	38.20
36" - 40"	S2@.544	Ea	34.60	13.80	48.40
44" - 48"	S2@.617	Ea	39.30	15.70	55.00
52" - 56"	S2@.758	Ea	48.20	19.20	67.40
60"	S2@.884	Ea	56.20	22.40	78.60

Description	Craft@Hrs	Unit	Material $	Labor $	Total $

24" wide galvanized steel rectangular duct drops and tap-in tees

Description	Craft@Hrs	Unit	Material $	Labor $	Total $
6"	S2@.231	Ea	14.70	5.87	20.57
8"	S2@.247	Ea	15.80	6.27	22.07
10"	S2@.262	Ea	16.70	6.65	23.35
12" - 16"	S2@.293	Ea	18.60	7.44	26.04
20" - 24"	S2@.354	Ea	22.60	8.99	31.59
28" - 32"	S2@.462	Ea	29.20	11.70	40.90
36" - 40"	S2@.581	Ea	37.10	14.80	51.90
44" - 48"	S2@.655	Ea	41.70	16.60	58.30
52" - 56"	S2@.798	Ea	50.70	20.30	71.00
60"	S2@.927	Ea	59.00	23.50	82.50

28" wide galvanized steel rectangular duct drops and tap-in tees

Description	Craft@Hrs	Unit	Material $	Labor $	Total $
6"	S2@.262	Ea	16.70	6.65	23.35
8"	S2@.278	Ea	17.60	7.06	24.66
10"	S2@.292	Ea	18.60	7.41	26.01
12" - 16"	S2@.324	Ea	20.70	8.23	28.93
20" - 24"	S2@.385	Ea	24.50	9.78	34.28
28" - 32"	S2@.497	Ea	31.60	12.60	44.20
36" - 40"	S2@.617	Ea	39.30	15.70	55.00
44" - 48"	S2@.694	Ea	44.10	17.60	61.70
52" - 56"	S2@.837	Ea	53.40	21.30	74.70
60"	S2@.969	Ea	61.90	24.60	86.50

32" wide galvanized steel rectangular duct drops and tap-in tees

Description	Craft@Hrs	Unit	Material $	Labor $	Total $
6"	S2@.356	Ea	22.70	9.04	31.74
8"	S2@.374	Ea	24.00	9.50	33.50
10"	S2@.393	Ea	25.00	9.98	34.98
12" - 16"	S2@.432	Ea	27.50	11.00	38.50
20" - 24"	S2@.506	Ea	32.20	12.80	45.00
28" - 32"	S2@.581	Ea	37.10	14.80	51.90
36" - 40"	S2@.655	Ea	41.70	16.60	58.30
44" - 48"	S2@.730	Ea	46.40	18.50	64.90
52" - 56"	S2@.877	Ea	56.00	22.30	78.30
60"	S2@1.02	Ea	64.60	25.90	90.50

Galvanized Steel Round Ductwork

Description	Craft@Hrs	Unit	Material $	Labor $	Total $
Galvanized steel round duct snap-lock with slip joints					
3"	S2@.080	LF	1.04	2.03	3.07
4"	S2@.090	LF	1.07	2.29	3.36
5"	S2@.095	LF	1.15	2.41	3.56
6"	S2@.100	LF	1.33	2.54	3.87
7"	S2@.110	LF	1.65	2.79	4.44
8"	S2@.120	LF	1.69	3.05	4.74
9"	S2@.140	LF	1.78	3.55	5.33
10"	S2@.160	LF	1.86	4.06	5.92
Galvanized steel round duct 90 degree adjustable elbow					
3"	S2@.250	Ea	10.70	6.35	17.05
4"	S2@.300	Ea	14.60	7.62	22.22
5"	S2@.350	Ea	17.90	8.89	26.79
6"	S2@.400	Ea	23.30	10.20	33.50
7"	S2@.450	Ea	26.70	11.40	38.10
8"	S2@.500	Ea	30.00	12.70	42.70
9"	S2@.600	Ea	32.40	15.20	47.60
10"	S2@.700	Ea	35.30	17.80	53.10
Galvanized steel round duct 45 degree adjustable elbow					
3"	S2@.250	Ea	6.72	6.35	13.07
4"	S2@.300	Ea	8.00	7.62	15.62
5"	S2@.350	Ea	10.20	8.89	19.09
6"	S2@.400	Ea	13.50	10.20	23.70
7"	S2@.450	Ea	15.20	11.40	26.60
8"	S2@.500	Ea	16.90	12.70	29.60
9"	S2@.600	Ea	18.40	15.20	33.60
10"	S2@.700	Ea	20.20	17.80	38.00

Standard fiberglass ductwork, or duct board, should not be used in systems where air velocities exceed 2,500 feet per minute, or where static pressures are higher than 2 inches of water gauge.

Additionally, it should not be used for the following applications:

1) Equipment rooms

2) Underground or outdoors

3) Final connections to air handling equipment

4) Plenums or casings

5) Fume, heat or moisture exhaust systems

The costs for manufacturing and installing fiberglass ductwork are based on the net areas of the duct and fittings.

Here are approximate current material costs, per square foot:

Fiberglass Type	Up to 2,500 SF Material ($)/SF	Over 2,500 SF Material ($)/SF
475	1.05	.95
800	1.15	1.05
1400	1.40	1.35

Additional costs: Approximately 25 cents per square foot should be added for the cost of tie rods, hangers, staples, tape and waste.

Type 475 fiberglass board is used almost exclusively for residential and small commercial air systems, while Types 800 and 1400 are used in larger commercial and industrial systems.

Fiberglass Ductwork Fabrication Labor* Size ranges (inches)						
	Up through 16		17 through 32		33 through 48	
Activity	SF/Hour Output	Labor $/SF	SF/Hour Output	Labor /SF	SF/Hour Output	Labor $/SF
Duct only	60	.42	58	.43	56	.46
Fittings only	40	.63	38	.66	35	.72
Duct & 15% fittings	55	.47	53	.48	50	.51
Duct & 25% fittings	53	.48	51	.50	48	.53
Duct & 35% fittings	51	.50	49	.52	46	.55

Labor rates are based on using a standard grooving machine and include all tie-rod reinforcing.

Labor correction factors:

1) Hand grooving in lieu of machine grooving: 1.45

2) Use of an auto-closer: .75

3) Tie-rods enclosed in conduit: 1.25

4) T-bar or channel reinforcing in lieu of tie-rods: 1.15

Fiberglass ductwork installation

12" (2.00 SF/LF)	S2@.070	LF	--	1.78	1.78
14" (2.35 SF/LF)	S2@.082	LF	--	2.08	2.08
16" (2.65 SF/LF)	S2@.093	LF	--	2.36	2.36
18" (3.00 SF/LF)	S2@.105	LF	--	2.67	2.67
20" (3.35 SF/LF)	S2@.117	LF	--	2.97	2.97
22" (3.65 SF/LF)	S2@.128	LF	--	3.25	3.25
24" (4.00 SF/LF)	S2@.140	LF	--	3.55	3.55
26" (4.35 SF/LF)	S2@.152	LF	--	3.86	3.86
28" (4.65 SF/LF)	S2@.163	LF	--	4.14	4.14
30" (5.00 SF/LF)	S2@.175	LF	--	4.44	4.44
32" (5.35 SF/LF)	S2@.187	LF	--	4.75	4.75
34" (5.65 SF/LF)	S2@.198	LF	--	5.03	5.03
36" (6.00 SF/LF)	S2@.210	LF	--	5.33	5.33
38" (6.35 SF/LF)	S2@.222	LF	--	5.64	5.64
40" (6.65 SF/LF)	S2@.232	LF	--	5.89	5.89
42" (7.00 SF/LF)	S2@.245	LF	--	6.22	6.22
44" (7.35 SF/LF)	S2@.257	LF	--	6.53	6.53
46" (7.65 SF/LF)	S2@.268	LF	--	6.80	6.80
48" (8.00 SF/LF)	S2@.280	LF	--	7.11	7.11
50" (8.35 SF/LF)	S2@.292	LF	--	7.41	7.41
52" (8.65 SF/LF)	S2@.303	LF	--	7.69	7.69
54" (9.00 SF/LF)	S2@.315	LF	--	8.00	8.00
56" (9.35 SF/LF)	S2@.327	LF	--	8.30	8.30
58" (9.65 SF/LF)	S2@.338	LF	--	8.58	8.58
60" (10.0 SF/LF)	S2@.350	LF	--	8.89	8.89
64" (10.6 SF/LF)	S2@.373	LF	--	9.47	9.47
68" (11.4 SF/LF)	S2@.397	LF	--	10.10	10.10
72" (12.0 SF/LF)	S2@.420	LF	--	10.70	10.70
76" (12.6 SF/LF)	S2@.443	LF	--	11.20	11.20
80" (13.4 SF/LF)	S2@.467	LF	--	11.90	11.90
84" (14.0 SF/LF)	S2@.490	LF	--	12.40	12.40
88" (14.6 SF/LF)	S2@.513	LF	--	13.00	13.00
92" (15.4 SF/LF)	S2@.537	LF	--	13.60	13.60
96" (16.0 SF/LF)	S2@.560	LF	--	14.20	14.20

For estimating purposes, when measuring the length of ductwork, include the fitting. Then apply the labor per run foot to the total.

1¼" thick wire reinforced flexible fiberglass duct with vinyl cover

4"	S2@.060	LF	1.74	1.52	3.26
5"	S2@.070	LF	1.80	1.78	3.58
6"	S2@.080	LF	1.98	2.03	4.01
8"	S2@.090	LF	2.50	2.29	4.79
9"	S2@.100	LF	2.68	2.54	5.22
10"	S2@.120	LF	2.80	3.05	5.85
12"	S2@.140	LF	3.62	3.55	7.17
14"	S2@.160	LF	4.33	4.06	8.39
16"	S2@.180	LF	5.03	4.57	9.60
18"	S2@.200	LF	5.72	5.08	10.80

Unitary or factory-packaged air conditioning units are often too small to meet the heating and cooling needs of a large single-zone space. In such cases, engineers design what's called a "built-up" system. They're custom fitted to the client's needs and the space. A typical "built-up" system includes the following equipment:

- centrifugal blowers or axial-flow fans
- damper assemblies
- coil banks
- filter banks

Once the system is assembled, it is enclosed in a shop-fabricated housing or casing. The illustration on page 262 shows a standard apparatus housing. The walls and roof are shop-built panels. These panels consist of a layer of thermal insulation covered on one or both sides with sheets of 18 gauge galvanized steel. An interior framework of steel angles or channels supports the panels. Seal all mating surfaces to minimize air leakage, using gaskets, caulking or sealant.

Apparatus Housings

There are two types of panels used in building apparatus housings: "double-skin" panels and "single-skin" panels. Double-skin panels have a layer of thermal insulation material sandwiched between two sheets of 18 gauge galvanized steel. Single-skin panels also have a layer of insulation. However, only the outside surface is clad in 18 gauge steel sheeting. On the inside the panel insulation is exposed.

The costs of constructing and installing apparatus housings vary greatly. This is due to the wide range of possible conditions and needs that you face on each job. Some of the many factors that influence costs here include the following:

- internal operating pressures
- external wind loads
- seismic requirements
- physical locations

That's why I recommend that you always prepare a detailed cost estimate. Then, you can take all the particulars into account. However, for budget estimating purposes only, use the following cost data:

Single-skin panels	$8.05 per SF*
Double-skin panels	$13.70 per SF*
Access doors (20" x 60")	$280.00 each

*Total area of walls and roof, less doors.

When you're estimating material costs for apparatus housings, remember to add in an extra 25 percent to cover the following costs: waste, seams, gaskets, sealants and miscellaneous assembly hardware.

Labor for Shop-Fabrication and Field Assembly of Apparatus Housings. The costs listed are manhours per square foot (SF) of panel. To find the total SF of panel, add up areas for all the panels.

Description	Shop Labor (MH per SF)	Field Labor (MH per SF)
Single-skin panels	.040	.250
Double-skin panels	.085	.300
Add for access doors	.400	240

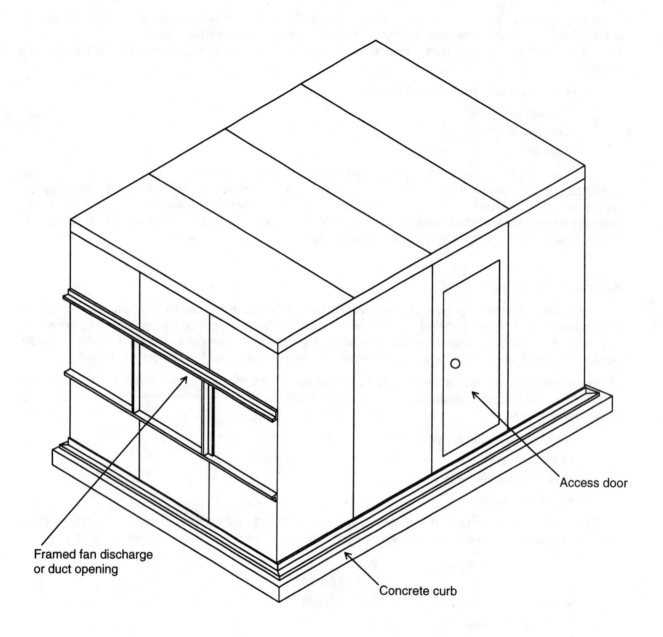

Framed fan discharge
or duct opening

Access door

Concrete curb

Figure 1 - Typical apparatus housing

Description	Craft@Hrs	Unit	Material $	Labor $	Total $
Supply register for residential ductwork					
8 x 4"	S2@.650	Ea	4.74	16.50	21.24
8 x 6"	S2@.650	Ea	4.92	16.50	21.42
10 x 4"	S2@.650	Ea	5.34	16.50	21.84
10 x 6"	S2@.650	Ea	9.10	16.50	25.60
10 x 8"	S2@.700	Ea	8.63	17.80	26.43
12 x 6"	S2@.700	Ea	9.19	17.80	26.99
12 x 8"	S2@.750	Ea	7.51	19.00	26.51
14 x 6"	S2@.750	Ea	7.51	19.00	26.51
14 x 8"	S2@.800	Ea	8.58	20.30	28.88
Return air grille for residential ductwork					
14 x 6"	S2@.750	Ea	4.08	19.00	23.08
14 x 8"	S2@.800	Ea	3.96	20.30	24.26
16 x 16"	S2@.850	Ea	11.90	21.60	33.50
20 x 10"	S2@.900	Ea	17.70	22.90	40.60
20 x 16"	S2@1.00	Ea	19.50	25.40	44.90
20 x 20"	S2@1.10	Ea	25.80	27.90	53.70
24 x 24"	S2@1.15	Ea	33.60	29.20	62.80
Supply register for commercial ductwork					
8 x 4"	S2@.650	Ea	15.20	16.50	31.70
8 x 6"	S2@.650	Ea	17.00	16.50	33.50
10 x 4"	S2@.650	Ea	19.70	16.50	36.20
10 x 6"	S2@.700	Ea	20.20	17.80	38.00
10 x 8"	S2@.750	Ea	21.60	19.00	40.60
12 x 6"	S2@.750	Ea	22.30	19.00	41.30
12 x 8"	S2@.800	Ea	25.10	20.30	45.40
12 x 10"	S2@.850	Ea	28.40	21.60	50.00
14 x 6"	S2@.850	Ea	28.40	21.60	50.00
14 x 8"	S2@.900	Ea	28.80	22.90	51.70
14 x 10"	S2@.950	Ea	30.70	24.10	54.80
16 x 8"	S2@.950	Ea	30.70	24.10	54.80
16 x 10"	S2@1.00	Ea	33.60	25.40	59.00
18 x 8"	S2@1.00	Ea	34.90	25.40	60.30
18 x 10"	S2@1.10	Ea	37.40	27.90	65.30
18 x 12"	S2@1.10	Ea	39.50	27.90	67.40

Ductwork Specialties

Description	Craft@Hrs	Unit	Material $	Labor $	Total $
Return register for commercial ductwork					
8 x 4"	S2@.650	Ea	13.90	16.50	30.40
8 x 6"	S2@.650	Ea	15.20	16.50	31.70
10 x 4"	S2@.650	Ea	16.80	16.50	33.30
10 x 6"	S2@.700	Ea	18.20	17.80	36.00
10 x 8"	S2@.750	Ea	19.40	19.00	38.40
12 x 6"	S2@.750	Ea	20.10	19.00	39.10
12 x 8"	S2@.800	Ea	22.60	20.30	42.90
12 x 10"	S2@.850	Ea	25.40	21.60	47.00
14 x 6"	S2@.850	Ea	25.40	21.60	47.00
14 x 8"	S2@.900	Ea	25.90	22.90	48.80
14 x 10"	S2@.950	Ea	27.40	24.10	51.50
16 x 8"	S2@.950	Ea	27.40	24.10	51.50
16 x 10"	S2@1.00	Ea	30.20	25.40	55.60
18 x 8"	S2@1.00	Ea	31.50	25.40	56.90
18 x 10"	S2@1.10	Ea	33.80	27.90	61.70
18 x 12"	S2@1.10	Ea	35.60	27.90	63.50
Ceiling diffuser for commercial ductwork					
6"	S2@.650	Ea	25.80	16.50	42.30
8"	S2@.650	Ea	33.60	16.50	50.10
10"	S2@.700	Ea	40.30	17.80	58.10
12"	S2@.750	Ea	47.40	19.00	66.40
14"	S2@.800	Ea	61.00	20.30	81.30
16"	S2@.900	Ea	68.30	22.90	91.20
18"	S2@1.00	Ea	116.00	25.40	141.40
20"	S2@1.20	Ea	139.00	30.50	169.50
Lay-in ceiling diffuser for commercial ductwork					
6"	S2@.520	Ea	34.70	13.20	47.90
8"	S2@.520	Ea	40.30	13.20	53.50
10"	S2@.560	Ea	46.90	14.20	61.10
12"	S2@.600	Ea	52.90	15.20	68.10
14"	S2@.640	Ea	64.50	16.20	80.70
16"	S2@.720	Ea	72.30	18.30	90.60

Description	Craft@Hrs	Unit	Material $	Labor $	Total $
Multi-blade damper for rectangular ductwork					
12 x 6"	S2@.600	Ea	26.20	15.20	41.40
12 x 10"	S2@.650	Ea	30.90	16.50	47.40
12 x 12"	S2@.700	Ea	38.00	17.80	55.80
18 x 12"	S2@.800	Ea	43.80	20.30	64.10
18 x 18"	S2@.850	Ea	48.60	21.60	70.20
24 x 12"	S2@.850	Ea	51.10	21.60	72.70
24 x 18"	S2@.900	Ea	55.70	22.90	78.60
24 x 24"	S2@1.00	Ea	62.80	25.40	88.20
30 x 12"	S2@.900	Ea	58.10	22.90	81.00
30 x 18"	S2@1.00	Ea	65.20	25.40	90.60
30 x 24"	S2@1.10	Ea	73.60	27.90	101.50
30 x 30"	S2@1.20	Ea	79.50	30.50	110.00
36 x 18"	S2@1.10	Ea	76.00	27.90	103.90
36 x 36"	S2@1.40	Ea	104.00	35.50	139.50
42 x 24"	S2@1.30	Ea	106.00	33.00	139.00
42 x 42"	S2@1.50	Ea	146.00	38.10	184.10
48 x 24"	S2@1.40	Ea	109.00	35.50	144.50
48 x 48"	S2@1.60	Ea	152.00	40.60	192.60
54 x 24"	S2@1.45	Ea	140.00	36.80	176.80
54 x 54"	S2@1.70	Ea	175.00	43.20	218.20
Single-blade damper for rectangular ductwork					
12 x 6"	S2@.600	Ea	17.30	15.20	32.50
12 x 8"	S2@.625	Ea	18.90	15.90	34.80
12 x 10"	S2@.650	Ea	21.20	16.50	37.70
12 x 12"	S2@.700	Ea	24.90	17.80	42.70
18 x 12"	S2@.800	Ea	27.60	20.30	47.90
18 x 18"	S2@.850	Ea	31.40	21.60	53.00
24 x 12"	S2@.850	Ea	33.20	21.60	54.80
24 x 18"	S2@.900	Ea	34.50	22.90	57.40
24 x 24"	S2@1.00	Ea	38.00	25.40	63.40
Single-blade damper for round ductwork					
6"	S2@.600	Ea	17.30	15.20	32.50
8"	S2@.550	Ea	16.20	14.00	30.20
10"	S2@.600	Ea	18.80	15.20	34.00
12"	S2@.650	Ea	21.30	16.50	37.80
14"	S2@.700	Ea	24.30	17.80	42.10
16"	S2@.765	Ea	26.90	19.40	46.30
18"	S2@.835	Ea	29.60	21.20	50.80
20"	S2@.900	Ea	33.30	22.90	56.20

Ductwork Specialties

Description	Craft@Hrs	Unit	Material $	Labor $	Total $
Curtain-type duct fire damper, 1½ hour rating, U.L. label					
12 x 6"	S2@1.00	Ea	27.50	25.40	52.90
12 x 8"	S2@1.00	Ea	29.60	25.40	55.00
12 x 10"	S2@1.05	Ea	32.40	26.70	59.10
12 x 12"	S2@1.10	Ea	35.00	27.90	62.90
18 x 12"	S2@1.20	Ea	48.90	30.50	79.40
18 x 18"	S2@1.30	Ea	52.90	33.00	85.90
24 x 12"	S2@1.40	Ea	59.10	35.50	94.60
24 x 18"	S2@1.50	Ea	67.20	38.10	105.30
24 x 24"	S2@1.60	Ea	80.20	40.60	120.80
30 x 12"	S2@1.50	Ea	68.10	38.10	106.20
30 x 18"	S2@1.80	Ea	75.60	45.70	121.30
30 x 24"	S2@2.00	Ea	99.80	50.80	150.60
30 x 30"	S2@2.30	Ea	108.00	58.40	166.40
36 x 18"	S2@2.00	Ea	86.10	50.80	136.90
36 x 36"	S2@2.40	Ea	129.00	60.90	189.90
42 x 24"	S2@2.40	Ea	112.00	60.90	172.90
42 x 42"	S2@2.90	Ea	162.00	73.60	235.60
48 x 24"	S2@2.90	Ea	121.00	73.60	194.60
48 x 48"	S2@3.50	Ea	202.00	88.90	290.90
54 x 24"	S2@3.00	Ea	131.00	76.20	207.20
54 x 54"	S2@3.80	Ea	234.00	96.50	330.50

Correction factors:
(1) For 22 gauge break-away connection, multiply material cost by 1.30
(2) For blades out of air stream cap, multiply material cost by 1.25

Description	Craft@Hrs	Unit	Material $	Labor $	Total $
Single skin 2" turning vanes for duct					
12 x 6"	S2@.280	Ea	3.20	7.11	10.31
12 x 8"	S2@.285	Ea	3.97	7.24	11.21
12 x 10"	S2@.290	Ea	5.14	7.36	12.50
12 x 12"	S2@.300	Ea	6.35	7.62	13.97
18 x 12"	S2@.375	Ea	9.56	9.52	19.08
18 x 18"	S2@.450	Ea	11.60	11.40	23.00
24 x 12"	S2@.450	Ea	12.70	11.40	24.10
24 x 18"	S2@.500	Ea	13.90	12.70	26.60
24 x 24"	S2@.525	Ea	15.70	13.30	29.00
30 x 12"	S2@.550	Ea	15.90	14.00	29.90
30 x 18"	S2@.750	Ea	23.00	19.00	42.00
30 x 24"	S2@.800	Ea	30.60	20.30	50.90
30 x 30"	S2@.900	Ea	38.70	22.90	61.60
36 x 18"	S2@.750	Ea	27.40	19.00	46.40
36 x 36"	S2@.800	Ea	55.50	20.30	75.80
42 x 24"	S2@.900	Ea	43.10	22.90	66.00
42 x 42"	S2@1.00	Ea	75.70	25.40	101.10
48 x 24"	S2@1.10	Ea	48.90	27.90	76.80
48 x 48"	S2@1.50	Ea	97.90	38.10	136.00
54 x 24"	S2@1.15	Ea	54.80	29.20	84.00
54 x 54"	S2@1.25	Ea	124.00	31.70	155.70
60 x 24"	S2@1.30	Ea	61.70	33.00	94.70
60 x 36"	S2@1.40	Ea	93.10	35.50	128.60
60 x 60"	S2@1.50	Ea	153.00	38.10	191.10
72 x 36"	S2@1.65	Ea	112.00	41.90	153.90
72 x 54"	S2@2.40	Ea	165.00	60.90	225.90

Note: Labor includes cost to cut, assemble and install vanes and rails. For 4-inch air-foil vanes, multiply costs by 1.80 and labor costs by .90

Description	Craft@Hrs	Unit	Material $	Labor $	Total $
Plain spin-in galvanized steel ductwork					
4"	S2@.200	Ea	2.45	5.08	7.53
5"	S2@.200	Ea	2.75	5.08	7.83
6"	S2@.200	Ea	3.04	5.08	8.12
7"	S2@.200	Ea	3.33	5.08	8.41
8"	S2@.250	Ea	3.86	6.35	10.21
9"	S2@.250	Ea	4.26	6.35	10.61
10"	S2@.300	Ea	4.61	7.62	12.23
12"	S2@.350	Ea	5.48	8.89	14.37
14"	S2@.400	Ea	6.58	10.20	16.78
16"	S2@.450	Ea	7.80	11.40	19.20
18"	S2@.500	Ea	9.73	12.70	22.43

Ductwork Specialties

Description	Craft@Hrs	Unit	Material $	Labor $	Total $
Spin-in galvanized steel duct with scoop and damper					
4"	S2@.250	Ea	8.25	6.35	14.60
5"	S2@.250	Ea	8.38	6.35	14.73
6"	S2@.250	Ea	8.49	6.35	14.84
7"	S2@.250	Ea	8.61	6.35	14.96
8"	S2@.300	Ea	8.79	7.62	16.41
9"	S2@.300	Ea	8.91	7.62	16.53
10"	S2@.350	Ea	9.08	8.89	17.97
12"	S2@.400	Ea	9.80	10.20	20.00
14"	S2@.450	Ea	10.50	11.40	21.90
16"	S2@.500	Ea	11.10	12.70	23.80
18"	S2@.550	Ea	11.90	14.00	25.90

Description	Craft@Hrs	Unit	Material $	Labor $	Total $
Duct-to-equipment flexible connection					
18 x 12"	S2@.650	Ea	7.49	16.50	23.99
18 x 18"	S2@.850	Ea	8.99	21.60	30.59
24 x 12"	S2@.850	Ea	8.99	21.60	30.59
24 x 18"	S2@1.00	Ea	10.50	25.40	35.90
24 x 24"	S2@1.30	Ea	12.00	33.00	45.00
30 x 12"	S2@1.00	Ea	10.50	25.40	35.90
30 x 18"	S2@1.30	Ea	12.00	33.00	45.00
30 x 30"	S2@1.60	Ea	15.00	40.60	55.60
36 x 12"	S2@1.30	Ea	12.00	33.00	45.00
36 x 24"	S2@1.60	Ea	15.00	40.60	55.60
36 x 36"	S2@1.80	Ea	17.90	45.70	63.60
48 x 24"	S2@1.80	Ea	17.90	45.70	63.60
48 x 48"	S2@2.00	Ea	23.70	50.80	74.50

Description	Craft@Hrs	Unit	Material $	Labor $	Total $

Roof-top air conditioning unit, gas heating

Description	Craft@Hrs	Unit	Material $	Labor $	Total $
2-T, 800 CFM	S2@5.00	Ea	2,780.00	127.00	2,907.00
2½-T, 1,000 CFM	S2@6.00	Ea	2,960.00	152.00	3,112.00
3-T, 1,200 CFM	S2@7.00	Ea	3,140.00	178.00	3,318.00
4-T, 1,600 CFM	S2@8.00	Ea	4,190.00	203.00	4,393.00
5-T, 2,000 CFM	S2@9.00	Ea	5,040.00	229.00	5,269.00
7½-T, 3,000 CFM	S2@10.0	Ea	9,020.00	254.00	9,274.00
10 T, 4,000 CFM	S2@12.0	Ea	11,900.00	305.00	12,205.00
12-T, 5,000 CFM	S2@14.0	Ea	14,700.00	355.00	15,055.00
15-T, 6,000 CFM	S2@15.0	Ea	17,400.00	381.00	17,781.00
20-T, 8,000 CFM	S2@16.0	Ea	22,600.00	406.00	23,006.00
25-T, 10,000 CFM	S2@17.0	Ea	27,900.00	432.00	28,332.00
30-T, 12,000 CFM	S2@19.0	Ea	32,800.00	482.00	33,282.00
40-T, 16,000 CFM	S2@22.0	Ea	43,000.00	559.00	43,559.00

Roof-top air conditioning unit, hot water coil Costs shown per ton (T) of rated capacity and cubic feet per minute (CFM)

Description	Craft@Hrs	Unit	Material $	Labor $	Total $
2-T, 800 CFM	S2@5.00	Ea	2,540.00	127.00	2,667.00
2½-T, 1,000 CFM	S2@6.00	Ea	2,710.00	152.00	2,862.00
3-T, 1,200 CFM	S2@7.00	Ea	2,880.00	178.00	3,058.00
4-T, 1,600 CFM	S2@8.00	Ea	3,850.00	203.00	4,053.00
5-T, 2,000 CFM	S2@9.00	Ea	4,630.00	229.00	4,859.00
7½-T, 3,000 CFM	S2@10.0	Ea	8,250.00	254.00	8,504.00
10-T, 4,000 CFM	S2@12.0	Ea	10,900.00	305.00	11,205.00
12-T, 5,000 CFM	S2@14.0	Ea	13,500.00	355.00	13,855.00
15-T, 6,000 CFM	S2@15.0	Ea	15,800.00	381.00	16,181.00
20-T, 8,000 CFM	S2@16.0	Ea	20,700.00	406.00	21,106.00
25-T, 10,000 CFM	S2@17.0	Ea	25,500.00	432.00	25,932.00
30-T, 12,000 CFM	S2@19.0	Ea	29,900.00	482.00	30,382.00
40-T, 16,000 CFM	S2@22.0	Ea	39,300.00	559.00	39,859.00

Residential gas-fired upflow heater Costs shown per thousand Btu per hour (MBH)

Description	Craft@Hrs	Unit	Material $	Labor $	Total $
50 MBH input	S2@1.20	Ea	428.00	30.50	458.50
75 MBH input	S2@1.50	Ea	563.00	38.10	601.10
100 MBH input	S2@1.80	Ea	654.00	45.70	699.70
125 MBH input	S2@2.00	Ea	741.00	50.80	791.80

Residential gas heater, electric cooler with exterior condenser Costs shown per thousand Btu per hour (MBH)

Description	Craft@Hrs	Unit	Material $	Labor $	Total $
30 MBH	S2@2.50	Ea	1,450.00	63.50	1,513.50
36 MBH	S2@2.75	Ea	1,730.00	69.80	1,799.80
42 MBH	S2@3.00	Ea	1,930.00	76.20	2,006.20
48 MBH	S2@3.50	Ea	2,260.00	88.90	2,348.90
60 MBH	S2@4.00	Ea	2,690.00	102.00	2,792.00

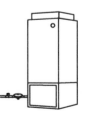

Description	Craft@Hrs	Unit	Material $	Labor $	Total $

Packaged air handler with chilled water and hot water/steam coil
Costs shown per ton (T) of rated capacity and cubic feet per minute (CFM)

Description	Craft@Hrs	Unit	Material $	Labor $	Total $
3-T, 1,200 CFM	S2@4.00	Ea	2,030.00	102.00	2,132.00
4-T, 1,600 CFM	S2@5.50	Ea	2,680.00	140.00	2,820.00
5-T, 2,000 CFM	S2@7.00	Ea	3,190.00	178.00	3,368.00
7½-T, 3,000 CFM	S2@9.00	Ea	4,490.00	229.00	4,719.00
10-T, 4,000 CFM	S2@11.0	Ea	5,700.00	279.00	5,979.00
12-T, 5,000 CFM	S2@13.0	Ea	6,730.00	330.00	7,060.00
15-T, 6,000 CFM	S2@14.0	Ea	7,510.00	355.00	7,865.00
20-T, 8,000 CFM	S2@15.0	Ea	8,970.00	381.00	9,351.00
25-T, 10,000 CFM	S2@16.0	Ea	9,720.00	406.00	10,126.00
30-T, 12,000 CFM	S2@18.0	Ea	11,300.00	457.00	11,757.00
40-T, 16,000 CFM	S2@21.0	Ea	14,000.00	533.00	14,533.00

Boilers. Shipped Factory Assembled, Skid Mounted. Costs shown include the boiler shell and structure, I-beam skids, combustion chamber, firetubes, crown sheet, tube sheets (for firetube), water tubes, mud drum(s), steam drum (for watertube boiler), finned watertubes and check valves (for hot water generators), refractory brick lining, explosion doors, combustion air inlets and stack outlet. Add the cost of the combustion train, boiler trim, fuel train piping, electrical wiring, refractory, stack, feedwater system, combustion controls, water treatment, expansion tanks (if hot water) or condensate return (if steam) and system start-up from the sections following the sections on the packaged boilers that follow. Costs shown are based on boiler horsepower rating (BHP). Note: Boiler horsepower (BHP) multiplied times 33.5 gives the approximate required 1000 Btu per hour input (MBH input)

Description	Craft@Hrs	Unit	Material $	Labor $	Total $

Hot water or steam cast iron gas boiler 30 PSIG Costs shown are based on boiler horsepower rating (BHP).

4.5 BHP	S2@4.50	Ea	2,160.00	114.00	2,274.00
5.9 BHP	S2@5.00	Ea	2,710.00	127.00	2,837.00
9.0 BHP	S2@5.50	Ea	4,040.00	140.00	4,180.00
15.0 BHP	S2@6.00	Ea	6,280.00	152.00	6,432.00
20.0 BHP	S2@6.50	Ea	8,310.00	165.00	8,475.00
30.0 BHP	S2@7.50	Ea	10,800.00	190.00	10,990.00
45.0 BHP	S2@8.50	Ea	15,600.00	216.00	15,816.00
60.0 BHP	S2@9.00	Ea	20,400.00	229.00	20,629.00

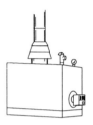

Hot water Scotch marine firetube gas boiler 30 PSIG Costs shown are based on boiler horsepower rating (BHP).

30 BHP	S2@7.50	Ea	15,500.00	190.00	15,690.00
40 BHP	S2@8.00	Ea	16,400.00	203.00	16,603.00
50 BHP	S2@8.50	Ea	17,100.00	216.00	17,316.00
60 BHP	S2@9.00	Ea	19,100.00	229.00	19,329.00
70 BHP	S2@9.50	Ea	20,600.00	241.00	20,841.00
80 BHP	S2@10.0	Ea	21,800.00	254.00	22,054.00
100 BHP	S2@12.0	Ea	26,100.00	305.00	26,405.00
125 BHP	S2@16.0	Ea	27,900.00	406.00	28,306.00
150 BHP	S2@16.0	Ea	29,900.00	406.00	30,306.00
200 BHP	S2@16.0	Ea	35,400.00	406.00	35,806.00
250 BHP	S2@32.0	Ea	39,000.00	812.00	39,812.00
300 BHP	S2@32.0	Ea	41,900.00	812.00	42,712.00
350 BHP	S2@32.0	Ea	47,400.00	812.00	48,212.00
400 BHP	S2@32.0	Ea	48,800.00	812.00	49,612.00
500 BHP	S2@32.0	Ea	58,200.00	812.00	59,012.00
600 BHP	S2@32.0	Ea	65,000.00	812.00	65,812.00
750 BHP	S2@64.0	Ea	68,600.00	1,620.00	70,220.00

Description	Craft@Hrs	Unit	Material $	Labor $	Total $

Packaged Scotch marine firetube oil or gas fired boiler 150 PSIG
Costs shown are based on boiler horsepower rating (BHP).

Description	Craft@Hrs	Unit	Material $	Labor $	Total $
60 BHP	S2@9.00	Ea	21,800.00	229.00	22,029.00
70 BHP	S2@9.50	Ea	23,600.00	241.00	23,841.00
80 BHP	S2@10.0	Ea	25,200.00	254.00	25,454.00
100 BHP	S2@12.0	Ea	30,000.00	305.00	30,305.00
125 BHP	S2@16.0	Ea	32,000.00	406.00	32,406.00
150 BHP	S2@16.0	Ea	34,400.00	406.00	34,806.00
200 BHP	S2@16.0	Ea	45,000.00	406.00	45,406.00
250 BHP	S2@32.0	Ea	48,200.00	812.00	49,012.00
300 BHP	S2@32.0	Ea	54,500.00	812.00	55,312.00
350 BHP	S2@32.0	Ea	56,100.00	812.00	56,912.00
400 BHP	S2@32.0	Ea	67,000.00	812.00	67,812.00
500 BHP	S2@32.0	Ea	74,700.00	812.00	75,512.00
600 BHP	S2@32.0	Ea	76,000.00	812.00	76,812.00
700 BHP	S2@64.0	Ea	79,000.00	1,620.00	80,620.00

Packaged hot water firebox oil or gas fired boiler
Costs shown are based on boiler horsepower rating (BHP).

Description	Craft@Hrs	Unit	Material $	Labor $	Total $
13 BHP	S2@8.00	Ea	8,240.00	203.00	8,443.00
16 BHP	S2@8.00	Ea	8,380.00	203.00	8,583.00
19 BHP	S2@8.00	Ea	8,670.00	203.00	8,873.00
22 BHP	S2@8.00	Ea	8,900.00	203.00	9,103.00
28 BHP	S2@8.00	Ea	9,250.00	203.00	9,453.00
34 BHP	S2@8.00	Ea	10,700.00	203.00	10,903.00
40 BHP	S2@8.00	Ea	11,600.00	203.00	11,803.00
46 BHP	S2@8.00	Ea	13,400.00	203.00	13,603.00
52 BHP	S2@8.00	Ea	15,700.00	203.00	15,903.00
61 BHP	S2@8.00	Ea	17,200.00	203.00	17,403.00
70 BHP	S2@8.00	Ea	18,400.00	203.00	18,603.00
80 BHP	S2@8.00	Ea	19,100.00	203.00	19,303.00
100 BHP	S2@8.00	Ea	21,400.00	203.00	21,603.00
125 BHP	S2@16.0	Ea	25,400.00	406.00	25,806.00
150 BHP	S2@16.0	Ea	27,500.00	406.00	27,906.00

Description	Craft@Hrs	Unit	Material $	Labor $	Total $

Packaged watertube industrial high pressure boilers Costs shown are based on boiler horsepower rating (BHP).

Description	Craft@Hrs	Unit	Material $	Labor $	Total $
480 BHP	S2@32.0	Ea	155,000.00	812.00	155,812.00
526 BHP	S2@32.0	Ea	160,000.00	812.00	160,812.00
572 BHP	S2@32.0	Ea	166,000.00	812.00	166,812.00
618 BHP	S2@32.0	Ea	170,000.00	812.00	170,812.00
664 BHP	S2@32.0	Ea	178,000.00	812.00	178,812.00
710 BHP	S2@64.0	Ea	181,000.00	1,620.00	182,620.00
747 BHP	S2@64.0	Ea	183,000.00	1,620.00	184,620.00
792 BHP	S2@64.0	Ea	189,000.00	1,620.00	190,620.00
838 BHP	S2@64.0	Ea	195,000.00	1,620.00	196,620.00
884 BHP	S2@64.0	Ea	201,000.00	1,620.00	202,620.00
930 BHP	S2@64.0	Ea	205,000.00	1,620.00	206,620.00
1,000 BHP	S2@64.0	Ea	207,000.00	1,620.00	208,620.00

Costs for Combustion Trains for Boilers. Costs shown include the burner, burner head, burner mount flange, associated refractory and gasketing, main fuel valve, draft damper and fan (for a power burner), or burner face plate (for an atmospheric or induced-draft burner). Note: Boiler horsepower (BHP) multiplied times 33.5 gives the approximate required 1000 Btu per hour input (MBH input).

Combustion train for Scotch marine firetube boilers Costs shown are based on boiler horsepower rating (BHP).

Description	Craft@Hrs	Unit	Material $	Labor $	Total $
20 BHP	S2@8.00	Ea	1,270.00	203.00	1,473.00
30 BHP	S2@8.00	Ea	1,730.00	203.00	1,933.00
40 BHP	S2@8.00	Ea	2,370.00	203.00	2,573.00
50 BHP	S2@8.00	Ea	2,960.00	203.00	3,163.00
60 BHP	S2@8.00	Ea	3,280.00	203.00	3,483.00
70 BHP	S2@8.00	Ea	3,570.00	203.00	3,773.00
80 BHP	S2@8.00	Ea	3,640.00	203.00	3,843.00
100 BHP	S2@8.00	Ea	3,980.00	203.00	4,183.00
125 BHP	S2@8.00	Ea	4,140.00	203.00	4,343.00
150 BHP	S2@8.00	Ea	4,580.00	203.00	4,783.00
200 BHP	S2@8.00	Ea	4,770.00	203.00	4,973.00
250 BHP	S2@8.00	Ea	5,250.00	203.00	5,453.00
300 BHP	S2@8.00	Ea	5,880.00	203.00	6,083.00
350 BHP	S2@8.00	Ea	6,020.00	203.00	6,223.00
400 BHP	S2@8.00	Ea	6,210.00	203.00	6,413.00
500 BHP	S2@16.0	Ea	6,740.00	406.00	7,146.00
600 BHP	S2@16.0	Ea	7,210.00	406.00	7,616.00
750 BHP	S2@32.0	Ea	7,650.00	812.00	8,462.00
800 BHP	S2@32.0	Ea	7,990.00	812.00	8,802.00

Description	Craft@Hrs	Unit	Material $	Labor $	Total $

Combustion train for package firebox boilers Costs shown are based on boiler horsepower rating (BHP).

Description	Craft@Hrs	Unit	Material $	Labor $	Total $
13.4 BHP	S2@8.00	Ea	665.00	203.00	868.00
16.4 BHP	S2@8.00	Ea	877.00	203.00	1,080.00
17.4 BHP	S2@8.00	Ea	1,040.00	203.00	1,243.00
22.4 BHP	S2@8.00	Ea	1,230.00	203.00	1,433.00
28.4 BHP	S2@8.00	Ea	1,330.00	203.00	1,533.00
34.4 BHP	S2@8.00	Ea	1,460.00	203.00	1,663.00
40.3 BHP	S2@8.00	Ea	1,580.00	203.00	1,783.00
46.3 BHP	S2@8.00	Ea	1,740.00	203.00	1,943.00
52.3 BHP	S2@8.00	Ea	1,940.00	203.00	2,143.00
61.1 BHP	S2@8.00	Ea	2,100.00	203.00	2,303.00
70.0 BHP	S2@8.00	Ea	2,150.00	203.00	2,353.00
80.0 BHP	S2@8.00	Ea	2,250.00	203.00	2,453.00
100.0 BHP	S2@8.00	Ea	2,370.00	203.00	2,573.00
125.0 BHP	S2@8.00	Ea	2,520.00	203.00	2,723.00
150.0 BHP	S2@8.00	Ea	2,760.00	203.00	2,963.00

Natural gas fuel train piping for firetube and watertube boilers Meets requirements of Underwriters' Laboratories. Shipped pre-assembled. Add the cost of venting and piping to a gas meter. Costs shown are based on boiler horsepower rating (BHP)

Description	Craft@Hrs	Unit	Material $	Labor $	Total $
20 BHP	S2@8.00	Ea	337.00	203.00	540.00
30 BHP	S2@8.00	Ea	462.00	203.00	665.00
40 BHP	S2@8.00	Ea	707.00	203.00	910.00
50 BHP	S2@8.00	Ea	843.00	203.00	1,046.00
60 BHP	S2@8.00	Ea	923.00	203.00	1,126.00
70 BHP	S2@8.00	Ea	979.00	203.00	1,182.00
80 BHP	S2@8.00	Ea	1,060.00	203.00	1,263.00
100 BHP	S2@8.00	Ea	1,230.00	203.00	1,433.00
125 BHP	S2@8.00	Ea	1,460.00	203.00	1,663.00
150 BHP	S2@8.00	Ea	1,730.00	203.00	1,933.00
200 BHP	S2@8.00	Ea	2,070.00	203.00	2,273.00
250 BHP	S2@8.00	Ea	2,550.00	203.00	2,753.00
300 BHP	S2@8.00	Ea	2,690.00	203.00	2,893.00
350 BHP	S2@8.00	Ea	2,890.00	203.00	3,093.00
400 BHP	S2@8.00	Ea	3,100.00	203.00	3,303.00
500 BHP	S2@16.0	Ea	3,490.00	406.00	3,896.00
600 BHP	S2@16.0	Ea	4,030.00	406.00	4,436.00
750 BHP	S2@16.0	Ea	4,350.00	406.00	4,756.00
800 BHP	S2@16.0	Ea	4,550.00	406.00	4,956.00

Description	Craft@Hrs	Unit	Material $	Labor $	Total $

Natural gas fuel train piping for package firebox boilers Costs shown are based on boiler horsepower rating (BHP).

Description	Craft@Hrs	Unit	Material $	Labor $	Total $
13 BHP	S2@8.00	Ea	451.00	203.00	654.00
16 BHP	S2@8.00	Ea	516.00	203.00	719.00
19 BHP	S2@8.00	Ea	554.00	203.00	757.00
22 BHP	S2@8.00	Ea	609.00	203.00	812.00
28 BHP	S2@8.00	Ea	642.00	203.00	845.00
34 BHP	S2@8.00	Ea	707.00	203.00	910.00
40 BHP	S2@8.00	Ea	843.00	203.00	1,046.00
45 BHP	S2@8.00	Ea	886.00	203.00	1,089.00
50 BHP	S2@8.00	Ea	1,060.00	203.00	1,263.00
60 BHP	S2@8.00	Ea	1,190.00	203.00	1,393.00
70 BHP	S2@8.00	Ea	1,360.00	203.00	1,563.00
80 BHP	S2@8.00	Ea	1,420.00	203.00	1,623.00
100 BHP	S2@8.00	Ea	1,690.00	203.00	1,893.00
125 BHP	S2@8.00	Ea	1,750.00	203.00	1,953.00
150 BHP	S2@8.00	Ea	1,910.00	203.00	2,113.00
Add for code approval:					
Factory Mutual	--	%	7.0	--	--
Industrial Risk	--	%	10.0	--	--

Oil Fuel Train Piping. Costs shown include recirculating pump, electric oil preheater, regulating valve, check valves, isolating pump stop cocks, oil line pressure relief valve and pressure gauges. Add the cost of concrete pump pad, electrical connections, relief line piping back to the oil tank and tank connecting piping. Add for number 4-6 fuel oil as shown.

Oil fuel train piping for boilers Costs shown are based on boiler horsepower rating (BHP) using number 2 fuel oil.

Description	Craft@Hrs	Unit	Material $	Labor $	Total $
20 to 30 BHP	S2@4.00	Ea	723.00	102.00	825.00
40 to 60 BHP	S2@4.00	Ea	802.00	102.00	904.00
70 to 80 BHP	S2@4.00	Ea	877.00	102.00	979.00
100 BHP	S2@8.00	Ea	1,130.00	203.00	1,333.00
125 to 150 BHP	S2@8.00	Ea	1,460.00	203.00	1,663.00
200 BHP	S2@8.00	Ea	1,930.00	203.00	2,133.00
250 BHP	S2@8.00	Ea	2,290.00	203.00	2,493.00
300 BHP	S2@16.0	Ea	2,450.00	406.00	2,856.00
350 BHP	S2@16.0	Ea	2,730.00	406.00	3,136.00
400 BHP	S2@16.0	Ea	3,270.00	406.00	3,676.00
500 BHP	S2@32.0	Ea	3,330.00	812.00	4,142.00
600 BHP	S2@32.0	Ea	3,480.00	812.00	4,292.00
750 to 800 BHP	S2@32.0	Ea	3,800.00	812.00	4,612.00
900 to 1000 BHP	S2@32.0	Ea	4,480.00	812.00	5,292.00
Add for #4-6 oil	--	%	10.0	--	--
Add for tie-in with:					
dual-fuel burner	--	%	5.0	--	--

Heating, Ventilating and Air Conditioning Equipment

Description	Craft@Hrs	Unit	Material $	Labor $	Total $

Costs for Electrical Service for Boilers, Subcontract. Costs shown include typical main power tie-in and fusing, circuit breaker panel, and main bus. Assumes controls, fan, blower switchgear and control console are pre-wired electrical service. These costs include the electrical subcontractor's overhead and profit.

Electrical service for firetube and watertube boilers Costs shown are based on boiler horsepower rating (BHP)

Description	Craft@Hrs	Unit	Material $	Labor $	Total $
20 to 100 BHP	--	Ea	--	--	612.00
125 to 200 BHP	--	Ea	--	--	655.00
250 to 350 BHP	--	Ea	--	--	753.00
400 to 500 BHP	--	Ea	--	--	856.00
500 to 600 BHP	--	Ea	--	--	1,240.00
750 to 800 BHP	--	Ea	--	--	2,230.00

Electrical service for package firebox boilers Costs shown are based on boiler horsepower rating (BHP)

Description	Craft@Hrs	Unit	Material $	Labor $	Total $
13 to 70 BHP	--	Ea	--	--	521.00
80 to 125 BHP	--	Ea	--	--	555.00
150 BHP	--	Ea	--	--	655.00

Costs for Refractory for Boilers, Subcontract. For package Scotch marine firetube, firebox and watertube boilers. Most new boilers are shipped from the factory with the refactory brick installed. If refractory brick is installed on the job site to repair shipping damage or to create firedoor seals, use the figures below. These costs include the refractory subcontractor's overhead and profit. Costs shown are per 100 pounds of refractory used.

Description	Craft@Hrs	Unit	Material $	Labor $	Total $
Repairs and seals	--	100lb	--	--	137.00

Costs for Boiler Feedwater Pumps. Costs shown are for the pump only. Add the cost of piping, pump actuating valves and concrete pump base.

Feedwater pumps for firetube and watertube boilers Costs shown are based on boiler horsepower rating (BHP)

Description	Craft@Hrs	Unit	Material $	Labor $	Total $
20 to 40 BHP	S2@8.00	Ea	4,430.00	203.00	4,633.00
50 to 80 BHP	S2@8.00	Ea	5,350.00	203.00	5,553.00
100 BHP	S2@8.00	Ea	5,920.00	203.00	6,123.00
125 to 200 BHP	S2@16.0	Ea	6,120.00	406.00	6,526.00
250 to 400 BHP	S2@32.0	Ea	6,730.00	812.00	7,542.00
500 BHP	S2@32.0	Ea	7,100.00	812.00	7,912.00
600 BHP	S2@32.0	Ea	7,380.00	812.00	8,192.00
750 BHP	S2@32.0	Ea	7,580.00	812.00	8,392.00
800 BHP	S2@32.0	Ea	10,300.00	812.00	11,112.00

Description	Craft@Hrs	Unit	Material $	Labor $	Total $

Feedwater pumps for package firebox boilers Costs shown are based on boiler horsepower rating (BHP)

13 to 35 BHP	S2@8.00	Ea	2,370.00	203.00	2,573.00
40 to 45 BHP	S2@8.00	Ea	2,790.00	203.00	2,993.00
50 to 60 BHP	S2@8.00	Ea	3,500.00	203.00	3,703.00
70 to 80 BHP	S2@8.00	Ea	4,170.00	203.00	4,373.00
100 BHP	S2@8.00	Ea	5,350.00	203.00	5,553.00
125 BHP	S2@16.0	Ea	5,350.00	406.00	5,756.00
150 BHP	S2@16.0	Ea	5,920.00	406.00	6,326.00

Costs for Natural Gas Combustion Controls for Boilers. Costs shown include automatic recycling or continuous pilot, spark igniter and coil (for electric ignition), timer to control the purge cycle, pilot, main burner (hi-lo-off) or main burner (proportional control), flame safeguard and sensor and combustion control panel.

Natural gas combustion controls for firetube and watertube boilers
Costs shown are based on boiler horsepower rating (BHP)

20 to 80 BHP	S2@4.00	Ea	1,820.00	102.00	1,922.00
100 BHP	S2@4.00	Ea	2,040.00	102.00	2,142.00
125 to 200 BHP	S2@8.00	Ea	2,040.00	203.00	2,243.00
250 BHP	S2@16.0	Ea	2,040.00	406.00	2,446.00
300 BHP	S2@16.0	Ea	2,450.00	406.00	2,856.00
350 or 400 BHP	S2@16.0	Ea	2,870.00	406.00	3,276.00
500 BHP	S2@32.0	Ea	2,870.00	812.00	3,682.00
600 BHP	S2@32.0	Ea	3,430.00	812.00	4,242.00
750 BHP	S2@64.0	Ea	3,990.00	1,620.00	5,610.00
800 BHP	S2@64.0	Ea	4,640.00	1,620.00	6,260.00

Natural gas combustion controls for firebox boilers Costs shown are
based on boiler horsepower rating (BHP)

13.4 to 80.0 BHP	S2@4.00	Ea	1,760.00	102.00	1,862.00
100.0 BHP	S2@4.00	Ea	2,040.00	102.00	2,142.00
126.9 BHP	S2@8.00	Ea	2,040.00	203.00	2,243.00
150.0 BHP	S2@8.00	Ea	3,790.00	203.00	3,993.00
Additive if burner controls are for:					
#2 oil-fired	--	%	--	--	+10.0
Heavy oil-fired	--	%	--	--	+20.0
Dual fuel burners	--	%	--	--	+70.0

Heating, Ventilating and Air Conditioning Equipment

Description	Craft@Hrs	Unit	Material $	Labor $	Total $

Costs for Water Softening Systems for Use with Boilers. Costs shown include the metering valve, tank, pump, and zeolite media. Add the cost of piping to connect the water softening system to the main feedwater tank and pump.

Water softening system for firetube and watertube boilers Costs shown are based on boiler horsepower rating (BHP)

Description	Craft@Hrs	Unit	Material $	Labor $	Total $
20 to 80 BHP	S2@4.00	Ea	3,020.00	102.00	3,122.00
100 BHP	S2@4.00	Ea	3,550.00	102.00	3,652.00
125 BHP	S2@8.00	Ea	4,110.00	203.00	4,313.00
150 or 200 BHP	S2@8.00	Ea	5,200.00	203.00	5,403.00
250 BHP	S2@8.00	Ea	6,140.00	203.00	6,343.00
300 or 350 BHP	S2@16.0	Ea	7,850.00	406.00	8,256.00
400 or 500 BHP	S2@16.0	Ea	9,800.00	406.00	10,206.00
600 BHP	S2@32.0	Ea	11,800.00	812.00	12,612.00
750 or 800 BHP	S2@64.0	Ea	14,000.00	1,620.00	15,620.00

Water softening system for package firebox boilers Costs shown are based on boiler horsepower rating (BHP)

Description	Craft@Hrs	Unit	Material $	Labor $	Total $
13.4 to 80 BHP	S2@4.00	Ea	3,560.00	102.00	3,662.00
100.0 BHP	S2@4.00	Ea	3,560.00	102.00	3,662.00
126.9 BHP	S2@8.00	Ea	4,170.00	203.00	4,373.00
150.0 BHP	S2@8.00	Ea	4,840.00	203.00	5,043.00

Description	Craft@Hrs	Unit	Material $	Labor $	Total $

Costs for Start-up, Shakedown, and Calibration of Boilers. Costs shown include minor changes and adjustments to meet requirements of the boiler inspector, adjustment of combustion efficiency at the burner by a control technician, pump and metering calibration testing and steam or hot water metering on start-up.

Start-up, shakedown & calibration of firetube and watertube boilers
Costs shown are based on boiler horsepower rating (BHP.) If a factory technician performs the start-up, add this cost at a minimum of $500 (8 hours at $62.50 per hour)

Description	Craft@Hrs	Unit	Material $	Labor $	Total $
13.4 to 100 BHP	S2@4.00	Ea	--	102.00	102.00
125 to 200 BHP	S2@8.00	Ea	--	203.00	203.00
250 to 400 BHP	S2@16.0	Ea	--	406.00	406.00
500 to 600 BHP	S2@24.0	Ea	--	609.00	609.00
750 to 800 BHP	S2@30.0	Ea	--	762.00	762.00

Costs for Stack Economizer Units for Boilers. Costs shown include piping connection to existing boiler, economizer fin tube coils, feedwater pump, check valves, thermostatic regulating valve, stack dampers, pressure relief valve, stack mount flanges and insulating enclosure, but no architectural modifications.

Stack waste heat recovery system for boilers
Costs shown are based on boiler horsepower rating (BHP)

Description	Craft@Hrs	Unit	Material $	Labor $	Total $
500 BHP	S2@40.0	Ea	17,100.00	1,020.00	18,120.00
750 BHP	S2@48.0	Ea	21,300.00	1,220.00	22,520.00
800 BHP	S2@48.0	Ea	22,900.00	1,220.00	24,120.00
1,000 BHP	S2@56.0	Ea	28,600.00	1,420.00	30,020.00

Costs for Condensate Receiver and Pumping Units for Boilers. Costs shown include a stainless steel factory assembled tank and pump with motor, float switch, condensate inlet, feedwater tank, connection, level gauge, thermometer, solenoid valve for direct boiler feed. No concrete mounting pad required.

Condensate receiver and pumping unit for boilers
Unit sizing and costs are based on storage for one minute at the designed steam return rate

Description	Craft@Hrs	Unit	Material $	Labor $	Total $
750 lbs per hour	S2@4.00	Ea	2,230.00	102.00	2,332.00
1,500 lbs per hour	S2@4.00	Ea	2,230.00	102.00	2,332.00
3,000 lbs per hour	S2@4.00	Ea	2,840.00	102.00	2,942.00
5,000 lbs per hour	S2@6.00	Ea	3,120.00	152.00	3,272.00
6,250 lbs per hour	S2@6.00	Ea	3,490.00	152.00	3,642.00

Heating, Ventilating and Air Conditioning Equipment

Description	Craft@Hrs	Unit	Material $	Labor $	Total $

Costs for Deaerator/Condenser Units for Boilers. Pressurized jet spray type. Costs shown include an ASME storage receiver, stand, jet spray deaerator head, makeup control valve, makeup controller, pressure control valve, level controls, pressure and temperature controls, pumped condensate inlet, pumps and motor, steam inlet and system accessories. Add for a concrete mounting pad and piping to the condensate return line (inlet) and piping to the boiler feedwater system (outlet).

Deaerator/condenser unit for boilers Costs shown are based on boiler horsepower rating (BHP)

Description	Craft@Hrs	Unit	Material $	Labor $	Total $
100 BHP	S2@8.00	Ea	22,900.00	203.00	23,103.00
200 BHP	S2@12.0	Ea	23,800.00	305.00	24,105.00
300 BHP	S2@14.0	Ea	25,300.00	355.00	25,655.00
400 BHP	S2@16.0	Ea	28,600.00	406.00	29,006.00
600 BHP	S2@24.0	Ea	30,600.00	609.00	31,209.00
800 BHP	S2@36.0	Ea	33,700.00	914.00	34,614.00
1,000 BHP	S2@48.0	Ea	39,300.00	1,220.00	40,520.00

Costs for Chemical Feed Duplex Pump, Packaged Units for Boilers. Costs shown include the chemical tanks, electronic metering, valve and stand. Add the cost of piping to the feedwater tank.

Chemical feed duplex pump for boilers Costs shown are based on boiler horsepower rating (BHP)

Description	Craft@Hrs	Unit	Material $	Labor $	Total $
20 to 100 BHP	S2@3.00	Ea	846.00	76.20	922.20
125 to 200 BHP	S2@3.00	Ea	1,100.00	76.20	1,176.20
200 to 400 BHP	S2@6.00	Ea	1,100.00	152.00	1,252.00
500 to 600 BHP	S2@6.00	Ea	2,490.00	152.00	2,642.00
700 to 800 BHP	S2@8.00	Ea	2,780.00	203.00	2,983.00
1,000 BHP	S2@10.0	Ea	3,770.00	254.00	4,024.00

Description	Craft@Hrs	Unit	Material $	Labor $	Total $

Costs for Packaged Boiler Feedwater Systems for Boilers. Costs shown include duplex pumps with heater assembly, thermal lining, pressure gauges, makeup feeder valve, ASME feedwater tank and stand, single phase electric motor and level gauge. Add the costs for a concrete mounting pad and piping to the boilers.

Packaged boiler feedwater systems Costs shown are based on boiler horsepower rating (BHP)

Description	Craft@Hrs	Unit	Material $	Labor $	Total $
15 to 50 BHP	S2@5.00	Ea	5,920.00	127.00	6,047.00
60 to 80 BHP	S2@5.00	Ea	6,510.00	127.00	6,637.00
100 BHP	S2@6.00	Ea	6,620.00	152.00	6,772.00
125 to 150 BHP	S2@6.00	Ea	7,100.00	152.00	7,252.00
200 BHP	S2@8.00	Ea	7,320.00	203.00	7,523.00
250 BHP	S2@8.00	Ea	7,640.00	203.00	7,843.00
300 BHP	S2@8.00	Ea	8,070.00	203.00	8,273.00
400 BHP	S2@8.00	Ea	8,340.00	203.00	8,543.00
500 BHP	S2@8.00	Ea	10,100.00	203.00	10,303.00
600 BHP	S2@12.0	Ea	10,600.00	305.00	10,905.00
750 BHP	S2@12.0	Ea	10,500.00	305.00	10,805.00

Costs for Continuous Blowdown Heat Recovery Systems. Costs shown include a flash receiver (50 PSI ASME rated tank), plate and frame heat exchanger, pneumatic level controller and valve, air filter regulator, water gauge, pressure gauge, backflush piping, pressure relief valve connected to mud drum and drain. Add for concrete pad mount, piping and electrical wiring.

Continuous blowdown heat recovery systems Costs shown by pounds per hour (lbs per hr) of rated makeup capacity

Description	Craft@Hrs	Unit	Material $	Labor $	Total $
5,000 lbs per hr	S2@8.00	Ea	19,900.00	203.00	20,103.00
10,000 lbs per hr	S2@8.00	Ea	39,300.00	203.00	39,503.00
25,000 lbs per hr	S2@8.00	Ea	49,700.00	203.00	49,903.00
50,000 lbs per hr	S2@8.00	Ea	79,100.00	203.00	79,303.00

Costs for Boiler Stacks. Costs shown assume a 20-foot stack height and include rigging, breeching, connection of stack sections, guy wires where required, vendor-supplied stack supports and insulation. Add the cost of a concrete base.

Boiler stack Costs shown assume a 20-foot stack height

Description	Craft@Hrs	Unit	Material $	Labor $	Total $
10" diameter	S2@12.0	Ea	1,410.00	305.00	1,715.00
12" diameter	S2@12.0	Ea	1,730.00	305.00	2,035.00
16" diameter	S2@16.0	Ea	2,100.00	406.00	2,506.00
20" diameter	S2@20.0	Ea	2,620.00	508.00	3,128.00
24" diameter	S2@24.0	Ea	3,460.00	609.00	4,069.00
30" diameter	S2@32.0	Ea	4,100.00	812.00	4,912.00
56" diameter	S2@40.0	Ea	8,090.00	1,020.00	9,110.00
Add for each 10' section (or portion thereof) beyond 20' high					
per 10' sect.	--	%	--	--	15.0

Heating, Ventilating and Air Conditioning Equipment

Description	Craft@Hrs	Unit	Material $	Labor $	Total $

Costs for Tankless Indirect Water Heater. Provides domestic hot water from boiler coil insert. Costs shown include heat exchanger, coil, circulating pump, thermostatically controlled valve, bypass piping and air bleeder valve. Add the cost of an equipment base, electrical wiring, piping and an insulated hot water storage tank.

Tankless indirect water heater Costs by gallons per hour (GPH)

Description	Craft@Hrs	Unit	Material $	Labor $	Total $
240 GPH	S2@8.00	Ea	8,100.00	203.00	8,303.00
360 GPH	S2@8.00	Ea	10,200.00	203.00	10,403.00
480 GPH	S2@10.0	Ea	10,400.00	254.00	10,654.00
840 GPH	S2@12.0	Ea	23,700.00	305.00	24,005.00
920 GPH	S2@16.0	Ea	26,600.00	406.00	27,006.00
1,240 GPH	S2@20.0	Ea	29,600.00	508.00	30,108.00

Costs for Boiler Trim. Costs shown include the pressure relief valve, level gauge (for steam systems), level controller & low water cutoff, pressure and temperature gauge, feedwater valve & actuator, main valve, blowdown valve, cleanout manholes and gaskets.

Boiler trim Costs shown are per boiler horsepower rating (BHP)

Description	Craft@Hrs	Unit	Material $	Labor $	Total $
20 to 100 BHP	S2@4.00	Ea	1,260.00	102.00	1,362.00
100 to 200 BHP	S2@6.00	Ea	1,430.00	152.00	1,582.00
200 to 400 BHP	S2@8.00	Ea	1,750.00	203.00	1,953.00
400 to 600 BHP	S2@8.00	Ea	2,690.00	203.00	2,893.00
600 to 800 BHP	S2@8.00	Ea	3,170.00	203.00	3,373.00
800 to 1000 BHP	S2@10.0	Ea	3,180.00	254.00	3,434.00

Description	Craft@Hrs	Unit	Material $	Labor $	Total $

Centrifugal air-foil wheel blower

Description	Craft@Hrs	Unit	Material $	Labor $	Total $
1/3 HP, 12" dia.	S2@2.50	Ea	710.00	63.50	773.50
1/2 HP, 15" dia.	S2@4.50	Ea	1,260.00	114.00	1,374.00
1 HP, 18" dia.	S2@7.00	Ea	1,430.00	178.00	1,608.00
2 HP, 18" dia.	S2@7.50	Ea	1,700.00	190.00	1,890.00
3 HP, 27" dia.	S2@9.00	Ea	2,690.00	229.00	2,919.00
5 HP, 30" dia.	S2@11.0	Ea	3,190.00	279.00	3,469.00
7½ HP, 36" dia.	S2@12.0	Ea	3,840.00	305.00	4,145.00
10 HP, 40" dia.	S2@14.0	Ea	5,360.00	355.00	5,715.00
15 HP, 44" dia.	S2@17.0	Ea	7,100.00	432.00	7,532.00
20 HP, 60" dia.	S2@20.0	Ea	8,110.00	508.00	8,618.00

Utility centrifugal blower

Description	Craft@Hrs	Unit	Material $	Labor $	Total $
1/3 HP, 10" dia.	S2@2.50	Ea	465.00	63.50	528.50
1/2 HP, 12" dia.	S2@4.50	Ea	610.00	114.00	724.00
1 HP, 15" dia.	S2@6.50	Ea	990.00	165.00	1,155.00
2 HP, 18" dia.	S2@7.50	Ea	1,130.00	190.00	1,320.00
3 HP, 24" dia.	S2@9.00	Ea	1,610.00	229.00	1,839.00
5 HP, 27" dia.	S2@11.0	Ea	1,850.00	279.00	2,129.00
7½ HP, 36" dia.	S2@12.0	Ea	3,000.00	305.00	3,305.00
10 HP, 40" dia.	S2@14.0	Ea	3,610.00	355.00	3,965.00
15 HP, 44" dia.	S2@17.0	Ea	3,910.00	432.00	4,342.00

Centrifugal chiller

Description	Craft@Hrs	Unit	Material $	Labor $	Total $
100 tons	S2@12.0	Ea	49,400.00	305.00	49,705.00
150 tons	S2@12.5	Ea	73,700.00	317.00	74,017.00
200 tons	S2@13.0	Ea	79,700.00	330.00	80,030.00
250 tons	S2@14.0	Ea	94,800.00	355.00	95,155.00
300 tons	S2@15.0	Ea	101,000.00	381.00	101,381.00
350 tons	S2@16.0	Ea	104,000.00	406.00	104,406.00
400 tons	S2@17.0	Ea	122,000.00	432.00	122,432.00
500 tons	S2@19.0	Ea	155,000.00	482.00	155,482.00

Reciprocating chiller

Description	Craft@Hrs	Unit	Material $	Labor $	Total $
20 tons	S2@10.0	Ea	15,100.00	254.00	15,354.00
25 tons	S2@10.5	Ea	17,100.00	267.00	17,367.00
30 tons	S2@12.0	Ea	18,800.00	305.00	19,105.00
40 tons	S2@13.0	Ea	22,600.00	330.00	22,930.00
50 tons	S2@14.0	Ea	31,900.00	355.00	32,255.00
60 tons	S2@15.0	Ea	41,200.00	381.00	41,581.00
75 tons	S2@16.0	Ea	44,000.00	406.00	44,406.00
100 tons	S2@16.5	Ea	47,000.00	419.00	47,419.00
125 tons	S2@17.0	Ea	56,300.00	432.00	56,732.00
150 tons	S2@17.5	Ea	65,500.00	444.00	65,944.00

Heating, Ventilating and Air Conditioning Equipment

Description	Craft@Hrs	Unit	Material $	Labor $	Total $

Air cooled condensing unit

Description	Craft@Hrs	Unit	Material $	Labor $	Total $
3 tons	S2@2.00	Ea	2,160.00	50.80	2,210.80
5 tons	S2@3.00	Ea	3,710.00	76.20	3,786.20
7½ tons	S2@4.00	Ea	4,800.00	102.00	4,902.00
10 tons	S2@4.50	Ea	6,400.00	114.00	6,514.00
12½ tons	S2@5.00	Ea	7,540.00	127.00	7,667.00
15 tons	S2@5.50	Ea	8,650.00	140.00	8,790.00
20 tons	S2@6.00	Ea	11,500.00	152.00	11,652.00
25 tons	S2@7.00	Ea	14,400.00	178.00	14,578.00
30 tons	S2@8.00	Ea	17,200.00	203.00	17,403.00
40 tons	S2@9.00	Ea	22,600.00	229.00	22,829.00
50 tons	S2@10.0	Ea	28,300.00	254.00	28,554.00

Galvanized steel cooling tower

Description	Craft@Hrs	Unit	Material $	Labor $	Total $
10 tons	S2@6.00	Ea	2,250.00	152.00	2,402.00
15 tons	S2@6.50	Ea	1,670.00	165.00	1,835.00
20 tons	S2@7.00	Ea	4,020.00	178.00	4,198.00
25 tons	S2@7.50	Ea	4,960.00	190.00	5,150.00
30 tons	S2@8.00	Ea	5,650.00	203.00	5,853.00
40 tons	S2@8.50	Ea	7,210.00	216.00	7,426.00
50 tons	S2@9.00	Ea	7,900.00	229.00	8,129.00
60 tons	S2@9.50	Ea	8,310.00	241.00	8,551.00
80 tons	S2@10.0	Ea	9,170.00	254.00	9,424.00
100 tons	S2@10.5	Ea	10,800.00	267.00	11,067.00
125 tons	S2@11.0	Ea	13,000.00	279.00	13,279.00
150 tons	S2@11.5	Ea	15,100.00	292.00	15,392.00
175 tons	S2@12.0	Ea	17,100.00	305.00	17,405.00
200 tons	S2@13.0	Ea	18,700.00	330.00	19,030.00
300 tons	S2@16.0	Ea	27,000.00	406.00	27,406.00
400 tons	S2@18.0	Ea	34,400.00	457.00	34,857.00
500 tons	S2@20.0	Ea	41,300.00	508.00	41,808.00

Redwood or fir induced-draft cooling tower

Description	Craft@Hrs	Unit	Material $	Labor $	Total $
100 tons	S2@26.0	Ea	6,400.00	660.00	7,060.00
200 tons	S2@32.0	Ea	10,800.00	812.00	11,612.00
300 tons	S2@40.0	Ea	15,400.00	1,020.00	16,420.00
400 tons	S2@45.0	Ea	17,600.00	1,140.00	18,740.00
500 tons	S2@50.0	Ea	20,400.00	1,270.00	21,670.00

Redwood or fir forced draft cooling tower

Description	Craft@Hrs	Unit	Material $	Labor $	Total $
100 tons	S2@26.0	Ea	8,760.00	660.00	9,420.00
200 tons	S2@32.0	Ea	16,700.00	812.00	17,512.00
300 tons	S2@40.0	Ea	19,500.00	1,020.00	20,520.00
400 tons	S2@45.0	Ea	25,800.00	1,140.00	26,940.00
500 tons	S2@50.0	Ea	31,400.00	1,270.00	32,670.00

Description	Craft@Hrs	Unit	Material $	Labor $	Total $
Vane-axial fan					
30 HP, 36" dia.	S2@16.0	Ea	9,900.00	406.00	10,306.00
50 HP, 42" dia.	S2@24.0	Ea	11,100.00	609.00	11,709.00
100 HP, 48" dia.	S2@32.0	Ea	13,900.00	812.00	14,712.00
200 HP, 54" dia.	S2@40.0	Ea	19,400.00	1,020.00	20,420.00
Tube-axial fan					
15 HP, 24" dia.	S2@10.0	Ea	2,940.00	254.00	3,194.00
25 HP, 30" dia.	S2@13.0	Ea	4,550.00	330.00	4,880.00
30 HP, 36" dia.	S2@16.0	Ea	6,050.00	406.00	6,456.00
75 HP, 40" dia.	S2@22.0	Ea	9,080.00	559.00	9,639.00
100 HP, 48" dia.	S2@32.0	Ea	1,950.00	812.00	2,762.00
Roof exhaust fan					
1/4 HP, 14" dia.	S2@2.50	Ea	772.00	63.50	835.50
1/3 HP, 22" dia.	S2@2.90	Ea	950.00	73.60	1,023.60
1/2 HP, 24" dia.	S2@3.50	Ea	1,080.00	88.90	1,168.90
1 HP, 30" dia.	S2@4.50	Ea	1,480.00	114.00	1,594.00
2 HP, 36" dia.	S2@5.00	Ea	2,140.00	127.00	2,267.00
3 HP, 40" dia.	S2@6.00	Ea	3,210.00	152.00	3,362.00
5 HP, 48" dia.	S2@7.00	Ea	4,440.00	178.00	4,618.00

Heating, Ventilating and Air Conditioning Equipment

Description	Craft@Hrs	Unit	Material $	Labor $	Total $

Ceiling-hung fan coil unit with cabinet

Description	Craft@Hrs	Unit	Material $	Labor $	Total $
200 CFM	S2@2.00	Ea	384.00	50.80	434.80
400 CFM	S2@2.50	Ea	455.00	63.50	518.50
600 CFM	S2@3.00	Ea	536.00	76.20	612.20
800 CFM	S2@3.25	Ea	676.00	82.50	758.50
1,000 CFM	S2@3.50	Ea	849.00	88.90	937.90

Less cabinet: Deduct 20% from material cost.
Floor-mounted: Add 10% to material cost.

2-pass shell and tube type heat exchanger

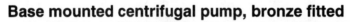

Description	Craft@Hrs	Unit	Material $	Labor $	Total $
4" dia. x 30"	S2@2.00	Ea	728.00	50.80	778.80
4" dia. x 60"	S2@2.50	Ea	856.00	63.50	919.50
6" dia. x 30"	S2@3.25	Ea	1,050.00	82.50	1,132.50
6" dia. x 60"	S2@3.50	Ea	1,300.00	88.90	1,388.90
8" dia. x 30"	S2@4.25	Ea	1,620.00	108.00	1,728.00
8" dia. x 60"	S2@4.50	Ea	2,100.00	114.00	2,214.00
10" dia. x 30"	S2@5.50	Ea	2,380.00	140.00	2,520.00
10" dia. x 60"	S2@5.75	Ea	3,170.00	146.00	3,316.00
10" dia. x 90"	S2@6.00	Ea	3,950.00	152.00	4,102.00
12" dia. x 60"	S2@6.50	Ea	4,150.00	165.00	4,315.00
12" dia. x 90"	S2@7.00	Ea	5,300.00	178.00	5,478.00
12" dia. x 120"	S2@7.25	Ea	6,430.00	184.00	6,614.00
14" dia. x 60"	S2@7.50	Ea	6,140.00	190.00	6,330.00
14" dia. x 90"	S2@8.00	Ea	7,620.00	203.00	7,823.00
14" dia. x 120"	S2@8.25	Ea	9,220.00	209.00	9,429.00
16" dia. x 60"	S2@8.50	Ea	7,530.00	216.00	7,746.00
16" dia. x 90"	S2@9.00	Ea	9,480.00	229.00	9,709.00
16" dia. x 120"	S2@9.25	Ea	10,800.00	235.00	11,035.00

Base mounted centrifugal pump, bronze fitted

Description	Craft@Hrs	Unit	Material $	Labor $	Total $
1/2 HP	S2@1.50	Ea	447.00	38.10	485.10
3/4 HP	S2@1.75	Ea	567.00	44.40	611.40
1 HP	S2@2.00	Ea	691.00	50.80	741.80
1½ HP	S2@2.40	Ea	835.00	60.90	895.90
2 HP	S2@2.70	Ea	966.00	68.60	1,034.60
3 HP	S2@3.00	Ea	1,230.00	76.20	1,306.20
5 HP	S2@3.50	Ea	1,400.00	88.90	1,488.90
7½ HP	S2@4.00	Ea	1,560.00	102.00	1,662.00
10 HP	S2@4.50	Ea	2,270.00	114.00	2,384.00
15 HP	S2@5.00	Ea	2,630.00	127.00	2,757.00
20 HP	S2@6.00	Ea	3,050.00	152.00	3,202.00
25 HP	S2@7.00	Ea	3,180.00	178.00	3,358.00
30 HP	S2@7.40	Ea	3,760.00	188.00	3,948.00
40 HP	S2@7.70	Ea	3,740.00	196.00	3,936.00
50 HP	S2@8.00	Ea	4,540.00	203.00	4,743.00

Description	Craft@Hrs	Unit	Material $	Labor $	Total $

Cast iron line-mounted centrifugal pump

Description	Craft@Hrs	Unit	Material $	Labor $	Total $
1/6 HP	S2@1.25	Ea	316.00	31.70	347.70
1/4 HP	S2@1.40	Ea	405.00	35.50	440.50
1/3 HP	S2@1.50	Ea	501.00	38.10	539.10
1/2 HP	S2@1.70	Ea	637.00	43.20	680.20
3/4 HP	S2@1.80	Ea	859.00	45.70	904.70
1 HP	S2@2.00	Ea	1,020.00	50.80	1,070.80

Duct-mounted flanged reheat coil

Description	Craft@Hrs	Unit	Material $	Labor $	Total $
12 x 6"	S2@1.50	Ea	247.00	38.10	285.10
12 x 8"	S2@1.60	Ea	261.00	40.60	301.60
12 x 10"	S2@1.70	Ea	275.00	43.20	318.20
12 x 12"	S2@1.80	Ea	289.00	45.70	334.70
18 x 6"	S2@1.90	Ea	307.00	48.20	355.20
18 x 12"	S2@2.00	Ea	322.00	50.80	372.80
18 x 18"	S2@2.10	Ea	343.00	53.30	396.30
24 x 12"	S2@2.30	Ea	370.00	58.40	428.40
24 x 18"	S2@2.50	Ea	394.00	63.50	457.50
24 x 24"	S2@2.75	Ea	446.00	69.80	515.80

Duct-mounted slip-in reheat coil

Description	Craft@Hrs	Unit	Material $	Labor $	Total $
12 x 6"	S2@1.10	Ea	296.00	27.90	323.90
12 x 8"	S2@1.20	Ea	336.00	30.50	366.50
12 x 10"	S2@1.30	Ea	411.00	33.00	444.00
12 x 12"	S2@1.40	Ea	446.00	35.50	481.50
18 x 6"	S2@1.50	Ea	494.00	38.10	532.10
18 x 12"	S2@1.60	Ea	536.00	40.60	576.60
18 x 18"	S2@1.70	Ea	570.00	43.20	613.20
24 x 12"	S2@1.90	Ea	605.00	48.20	653.20
24 x 18"	S2@2.10	Ea	639.00	53.30	692.30
24 x 24"	S2@2.40	Ea	788.00	60.90	848.90

Ceiling suspended gas-fired unit heater

Description	Craft@Hrs	Unit	Material $	Labor $	Total $
200 CFM	S2@2.00	Ea	523.00	50.80	573.80
400 CFM	S2@2.50	Ea	612.00	63.50	675.50
600 CFM	S2@3.00	Ea	739.00	76.20	815.20
800 CFM	S2@3.25	Ea	804.00	82.50	886.50
1,000 CFM	S2@3.50	Ea	866.00	88.90	954.90
1,500 CFM	S2@3.75	Ea	1,010.00	95.20	1,105.20
2,000 CFM	S2@4.00	Ea	1,130.00	102.00	1,232.00

Heating, Ventilating and Air Conditioning Equipment

Description	Craft@Hrs	Unit	Material $	Labor $	Total $
Variable-air volume cooling unit					
200- 400 CFM	S2@2.00	Ea	314.00	50.80	364.80
400- 600 CFM	S2@2.40	Ea	373.00	60.90	433.90
600- 800 CFM	S2@2.80	Ea	431.00	71.10	502.10
800-1,000 CFM	S2@3.20	Ea	483.00	81.20	564.20
1,000-1,500 CFM	S2@3.60	Ea	536.00	91.40	627.40
1,500-2,000 CFM	S2@4.00	Ea	600.00	102.00	702.00
Variable-air volume cooling and reheat unit					
200- 400 CFM	S2@2.50	Ea	455.00	63.50	518.50
400- 600 CFM	S2@2.90	Ea	523.00	73.60	596.60
600- 800 CFM	S2@3.30	Ea	634.00	83.80	717.80
800-1,000 CFM	S2@3.70	Ea	699.00	93.90	792.90
1,000-1,500 CFM	S2@4.10	Ea	856.00	104.00	960.00
1,500-2,000 CFM	S2@4.50	Ea	943.00	114.00	1,057.00

Description	Craft@Hrs	Unit	Material $	Labor $	Total $

Closed Cell Elastomeric Pipe and Tubing Insulation. Semi split, by nominal pipe or tube diameter with insulation wall thickness as shown, no cover. R factor equals 3.58 at 220 degrees F. Manufactured by Rubatex. These costs do not include scaffolding, add for same, if required. Also see fittings, flanges and valves at the end of this section.

Description	Craft@Hrs	Unit	Material $	Labor $	Total $
1/4", 1/2" thick	P2@.039	LF	.31	.96	1.27
3/8", 1/2" thick	P2@.039	LF	.33	.96	1.29
1/2", 1/2" thick	P2@.039	LF	.37	.96	1.33
3/4", 1/2" thick	P2@.039	LF	.42	.96	1.38
1", 1/2" thick	P2@.042	LF	.47	1.03	1.50
1-1/4", 1/2" thick	P2@.042	LF	.53	1.03	1.56
1-1/2", 1/2" thick	P2@.042	LF	.62	1.03	1.65
2", 1/2" thick	P2@.046	LF	.78	1.13	1.91
2-1/2", 3/8" thick	P2@.056	LF	.62	1.38	2.00
2-1/2", 1/2" thick	P2@.056	LF	.83	1.38	2.21
2-1/2", 3/4" thick	P2@.056	LF	1.08	1.38	2.46
3", 3/8" thick	P2@.082	LF	.71	2.02	2.73
3", 1/2" thick	P2@.082	LF	.94	2.02	2.96
3", 3/4" thick	P2@.082	LF	1.22	2.02	3.24
4", 3/8" thick	P2@.095	LF	.83	2.34	3.17
4", 1/2" thick	P2@.095	LF	1.10	2.34	3.44
4", 3/4" thick	P2@.095	LF	1.43	2.34	3.77

Pipe fittings and flanges
For each fitting or flange use the cost for 3 LF of pipe or tube of the same size)

Valves
Body only (for each valve body use the cost for 5 LF of pipe or tube of the same size)
Body and bonnet or yoke (for each valve use the cost for 10 LF of pipe or tube of the same size)
Flanged valves (add the cost for insulation of flanges per above)

Fiberglass Pipe Insulation with AP-T Plus Jacket

Description	Craft@Hrs	Unit	Material $	Labor $	Total $

Fiberglass Pipe Insulation with AP-T Plus (all purpose self sealing) Jacket. By nominal pipe diameter. R factor equals 2.56 at 300 degrees F. These costs do not include scaffolding, add for same, if requiredAlso see fittings, flanges and valves at the end of this section.

1/2" diameter pipe

Description	Craft@Hrs	Unit	Material $	Labor $	Total $
1/2" thick	P2@.038	LF	.77	.94	1.71
1" thick	P2@.038	LF	.91	.94	1.85
1-1/2" thick	P2@.040	LF	1.86	.99	2.85

3/4" diameter pipe

Description	Craft@Hrs	Unit	Material $	Labor $	Total $
1/2" thick	P2@.038	LF	.84	.94	1.78
1" thick	P2@.038	LF	1.04	.94	1.98
1-1/2" thick	P2@.040	LF	1.93	.99	2.92

1" diameter pipe

Description	Craft@Hrs	Unit	Material $	Labor $	Total $
1/2" thick	P2@.040	LF	.90	.99	1.89
1" thick	P2@.040	LF	.91	.99	1.90
1-1/2" thick	P2@.042	LF	2.05	1.03	3.08

1-1/4" diameter pipe

Description	Craft@Hrs	Unit	Material $	Labor $	Total $
1/2" thick	P2@.040	LF	1.00	.99	1.99
1" thick	P2@.040	LF	1.21	.99	2.20
1-1/2" thick	P2@.042	LF	2.18	1.03	3.21

1-1/2" diameter pipe

Description	Craft@Hrs	Unit	Material $	Labor $	Total $
1/2" thick	P2@.042	LF	1.07	1.03	2.10
1" thick	P2@.042	LF	1.36	1.03	2.39
1-1/2" thick	P2@.044	LF	2.35	1.08	3.43

2" diameter pipe

Description	Craft@Hrs	Unit	Material $	Labor $	Total $
1/2" thick	P2@.044	LF	1.14	1.08	2.22
1" thick	P2@.044	LF	1.56	1.08	2.64
1-1/2" thick	P2@.046	LF	2.52	1.13	3.65

2-1/2" diameter pipe

Description	Craft@Hrs	Unit	Material $	Labor $	Total $
1" thick	P2@.047	LF	.73	1.16	1.89
1-1/2" thick	P2@.048	LF	.85	1.18	2.03
2" thick	P2@.050	LF	.92	1.23	2.15

3" diameter pipe

Description	Craft@Hrs	Unit	Material $	Labor $	Total $
1" thick	P2@.051	LF	.84	1.26	2.10
1-1/2" thick	P2@.054	LF	.95	1.33	2.28
2" thick	P2@.056	LF	1.06	1.38	2.44

Description	Craft@Hrs	Unit	Material $	Labor $	Total $
4" diameter pipe					
1" thick	P2@.063	LF	1.04	1.55	2.59
1-1/2" thick	P2@.066	LF	1.18	1.63	2.81
2" thick	P2@.069	LF	1.34	1.70	3.04
6" diameter pipe					
1" thick	P2@.077	LF	1.25	1.90	3.15
1-1/2" thick	P2@.080	LF	1.45	1.97	3.42
2" thick	P2@.085	LF	1.63	2.09	3.72
8" diameter pipe					
1" thick	P2@.117	LF	1.81	2.88	4.69
1-1/2" thick	P2@.123	LF	2.08	3.03	5.11
2" thick	P2@.129	LF	2.36	3.18	5.54
10" diameter pipe					
1" thick	P2@.140	LF	2.16	3.45	5.61
1-1/2" thick	P2@.147	LF	2.53	3.62	6.15
2" thick	P2@.154	LF	2.35	3.79	6.14
12" diameter pipe					
1" thick	P2@.155	LF	2.59	3.82	6.41
1-1/2" thick	P2@.163	LF	2.98	4.02	7.00
2" thick	P2@.171	LF	3.36	4.21	7.57
Add for .016" aluminum jacket					
Per SF of surface	P2@.018	SF	.19	.44	.63

Pipe fittings and flanges

For each fitting or flange use the cost for 3 LF of pipe of the same pipe size

Valves

Body only

for each valve body use the cost for 5 LF of pipe of the same pipe size

Body and bonnet or yoke

for each valve use the cost for 10 LF of pipe of the same pipe size

Flanged valves

add the cost for of flanges per above

Calcium Silicate Pipe Insulation with 0.16" Aluminum Jacket

Description	Craft@Hrs	Unit	Material $	Labor $	Total $

Calcium Silicate Pipe Insulation with .016" Aluminum Jacket. By nominal pipe diameter. R factor equals 2.20 at 300 degrees F. Manufactured by Pabco. These costs do not include scaffolding, add for same, if required. Also see fittings, flanges and valves at the end of this section.

6" diameter pipe

Description	Craft@Hrs	Unit	Material $	Labor $	Total $
2" thick	P2@.183	LF	7.60	4.51	12.11
4" thick	P2@.218	LF	8.70	5.37	14.07
6" thick	P2@.288	LF	10.60	7.10	17.70

8" diameter pipe

Description	Craft@Hrs	Unit	Material $	Labor $	Total $
2" thick	P2@.194	LF	8.25	4.78	13.03
4" thick	P2@.228	LF	10.40	5.62	16.02
6" thick	P2@.308	LF	12.40	7.59	19.99

10" diameter pipe

Description	Craft@Hrs	Unit	Material $	Labor $	Total $
2" thick	P2@.202	LF	9.75	4.98	14.73
4" thick	P2@.239	LF	12.80	5.89	18.69
6" thick	P2@.337	LF	16.00	8.30	24.30

12" diameter pipe

Description	Craft@Hrs	Unit	Material $	Labor $	Total $
2" thick	P2@.208	LF	11.10	5.13	16.23
4" thick	P2@.261	LF	14.50	6.43	20.93
6" thick	P2@.388	LF	19.10	9.56	28.66

Pipe fittings and flanges
  For each fitting or flange use the cost for 3 LF of pipe of the same pipe size

Valves
  Body only
    for each valve body use the cost for 5 LF of pipe of the same pipe size)
  Body and bonnet or yoke
    for each valve use the cost for 10 LF of pipe of the same pipe size
  Flanged valves
    add the cost for insulation of flanges per above

Description	Craft@Hrs	Unit	Material $	Labor $	Total $

Fiberglass blanket duct insulation

Description	Craft@Hrs	Unit	Material $	Labor $	Total $
1"	S2@.020	SF	.60	.51	1.11
1½"	S2@.023	SF	.80	.58	1.38
2"	S2@.025	SF	1.00	.63	1.63

Fiberglass board duct insulation

Description	Craft@Hrs	Unit	Material $	Labor $	Total $
1"	S2@.067	SF	.90	1.70	2.60

Thermal fiberglass board duct insulation vapor barrier

Description	Craft@Hrs	Unit	Material $	Labor $	Total $
1"	S2@.063	SF	.70	1.60	2.30

Thermal fiberglass blanket duct lining

Description	Craft@Hrs	Unit	Material $	Labor $	Total $
1/2"	S2@.032	SF	.65	.81	1.46
3/4"	S2@.035	SF	.75	.89	1.64

Note: Material prices include all necessary tape, pins, mastic and 15% waste.

Trenching

Because of varying soil compositions and densities, trenching costs can sometimes be difficult to accurately estimate. The tables in this section are based on average soil conditions, which are loosely defined as dirt, soft clay or gravel mixed with rocks. The costs for trenching through limestone, sandstone or shale can increase by as much as 1,000 percent. For these conditions, it is best to consult with experts or to subcontract the work to specialists.

The primary consideration in excavations is to ensure the safety of the pipe layers. Current OSHA statutes require all excavations over 5 feet deep to be sloped, shored, sheeted or braced to prevent cave-ins. The use of shoring minimizes trenching costs by eliminating the need to slope the trench walls, but increases pipe-laying costs because of the difficulty in working between cross-bracing.

The estimator should carefully compare the costs for installing and removing shoring, and reduced pipe installation productivity, against the additional excavation costs for sloping the walls of the trench. For average soil conditions, not exceeding 6 to 8 feet in depth, it is usually more economical to slope the trench walls.

The current average cost for a backhoe with operator is $67.50 per hour. Some trenching contractors additionally charge $25.00 move-on and $25.00 move-off costs, particularly for small projects.

Total trenching, backfill and compaction costs are calculated by multiplying the backhoe hours from the following charts by $67.50 per hour, and adding move-on and move-off charges, if applicable.

Backhoe Trenching Trench walls at a 90 degree angle, based on average soil, including trenching, backfill and compaction. Hours per 100 linear feet of trench						
Trench depth (feet)	Trench width (feet)					
	1	2	3	4	5	6
2	2.6	5.1	7.7	10.2	12.8	15.3
3	3.8	7.7	11.5	15.3	19.2	23.0
4	5.1	10.2	15.3	20.4	25.6	30.7
5	6.4	12.8	19.2	25.6	32.0	38.3
6	7.7	15.3	23.0	30.7	38.3	46.0
7	8.9	17.9	26.8	35.8	44.7	53.7
8	10.2	20.4	30.7	40.9	51.1	61.3
9	11.5	23.0	34.5	46.0	57.5	69.0
10	12.8	25.5	38.3	51.1	63.9	76.6

Notes:
1) Compaction is based on the use of a compactor wheel.
2) Additional labor costs must be added if watering-in, stone removal or excess spoil removal is required.

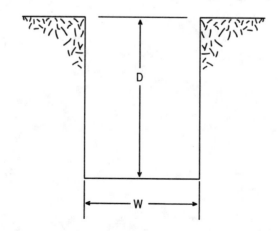

Backhoe Trenching Trench walls at a 45 degree angle, based on average soil including trenching, backfill and compaction. Hours per 100 linear feet of trench						
Trench depth (feet)	**Trench width (feet)**					
	1	**2**	**3**	**4**	**5**	**6**
2	7.6	10.3	12.6	15.5	17.9	20.3
3	15.2	22.6	22.6	26.6	30.7	34.5
4	25.5	30.7	35.5	40.7	45.9	51.0
5	38.3	44.5	51.0	57.6	63.8	70.0
6	53.4	61.0	69.0	76.6	84.1	91.7
7	71.4	80.3	89.3	98.3	107.2	115.9
8	92.1	102.1	112.4	122.4	132.8	143.1
9	114.8	126.6	137.9	149.3	161.0	172.4
10	140.3	153.1	165.9	178.6	191.4	204.5

Notes:

1) Compaction is based on the use of a compactor wheel.
2) Additional labor costs must be added if watering-in, stone removal or excess spoil removal is required.

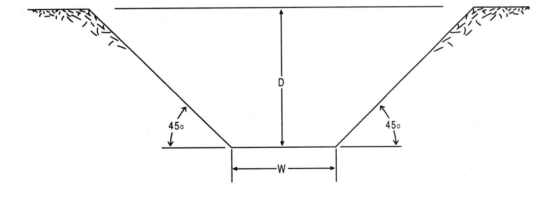

Equipment Rental

Equipment Rental			
Description	**Per day ($)**	**Per week ($)**	**Per month ($)**
Air compressor, gas or diesel, 100-125 CFM with 100' hose, air tool and 4 bits	42.00	260.00	1,190.00
Air compressor, gas or diesel, 165-185 CFM with 100' hose, air tool and 4 bits	50.00	284.00	1,260.00
Air compressor, gas or diesel, 100-125 CFM, no accessories	60.00	193.00	554.00
Air compressor, gas or diesel, 165-185 CFM, no accessories	85.00	261.00	750.00
Backhoe with bucket, 55 HP	219.00	727.00	2,070.00
Backhoe with bucket, 100 HP	362.00	1,150.00	3,360.00
Backhoe compaction wheel, 60 HP	153.00	231.00	772.00
Bender, hydraulic	35.00	105.00	320.00
Bevel machine with torch	15.00	45.00	135.00
Boom lift, 31' - 40'	245.00	690.00	1,960.00
Boom lift, 51' - 60'	375.00	1,090.00	3,120.00
Chain hoist, to 9.9 tons	31.00	85.00	200.00
Chain-hoist, 10 tons & over	126.00	335.00	888.00
Come-along, 1/2-ton	10.00	20.00	60.00
Come-along, 1-ton	12.00	40.00	85.00
Come-along, 4-ton	24.00	60.00	145.00
Compactor, soil, gas or diesel, 21" x 24" plate	65.00	206.00	588.00
Crane, truck-mounted, 6-ton capacity	269.00	800.00	2,380.00
Crane, truck-mounted, 10-ton capacity	357.00	1,070.00	3,110.00
Forklift, 2,000 pound capacity	150.00	450.00	1,300.00
Forklift, 4,000 pound capacity	184.00	554.00	1,620.00
Forklift, 8,000 pound capacity	229.00	706.00	2,090.00
Front-end loader, 1½ yard capacity	270.00	851.00	2,500.00
Front-end loader, 2 yard capacity	315.00	1,070.00	2,940.00
Front-end loader, 3 yard capacity	435.00	1,410.00	4,120.00
Pipe machine, up to 2" capacity	30.00	110.00	325.00
Pipe machine, 2½" to 4" capacity	50.00	175.00	535.00
Roustabout, small	26.00	94.00	285.00
Roustabout, medium	37.00	125.00	375.00
Roustabout, large	42.00	155.00	460.00
Scissors-lift, to 20'	103.00	285.00	778.00
Scissors-lift, 21' - 30'	135.00	385.00	1,200.00
Skip-loader	150.00	625.00	1,920.00
Trailer, office, 25'	---	---	150.00
Truck, 3/4 ton pickup	55.00	285.00	800.00
Truck, 1 ton pickup	60.00	295.00	820.00
Truck, 1 ton stakebed	70.00	315.00	850.00
Van, storage, 26'	---	---	225.00
Van, storage, 40'	---	---	275.00
Welding machine, gas, 200 amp, trailer-mounted with 50' bare lead	51.00	161.00	442.00
Welding machine, diesel, 400 amp, trailer-mounted with 50' bare lead	60.00	195.00	550.00

Description	Craft@Hrs	Unit	Material $	Labor $	Total $
1½" brass valve tags with black filled numbers					
--	P2@.150	Ea	1.50	3.70	5.20
1½" blank brass valve tags					
--	P2@.150	Ea	.75	3.70	4.45
1½" plastic valve tags with engraved numbers					
--	P2@.150	Ea	1.25	3.70	4.95
Engraved plastic equipment nameplates					
--	P2@.400	Ea	3.25	9.86	13.11
Pipe and duct markers					
--	P2@.250	Ea	1.85	6.16	8.01
As-built drawings					
--	P2@.500	Ea	--	12.30	12.30
Instructing owner's operating personnel					
--	P2@8.00	Ea	--	197.00	197.00

Budget Estimates for Plumbing and HVAC Work

General contractors and owners often ask mechanical subcontractors to prepare budget cost estimates for use in planning projects. A budget estimate isn't a bid for the contract and you probably won't be paid for preparing the estimate. That's why you won't want to spend too much time on these estimates. Accuracy within 8 to 10 percent of the final bid price will usually be acceptable. Normally you'll have only preliminary architectural drawings and possibly a guide specification to work from. The tables that follow will help you prepare budget estimates for nearly any plumbing or HVAC job.

The costs in these tables are averages and will vary considerably with design requirements and by geographic area. Make it clear to the owner or architect requesting the figures that these are budget estimates and will be revised when final plans and specifications are available.

Budget Estimates for Plumbing: First, count the number of plumbing fixtures. Multiply the number of fixtures of each type by the price per fixture shown in the table below. These prices include all the usual labor and material for a complete average plumbing installation. Prices are based on branch piping lengths averaging 35 feet between the fixture and the main. Typical main piping is also included in the prices.

Description	Craft@Hrs	Unit	Material $	Labor $	Total $
Budget plumbing estimate, cast iron DWV, copper supply pipe					
Bathtubs	P2@15.0	Ea	1,050.00	370.00	1,420.00
Showers	P2@9.00	Ea	575.00	222.00	797.00
Lavatories	P2@11.0	Ea	990.00	271.00	1,261.00
Kitchen sinks	P2@12.0	Ea	1,030.00	296.00	1,326.00
Service sinks	P2@10.0	Ea	895.00	246.00	1,141.00
Bar sinks	P2@9.00	Ea	750.00	222.00	972.00
Floor sinks	P2@7.00	Ea	465.00	172.00	637.00
Water closets					
Tank-type	P2@11.0	Ea	750.00	271.00	1,021.00
Flush valve	P2@12.0	Ea	850.00	296.00	1,146.00
Urinals	P2@8.00	Ea	1,080.00	197.00	1,277.00
Drinking fountains					
(refrigerated)	P2@12.0	Ea	1,250.00	296.00	1,546.00
Wash fountains	P2@25.0	Ea	2,700.00	616.00	3,316.00
Can washers	P2@18.0	Ea	1,370.00	444.00	1,814.00
Floor drains	P2@8.00	Ea	425.00	197.00	622.00
Area drains	P2@6.00	Ea	365.00	148.00	513.00
Roof drains	P2@18.0	Ea	960.00	444.00	1,404.00
Overflow drains	P2@18.0	Ea	960.00	444.00	1,404.00
Deck drains	P2@4.00	Ea	300.00	98.60	398.60
Cleanouts	P2@5.00	Ea	325.00	123.00	448.00
Trap primers	P2@2.00	Ea	165.00	49.30	214.30
Sump pumps	P2@6.00	Ea	1,600.00	148.00	1,748.00
Water heaters					
Gas, 40 gal.	P2@16.0	Ea	1,900.00	394.00	2,294.00
Gas, 80 gal.	P2@17.0	Ea	2,100.00	419.00	2,519.00
Gas, 120 gal.	P2@18.0	Ea	2,450.00	444.00	2,894.00

Description	Craft@Hrs Unit		Material $	Labor $	Total $

Budget plumbing estimate, plastic DWV and plastic supply pipe

Description	Craft@Hrs	Unit	Material $	Labor $	Total $
Bathtubs	P2@12.0	Ea	950.00	296.00	1,246.00
Showers	P2@7.00	Ea	460.00	172.00	632.00
Lavatories	P2@9.00	Ea	890.00	222.00	1,112.00
Kitchen sinks	P2@10.0	Ea	980.00	246.00	1,226.00
Service sinks	P2@8.00	Ea	790.00	197.00	987.00
Bar sinks	P2@7.00	Ea	660.00	172.00	832.00
Floor sinks	P2@6.00	Ea	400.00	148.00	548.00
Water closets					
Tank-type	P2@8.00	Ea	620.00	197.00	817.00
Flush valve	P2@9.00	Ea	740.00	222.00	962.00
Urinals	P2@6.00	Ea	1,000.00	148.00	1,148.00
Drinking fountains					
(refrigerated)	P2@10.0	Ea	1,150.00	246.00	1,396.00
Wash fountains	P2@22.0	Ea	2,650.00	542.00	3,192.00
Can washers	P2@15.0	Ea	1,300.00	370.00	1,670.00
Floor drains	P2@7.00	Ea	330.00	172.00	502.00
Area drains	P2@5.00	Ea	270.00	123.00	393.00
Roof drains	P2@16.0	Ea	920.00	394.00	1,314.00
Overflow drains	P2@16.0	Ea	920.00	394.00	1,314.00
Deck drains	P2@3.00	Ea	230.00	73.90	303.90
Cleanouts	P2@4.00	Ea	280.00	98.60	378.60
Trap primers	P2@2.00	Ea	140.00	49.30	189.30
Sump pumps	P2@5.00	Ea	1,600.00	123.00	1,723.00
Water heaters					
Gas, 40 gal.	P2@15.0	Ea	1,950.00	370.00	2,320.00
Gas, 80 gal.	P2@16.0	Ea	2,150.00	394.00	2,544.00
Gas, 120 gal.	P2@17.0	Ea	2,500.00	419.00	2,919.00

Budget Estimating

Budget Estimates for HVAC Work: Budget estimates for heating, ventilating and air conditioning work are figured by the square foot of floor. Multiply the area of conditioned space by the cost per square foot in the table that follows. The result is the estimated cost including a complete heating, ventilating and air conditioning system and 20 percent overhead and profit. The table also shows average heating and cooling requirements for various types of buildings.

Budget HVAC estimate, *subcontract basis*

Description	Cooling (Btu/SF)	Heating (Btu/SF)	Total Selling Price ($) Per SF	Per Ton
Apartments and condominiums	25	25	4.90	2,350.00
Auditoriums	40	40	12.50	3,750.00
Banks	45	30	13.80	3,680.00
Barber shops	40	40	12.90	3,870.00
Beauty shops	60	30	15.20	3,040.00
Bowling alleys	40	40	10.00	3,000.00
Churches	35	35	10.80	3,700.00
Classrooms	45	40	19.60	5,230.00
Cocktail lounges	70	30	13.70	2,350.00
Computer rooms	145	20	27.90	2,300.00
Department stores	35	30	8.80	3,010.00
Laboratories	60	45	24.30	4,860.00
Libraries	45	35	14.50	3,860.00
Manufacturing plants	40	35	5.60	1,690.00
Medical buildings	35	35	9.60	3,290.00
Motels	30	30	6.70	2,680.00
Museums	35	40	10.70	3,670.00
Nursing homes	45	35	12.40	3,310.00
Office buildings, low-rise	35	35	9.80	3,350.00
Office buildings, high-rise	40	30	13.20	3,970.00
Residences	20	30	3.00	1,800.00
Retail shops	45	30	10.00	2,660.00
Supermarkets	30	30	5.20	2,080.00
Theaters	40	40	9.80	2,950.00

The following pages contain business forms and letters most plumbing and HVAC contractors can use.

- **Change Orders**

  Change Estimate Takeoff

  Change Estimate Worksheet

  Change Estimate Summary

  Change Order Log

- **Subcontract Forms**

  Standard Form Subcontract

  Subcontract Change Order

- **Purchase Order**

- **Construction Schedule**

- **Letter of Intent**

- **Submittal Data Cover Sheet**

- **Submittal Index**

- **Billing Breakdown Worksheet**

- **Monthly Progress Billing**

- **Warranty Letter**

An explanation is provided on how the forms function in the contractor's office, how to fill them out, and a sample provided. Where appropriate, a blank copy of the form is provided for your use.

Permission is hereby given by the publisher for the duplication of any of the blank forms on a copy machine or at a printer, provided they are for your own use and not to be sold at a cost exceeding the actual printing and paper cost.

Space has been provided at the top of the forms so you can add your own letterhead or logo.

Change Estimates

Very few projects are completed without some changes to the original plans. Most changes in the scope of work are a result of either errors by the designer or requests by the owner.

Here's the usual sequence in handling changes.

1) The project architect usually initiates a change by sending a Request for Quotation (R.F.Q.) to the general contractor with a detailed description of the work to be modified.

2) The general contractor prepares a Change Estimate (C.E.) if the work is to be done by the general

3) The subcontractor prepares a change estimate which shows the work to be added to or deleted from the contract and the cost of changes, including markup. The general contractor reviews the change estimate and forwards a summary to the architect for decision.

4) If the decision is to go ahead with the change, the architect issues a Change Order (C.O.) to the general contractor. This change order is authorization to proceed with the change at the price quoted.

Your costs for making changes will normally be higher than your costs for similar items on the original estimate. Here's why:

- Equipment and materials for changes are almost always purchased in smaller quantities. Volume discounts are seldom available.

- Special handling and delivery of small orders will further increase equipment and material costs.

- Scheduled work will be disrupted.

- Overhead costs will be higher because special purchase orders or subcontracts will be required.

- Vendor restocking charges for deleted items will range from 20 to 50 percent of the original cost.

- Large change orders may require additional manpower, which usually results in decreased efficiency.

- The cumulative effect of change orders will be to increase overhead costs if the project completion date is extended.

To compensate for these higher costs, most estimators use a higher markup on change estimates:

- Add an extra 10 percent markup to equipment and material costs to cover increased over-head and handling costs, prior to adding final Overhead and Profit markup.

- Increase competitive labor units (manhours required to perform a specific task) by 15 percent to compensate for inefficiencies such as disruptions in the work schedule, extra handling of equipment and material, and additional manpower.

- Mark up subcontractor prices by 10 percent.

- Figure overhead and profit markup at 15/10 or 10/10.

By following these suggestions, the mathematical gross profit will appear to be 40 to 45 percent, but the true gross will be closer to 25 percent because your costs for change order work are higher, as listed above.

Change Estimate Example

Let's assume that the general contractor on your job has sent a R.F.Q. to your attention. The extra work is a run of hot water piping that will connect to equipment being furnished and set by others. The first step is to list all material and labor required to do the work. Notice the items entered on the Change Estimate Take-off form on page 304.

When the take-off is complete, transfer the manhour total to the Change Estimate Worksheet on page 305. To complete this form:

1) Multiply manhours by the cost per hour.

2) Add the cost of supervision.

3) Total direct labor cost.

4) Add the cost of preparing the change order estimate (C.E.).

5) Add equipment and tool costs.

6) Add subcontract costs.

Then bring totals forward to the Change Estimate Summary on page 306:

1) Enter the material cost total from the Change Estimate Take-off form on page 304.

2) Enter direct and indirect labor cost totals from the Change Estimate Worksheet.

3) Enter equipment, tool and subcontract cost totals from the Change Estimate Worksheet.

Notice the additional bond charge on line M of the Change Estimate Summary. This is the extra premium charged by the bonding company to cover the increase in contract value. The next line, *Service Reserve,* is the allowance for service work (if any is required) during the warranty period.

A sample Change Order Log, used to record the status of changes, is also included on page 307.

ACME

Mechanical Contractors

7600 Oak Avenue - Anytown, U.S.A. 12345-6789

Voice 1-234-5678 -- FAX 1-234-5680

Change Estimate Take-Off

Project Chancey Bank Building

C. E. Number One

Take-off by MA

Job No 7777

Date 2-3-97

Item	Equipment and Material Description	Quantity	Material $ Unit	Total	Labor Manhours Unit	Total
1	1" Type L hard copper pipe	40 LF	2.14	85.60	.038	1.52
2	1" 90 degree copper ell	6 Ea	2.11	12.66	.23	1.38
3	1" copper coupling	4 Ea	1.60	6.40	.24	.96
4	1" bronze ball valve	2 Ea	12.40	24.80	.36	.72
5	2" X 2" x 1" reducing copper tee	2 Ea	12.10	24.20	.40	.80
6	1" copper union	2 Ea	8.52	17.04	.26	.52
7	1" pipe hanger and inserts	8 Ea	2.20	17.60	.45	3.60
8	Miscellaneous material	1 Ea	LS	12.50	1.78	1.78
		Totals		200.80		9.50

ACME

Mechanical Contractors

7600 Oak Avenue - Anytown, U.S.A. 12345-6789

Voice 1-234-5678 -- FAX 1-234-5680

Change Estimate Worksheet

Estimated by MA

Date 2-3-97

C. E. Number One

Job No.: 7777

Direct Labor Expense (including benefits)

Plumber or pipefitter journeyman	9.50 manhours at $ 28.78 per hour =	$	270.18
Plumber or pipefitter supervisor	1.00 manhours at $ 38.78 per hour =	$	38.78
Sheet metal journeyman	_____ manhours at $ _____ per hour =	$	_____
Sheet metal supervisor	_____ manhours at $ _____ per hour =	$	_____
Laborer journeyman	_____ manhours at $ _____ per hour =	$	_____
	_____ manhours at $ _____ per hour =	$	_____
	Total Direct Labor Cost	$	308.96

Indirect Costs

Travel		$	_____
Subsistence		$	_____
Labor	Cost Estimate Preparation (1 Hour)	$	33.50
	Total Indirect Cost	$	33.50

Equipment and Tool Costs

	Ramset rental	$	23.50
	Propane	$	2.50
	Total Equipment and Tool Cost	$	26.00

Subcontract Costs

	Quikrap Insulation Co.	$	65.00
		$	_____
	Total Subcontract Cost	$	**65.00**

ACME
Mechanical Contractors

7600 Oak Avenue - Anytown, U.S.A. 12345-6789

Voice 1-234-5678 -- FAX 1-234-5680

Change Estimate Summary

Quotation to ABC General Contractors, Inc.
Address 380 First Street
Address _____
City, ST., ZIP Anytown, CA 63876
Attention: Mr. J. H. Smith
Job Name: Chancey Bank Building
Reference: Your R. F. O. No. 7 Labor and material to furnish and install
          additional hot water piping.

Date 2-3-97
Change Order Number One
Job Number 7777

_____
_____
_____

A.	Materials and equipment:	$ 200.80			
B.	Sales tax: 7%	$ 14.06			
C.	Direct labor:	$ 308.96			
D.	Indirect costs:	$ 33.50			
E.	Equipment and tools:	$ 26.00			
F.			**Subtotal:**	$ 583.32	
G.	Overhead at 15% of line F:	$ 87.50			
H.	Subcontracts:	$ 65.00			
I.	Overhead at 10% of line H:	$ 6.50			
J.			**Subtotal:**	$ 742.32	
K.	Profit at 10% of line J:	$ 74.23			
L.			**Subtotal:**	$ 816.55	
M.	Bond premium at 1% of line L:	$ 8.17			$ 8.17
N.	Service reserve at 0.5% of line L:	$ 4.08			$ 4.08
O.	**Total cost estimate, lines L thru N:**		Add Deduct		$ 828.80

P.   Exclusions from this estimate: Per Base Contract

Q.   This quotation is valid for 30 days.

R.   We require 1 days extension of the contract time.

S.   We are are not proceeding with this work per pending your authorization.

T.   Please forward your confirming change order and authorization to proceed.

Signed by: _____
                   Project Manager

Change Order Log

Job Name   Job Number

Change Number	Customer Number	Date	Amount	Date Approved	Amount Approved	Remarks

Change Estimate Take-Off

Project _____ C. E. Number _____

Take-off by _____ Job No. _____

Item	Description	Quantity	Material $ Unit	Material $ Total	Labor Manhours Unit	Labor Manhours Total

Change Estimate Worksheet

Estimated by _____   C. E. Number _____
Date _____   Job No.: _____

Direct Labor Expense (including benefits)

Plumber or pipefitter journeyman _____ manhours at $_____ per hour = $ _____
Plumber or pipefitter supervisor _____ manhours at $_____ per hour = $ _____
Sheet metal journeyman _____ manhours at $ ____ per hour = $ _____
Sheet metal supervisor _____ manhours at $_____ per hour = $ _____
Laborer journeyman _____ manhours at $_____ per hour = $ _____
_____ _____ manhours at $_____ per hour = $ _____

Total Direct Labor Cost $ _____

Indirect Costs

Travel _____ $ _____
Subsistence _____ $ _____
Labor _____ $ _____

Total Indirect Cost $ _____

Equipment and Tool Costs

_____ $ _____
_____ $ _____

Total Equipment and Tool Cost $ _____

Subcontract Costs

_____ $ _____
_____ $ _____

Total Subcontract Cost $ _____

Change Estimate Summary

Quotation to _____ Date _____
Address _____ Change Order Number _____
Address _____ Job Number _____
City, ST, ZIP _____ _____
Attention _____ _____
Job Name _____ _____
Reference _____ _____

_____

A. Materials and equipment: $ _____

B. Sales tax: $ _____

C. Direct labor: $ _____

D. Indirect costs: $ _____

E. Equipment and tools: $ _____

F.   **Subtotal:** .... $ _____

G. Overhead at _____% of line F: $ _____

H. Subcontracts: $ _____

I. Overhead at _____% of line H: $ _____

J.   **Subtotal:** .... $ _____

K. Profit at _____% of line J: $ _____

L.   **Subtotal:** .... $ _____

M. Bond premium at _____% of line L: $ _____

N. Service reserve at _____% of line L: $ _____

O. **Total cost estimate, lines L thru N:**   ☐ Add ☐ Deduct $ _____

P. Exclusions from this estimate: _____

_____

Q. ☐   This quotation is valid for _____ days.

R. ☐   We require _____ days extension of the contract time.

S. ☐   We are proceeding with this work per your authorization.

T. ☐   Please forward your confirming change order.

Signed by: _____
                        Project Manager

Another term for the general contractor is *prime contractor*. Subcontractors working directly for the prime contractor are called *second-tier subcontractors*. Contractors working for these second-tier subs are usually known as *third-tier subcontractors*. On a large job, many second-tier subs may hire third-tier subs to perform specialized work such as insulation, fireproofing, test and balance, water treatment, instrumentation and trenching.

If a portion of your work will be subcontracted to others, I recommend drawing up a subcontract agreement for that work. Any subcontractors who furnish both material and labor for a job should have a written contract.

Most larger plumbing, heating and A/C subcontractors use a two-part *Standard Form Subcontract* published by the Associated General Contractors of America. This two-part form is used to write the base subcontract. The A.G.C. also publishes a *Subcontract Change Order* form which is used when adding or deleting items after the base subcontract is signed. These forms are sold at modest cost at most A.G.C. offices.

The first section of page one of the Standard Form Subcontract identifies the two contractors entering into the agreement. The party issuing the subcontract is the *contractor*. Recipient of the agreement is the *subcontractor*.

Under *Recitals,* the contractor must affirm the right to issue a subcontract by declaring that a legal contract exists with the *owner* (general contractor).

Section 1 - Entire Contract describes the contract documents: the project plans, specifications and addenda. This list of documents should always be the same as the documents identified in the second-tier subcontractor's contract with the general contractor.

Section 2 - Scope describes the specific scope of work to be performed.

Section 3 - Contract Price. Complete this section by inserting the agreed-upon price for doing the work described.

Section 4 - Payment Schedule on page two should show the same payment terms as your contract with the general contractor.

Section 7 - Special Provisions is used to list any special agreements or exclusions.

When the contract is typed, do not sign it. Instead, forward it to the third-tier subcontractor with instructions to sign it and return it intact. When the signed contract is received, sign it and send a copy to the third-tier subcontractor.

Standard Form Subcontract

Subcontract No. <u>7777-1</u>

THIS AGREEMENT, made and entered into at Anytown, CA. this 3rd day of May, 19, 1997, by and between Acme Mechanical Contractors hereinafter called **CONTRACTOR,** with principal office at 7600 Oak Avenue, Anytown, California, and Quikrap Insulation Co., 320 Maple Blvd., Anytown, CA hereinafter called a **SUBCONTRACTOR.**

RECITALS

On or about the _____ 2nd _____ day of _____ January _____, 19 <u>97</u>, CONTRACTOR entered into a prime contract with

ABC General Contractors, Inc. _____ hereinafter called OWNER, whose address

is 380 First Street, Anytown, California

to perform the following construction work: The installation of all mechanical and plumbing systems for the proposed Mile-Hi office building to be located at 2600 Second, North, Anytown, California. Said work is to be performed in accordance with the prime contract and the plans and specifications. Said plans and specifications have been prepared by or on behalf of <u>Quik-Draw and Associates,</u> <u>Anytown, California</u>_____, ARCHITECT

SECTION 1 - ENTIRE CONTRACT

SUBCONTRACTOR certifies and agrees that he is fully familiar with all of the terms, conditions and obligations of the Contract Documents, as hereinafter defined, the location of the job site, and the conditions under which the work is to be performed, and that he enters into this Agreement based upon his investigation of all of such matters and is in no way relying upon any opinions or representations of CONTRACTOR. It is agreed that this Agreement represents the entire agreement. It is further agreed that the Contract Documents are incorporated in this Agreement by this reference, with the same force and effect as if the same were set forth at length herein, and that SUBCONTRACTOR and his subcontractors will be and are bound by any and all of said Contract Documents insofar as they relate in any part or in any way, directly or indirectly to the work covered by this Agreement. SUBCONTRACTOR agrees to be bound to CONTRACTOR in the same manner and to the same extent as CONTRACTOR is bound to OWNER under the Contract Documents, to the extent of the work provided for in this Agreement, and that where, in the Contract Documents, reference is made to CONTRACTOR and the work or specification therein pertains to SUBCONTRACTOR'S trade, craft, or type of work, then such work or specification shall be interpreted to apply to SUBCONTRACTOR instead of CONTRACTOR. The phrase "Contract Documents" is defined to mean and include:

> Drawing Nos. A-1 thru A-17, C-1 thru C-4, S-1 thru S-7, E-1 thru E-6 and M-1 thru M-7, all dated 10-4-1995, Specifications dated 10-2-95 and Addendum No. 1 dated 10-18-95.

SECTION 2 - SCOPE

SUBCONTRACTOR agrees to furnish all labor, services, materials, installation, cartage, hoisting, supplies, insurance, equipment, scaffolding, tools and other facilities of every kind and description required for the prompt and efficient execution of the work described herein and to perform the work necessary or incidental to complete <u>thermal insulation for all plumbing, HHW, CHW and</u> <u>condensate piping systems, including pumps P-1, P-2 and P-3</u> for the project in strict accordance with the Contract Documents and as more particularly, though not exclusively, specified in:

> Division 15, Section 400 entitled "Thermal Insulation systems" and General Conditions, as applicable.

SECTION 3 - CONTRACT PRICE

CONTRACTOR agrees to pay SUBCONTRACTOR for the strict performance of his work, the sum of:
<u>Forty-two thousand, three hundred dollars</u>_____($42,300.00_____), subject to additions and deductions for changes in the work as may be agreed upon, and to make payment in accordance with the Payment Schedule, Section 4.

SECTION 4 - PAYMENT SCHEDULE

CONTRACTOR agrees to pay SUBCONTRACTOR in monthly payments of ____90____ % of labor and materials which have been placed in position and for which payment has been made by OWNER to CONTRACTOR. The remaining ____10____% shall be retained by CONTRACTOR until he receives final payment from OWNER, but not less than thirty-five days after the entire work required by the prime contract has been fully completed in conformity with the Contract Documents and has been delivered and accepted by OWNER, ARCHITECT, and CONTRACTOR. Subject to the provisions of the next sentence, the retained percentage shall be paid SUBCONTRACTOR promptly after CONTRACTOR receives his final payment from OWNER. SUBCONTRACTOR agrees to furnish, if and when required by CONTRACTOR, payroll affidavits, receipts, vouchers, release of claims for labor, material and subcontractors performing work or furnishing materials under this Agreement, all in form satisfactory to CONTRACTOR, and it is agreed that no payment hereunder shall be made, except at CONTRACTOR'S option, until and unless such payroll affidavits, receipts, vouchers or release; or any or all of them, have been furnished. And payment made hereunder prior to completion and acceptance of the work, as referred to above, shall not be construed as evidence of acceptance of any part of SUBCONTRACTOR'S work.

REVISED 5-81   Published by AGC San Diego
Page 1 of 2

SECTION 5 - GENERAL SUBCONTRACT PROVISIONS

General Subcontract Provisions on back of Pages 1 and 2 are an integral part of this Agreement.

SECTION 6 - GENERAL PROVISIONS

1. *SUBCONTRACTOR* agrees to begin work as soon as instructed by the CONTRACTOR, and shall carry on said work promptly, efficiently and at a speed that will not cause delay in the progress of the CONTRACTOR'S work or work of other subcontractors. If, in the opinion of the CONTRACTOR, the SUBCONTRACTOR falls behind in the progress of the work, the CONTRACTOR may direct the SUBCONTRACTOR to take such steps as the CONTRACTOR deems necessary to improve the rate of progress, including, without limitation, requiring the SUBCONTRACTOR to increase the number of shifts, personnel, overtime operations, days of work, equipment, amount of plant, or other remedies and to submit to CONTRACTOR for CONTRACTOR'S approval an outline schedule demonstrating the manner in which the required rate of progress will be regained, without additional cost to the CONTRACTOR. CONTRACTOR may require SUBCONTRACTOR to prosecute, in preference to other parts of the work, such part or parts of the work as CONTRACTOR may specify.

The SUBCONTRACTOR shall complete the work as required by the progress schedule prepared by the CONTRACTOR, which may be amended from time to time. The progress schedule may be reviewed in the office of the CONTRACTOR and sequence of construction will be as directed by the CONTRACTOR.

The SUBCONTRACTOR agrees to have an acceptable representative (an officer of SUBCONTRACTOR if requested by the CONTRACTOR) present at all job meetings and to submit weekly progress reports in writing if requested by the CONTRACTOR. Any job progress schedules are hereby made a part of and incorporated herein by reference.

2. *Reserved Gate Usage*

SUBCONTRACTOR shall notify in writing, and assign its employees, material men and suppliers, to such gates or entrances as may be established for their use by CONTRACTOR and in accordance with such conditions and at such times as may be imposed by CONTRACTOR. Strict compliance with CONTRACTOR'S gate usage procedures shall be required by the SUBCONTRACTOR who shall be responsible for such gate usage by its employees, material men, suppliers, subcontractors, and their material men and suppliers.

3. *Staggered Days and Hours of Work and for Deliveries*

SUBCONTRACTOR shall schedule the work and the presence of its employees at the jobsite and any deliveries of supplies or materials by its material men and suppliers to the jobsite on such days, and at such times and during such hours, as may be directed by CONTRACTOR. SUBCONTRACTOR shall assume responsibility for such schedule compliance not only for its employees but for all its material men, suppliers and subcontractors, and their material men and suppliers.

SECTION 7 - SPECIAL PROVISIONS

```
        1. Pipe hanger inserts will be furnished and installed by Contractor.
        2. All trash pickup by Subcontractor; haul-away will be by General Contractor.
```

Contractors are required by law to be licensed and regulated by the Contractors' State License Board. Any questions concerning a contractor may be referred to the registrar of the board whose address is:

Contractors' State License Board -- P. O. Box 26000, Sacramento, California 95826

IN WITNESS WHEREOF: The parties hereto have executed this Agreement for themselves, their heirs, executors, successors, administrators, and assignees on the day and year first above written.

SUBCONTRACTOR	CONTRACTOR
Quikrap Insulation Company	Acme Mechanical Contractors
By _____	By _____
Name Title	Name Title
☐ Corporation ☐ Partnership ☐ Proprietorship	
(Seal)	
Contractor's State License No. _____	Contractor's State License No. _____

Revised 5-81   Published by AGC San Diego
Page 2 of 2

Subcontract Change Order

Number   One

Date: 6-11-97		Subcontract Number: 7777-1
To: Quikrap Insulation Co.		Our Job Number: 7777
320 Maple Blvd.		Our Proposal Number:
Anytown, California		Architect's C.O. Number:
		Effective Date of Change: 6-17-97

Subject to all the provisions of this Change Order, you are hereby directed to make the following change(s)

Delete HHW and CHW piping insulation located above Room No. 322

per ABC General Contractor's Change Order No. 18, dated 6-5-97.

The following change(s) will alter the price provided in your subcontract by the deduction *of $* 465.00

*Surety Consent Date:_____	Adjusted Subcontract Price Through C.O. Number: 0	$ 42,300.00
This change is approved		
Name _____	Amount this C.O. Number:	$ (465.00)
By: _____		
Title: _____	Current Adjusted Subcontract Price:	$ 41,835.00

When this Change Order is signed by both parties (and by Subcontractor's surety if subcontract is bonded), it constitutes their agreement:

(A) That the subcontract price is adjusted as shown above and that no further adjustment in that price by reason of the change(s) provided herein shall be made; and (B) That all the terms and conditions of the subcontract, except as modified by this and any previous changes, shall remain in full force and effect and apply to the work as so changed.

Accepted and Agreed: Date _____	
By:	**By:**
Subcontractor	Authorized Signature - Contractor
Title:	**Title:**

*Subcontractor: Sign pink copy and return immediately. If subcontract is bonded, obtain consent of surety endorsed thereon. No payments on account of this Change Order will be made until you have complied with the foregoing.

Published by San Diego Chapter A.G.C.

A purchase order is a legal document used to order equipment and materials from a vendor, with the implied promise to pay for these goods within a specified time after receipt of the items. Purchase orders are sometimes used to write subcontracts whose value does not exceed one or two thousand dollars. Refer to the sample purchase order on page 316.

While most of the purchase order is self-explanatory, certain items need to be emphasized:

"Description" should include all pertinent data such as model numbers, arrangements, performance requirements, applicable standards, electrical characteristics and so on.

"Tag" is used to identify a particular piece of equipment to facilitate unloading it at the proper location at the jobsite. Tag numbers for all major equipment can usually be found on the contract drawings.

"48 hour delivery notice": This can be very important if the material or equipment being delivered is too heavy to be unloaded by hand. Time will be required to arrange for a crane or forklift to unload the truck.

"Sales tax" is the current state sales tax which must be included unless the goods are to be delivered and installed at a project located out-of-state.

"Terms": The usual payment terms are "Net 30" which means the entire amount of the purchase order must be paid within thirty days after receipt of all goods, undamaged. If items are received damaged, payment should be withheld until satisfactory repair or replacement is made.

ACME

Mechanical Contractors

7600 Oak Avenue - Anytown, U.S.A. 12345-6789
Voice 123-456-7890 -- Fax 123-456-0987

P. O. Number: __7777-36_____
Job: _Chancey Bank Building
Job No: 7777   Date: 1-6-97

Purchase Order

Seller: Anytown Industrial Sales Co.	Ship to: Acme Mechanical Contractors
3657 Third Avenue Northwest	c/o Chancey Bank Building
Anytown, CA 63876	465 Commercial Street
	Anytown, CA 63876

By: M. D. Kiley

Attn: Mr. L. H. Seller

Delivery date: 2-28-97   ☐ 48 Hr. Delivery Notice Required: Yes   Payment terms: Net 30

Quantity	Description	Unit	Total
Two	Big-Blo Model No. 1JN 86-13 centrifugal fans	$443.00	$886.00
	with 480/3/60 one-half H.P., O.D.P. motor.		
	3250 CFM @ 0.50 T.S.P.		
	Tag one EF-13 and one EF-23		

Number of copies of operation and maintenance manuals required 2	Total before tax		$886.00
	Sales Tax 7%		$62.02
	Freight		Included
All shipments are to be F.O.B. jobsite or shop unless otherwise noted.	Total order		$948.02
	Buyer: M. D. Kiley		

Acceptance by seller: Date:

Sample Purchase Order

The general contractor's construction schedule may range from a simple bar chart to a complicated computer-generated CPM (Critical Path Method) diagram. More complex jobs require more complete and detailed schedules. On larger jobs, lower-tier subcontractors may be required to prepare a bar chart showing their construction schedule.

If you have to prepare a construction schedule, make conservative estimates of the time required to complete each part of the job. A slower schedule makes it possible to use smaller crews which are usually more efficient.

You can't schedule your own work until the general contractor has supplied a schedule for the balance of the project. When you have that schedule, prepare a list of tasks your crews will perform. Show this list to the general contractor to be certain there are no conflicts with the master schedule.

When the list of tasks to be performed has been approved, begin recording the manhours required for each task. Take these manhours from your estimate. When the duration for each task has been determined, figure the crew sizes required.

For example, suppose the estimate shows that it will take 1,040 manhours to install the underground plumbing for a project. The work must be completed in two months. Assuming one worker averages 173 hours of work in one month, how many workers will be required to meet the schedule? Here's the solution:

$$\frac{1{,}040 \text{ manhours}}{173 \text{ MH/Mo. x 2 months}} = 3 \text{ workers}$$

The first of the two bar charts on the following pages shows the schedule a general contractor might expect to receive from a plumbing and HVAC subcontractor. The chart on page 318 is intended for use by the subcontractor. It includes information a mechanical subcontractor needs to determine proper crew sizes for each task. The *Cumulative Manhours* line shows the budgeted manhours for each month of the project. If labor costs are to be kept within budget, actual hours should not exceed estimated hours. Monitor labor costs each month by comparing the line *Actual MH Used* (actual manhours used) with the line *Cumulative MH* (estimated cumulative manhours).

Construction Schedule

Company: Acme Mechanical Contractors   **Project:** Mile-Hi Office Bldg   **Job Number:** 7777   **Date:** 6-8-97

Activity	1996 Jul	Aug	Sep	Oct	Nov	Dec	1997 Jan	Feb	Mar	Apr	May	Crew Sizes
Submittals	▓											
Purchasing		▓	▓									
U/G Plumbing			▓	▓								
A/G Plumbing				▓	▓	▓	▓	▓				
Fin. Plumbing								▓	▓			
HVAC Piping					▓	▓	▓	▓				
HVAC Ducting						▓	▓	▓	▓			
Equipment									▓			
Insulation									▓			
Temp. Controls					▓	▓	▓	▓	▓			
Start-up										▓		

Construction Scheduling Form

318

Construction Schedule

Company: Acme Mechanical Cont. **Project:** Mile-Hi Office Building   **Job No:** 7777   **Date:** 6-8-97

Activity	1996 Jul	Aug	Sep	Oct	Nov	Dec	1997 Jan	Feb	Mar	Apr	May	CREW SIZES
Submittals	█											N/A
Purchasing		█	█									N/A
U/G Plumbing			█	█								1040 MH ÷ (173 x 2) = 3
A/G Plumbing				█	█	█	█					2075 MH ÷ (173 x 4) = 3
Fin. Plumbing								█				130 MH ÷ (173 x .5) = 1.5
HVAC Piping					█	█	█	█				2790 MH ÷ (173 x 4) = 4
HVAC Ducting					█	█		█				2760 MH ÷ (173 x 4) = 4
Equipment								█	█			690 MH ÷ (173 x 2) = 2
Insulation								█	█			Subcontract
Temp. Controls					█	█	█	█	█			Subcontact
Start-up										█		85 MH ÷ (173 x 2) = 2
Crew size/month			3	6	11	11	11	10.75	2	.5		
Cumulative MH*			520	1559	3466	5373	7280	9143	0499	0583		MH estimated = 9570
Actual MH used												
*Per estimate												

Construction Scheduling Form

Letter of Intent

On most larger projects the contractor is required to submit drawings and technical data to the owner's representative for approval before work actually begins. Most of these submittals will be prepared by your suppliers and third-tier subs. Vendors and subs may be reluctant to prepare these documents without some assurance that they've been selected to do the work. That's the purpose of a letter of intent. It notifies a proposed vendor or subcontractor that you plan to contract with them when the owner approves the information submitted.

A letter of intent isn't a contract. It's an expressed intention to make a contract, which isn't a contract at all. But a vendor who gets your letter of intent and supplies all the information requested should get the contract. If you place the order with a different vendor, obviously, your letter of intent was worthless. That first vendor may be very reluctant to cooperate in the future.

A sample letter of intent follows this section.

ACME

Mechanical Contractors

7600 Oak Avenue - Anytown, U.S.A. 12345-6789
Voice 123-456-7890 -- Fax 123-456-0987

December 20, 1996

Quikrap Insulation Company
320 Maple Boulevard
Anytown, U.S.A. 12345-6789

Attn: Mr. R. S. Quikrap

Dear Mr. Quikrap:

Acme Mechanical Contractors has been issued a subcontract to furnish and install the mechanical and plumbing systems for the proposed Mile-Hi Office Building to be located at 2600 Second Street in Anytown.

It is my intention to award your firm a subcontract to provide the thermal insulation for these systems in accordance with your proposal dated 12-13-96 in the amount of $42,300.00.

Please prepare ten copies of your submittal data and forward them to me no later than January 15, 1997. Upon approval of your submittal, I will mail you the subcontract for your review and signature.

Thank you,

M. D. Kiley
Project Manager

CC: Project file

Letter of Intent

Submittal Data

The construction specifications may require that you submit manufacturer's technical data on certain equipment you plan to install. You may have to submit six to ten copies of this technical data for approval before buying the equipment or contracting with third-tier subcontractors.

First, find out how many submittal copies will be required. Request that number of copies plus at least two additional sets for your use. A subcontractor who provides only services (no materials) may have to submit a detailed written explanation of the work to be done.

The second step is to label each submittal with the paragraph number of the specification where that equipment, material or service is described. Type or write this number in the upper right corner of each submittal sheet. From this set of numbered submittals, prepare a submittal index, a list of submittals identified by title and paragraph number. This index should be arranged in specification paragraph order. A blank submittal index form follows this section.

I prefer to use a three-ring binder to hold submittals. That makes it easier to add and delete sections as needed. Be sure to include a cover sheet, usually on company letterhead, which identifies the project, job number and scope of work being submitted. A sample submittal cover sheet follows this section.

Purchase orders and subcontracts should not be written until an approved copy of the submittal brochure has been received.

ACME Mechanical Contractors

Division 15
Heating, Ventilating, Air Conditioning
and Plumbing Systems

Submittal Data

for
Mile-Hi Office Building
2600 Second Street
Anytown, U.S.A.

7600 Oak Avenue - Anytown, U.S.A. 12345

Submittal Data Cover Sheet

Submittal Index

Specification Section	Specification Paragraph	Item Description	Proposed Vendor or Sub	As Specified	See Submittal Data Page

Submittal Index Form

The format to be used for monthly progress billings is seldom specified in the contract documents. In any case, your bill has to include enough detail to satisfy the owner and the lender.

Before preparing your bill, ask the general contractor how much detail is required. Some general contractors will want detailed cost breakdowns showing costs for each system on each building floor. Others will require much less detail.

Find out if billings can include the cost of equipment and materials delivered and stored either on site or off site but not yet installed. If billings can include materials stored off site, do these materials have to be stored in a bonded warehouse? Do vendor's invoices have to accompany progress billings?

When you understand what's required for monthly progress billings, prepare a billing breakdown worksheet. A sample worksheet follows this section.

Column one, Activity Lists each work item in the order work will be performed. The first item listed will usually be mobilization. This includes the cost of the job site trailer, electrical and telephone hookups, office furniture and supplies and initial labor costs. Billings for mobilization costs are sometimes denied by the lender. Many contractors minimize these costs to increase the chance of approval. It's better to receive partial payment for these costs than none at all.

Column two, Cost ($) Lists the actual contractor's costs for each activity, including labor, material, equipment and sales taxes. Markup for overhead and profit are not included.

Column three, Factor This column shows the markup assigned to each activity. Activities scheduled to be completed first are usually assigned the highest markups. This is called *front loading* and is common in the construction industry. Most bills won't be paid for 30 to 60 days. When the bill is paid, the amount received probably won't include the percentage allowed for *retention*. Generally, retention is 10 percent and isn't released until the project is complete and accepted by the owner.

Front loading helps contractors carry the financial burden of work in progress. Few subcontractors have enough cash to pay all their bills when due and still wait months to collect for work that's been completed. Employees have to be paid in full and on time. Assigning a higher markup to work completed first accelerates the payment schedule and helps spread payments more evenly over the entire project. Front loading is routine with subcontractors and unpopular with owners and lenders. In practice, subcontractors have no choice but to place a higher price or higher markup on work completed first.

Note that figures in the *Factor* column are less than 1.0 for work done near the end of the project. That's because front loading doesn't change the contract price. It changes only when that money is received.

Column four, Sell ($) Values for this column are the product of the cost and factor columns. Use the figures in this column when submitting monthly progress invoices.

ACME

Mechanical Contractors

7600 Oak Avenue - Anytown, U.S.A. 12345-6789
Voice 123-456-7890 -- Fax 123-456-0987

Billing Breakdown Worksheet

(Chancey Bank Building - Job No. 7777)

Activity	Cost ($)	Factor	Sell ($)
Mobilization	2,000	1.50	3,000
Underground chilled and heated water piping	120,000	1.35	162,000
Underground plumbing & piping	87,000	1.35	117,450
Above ground chilled and heating water piping	115,000	1.20	138,000
Above ground plumbing & piping	87,000	1.20	104,400
HVAC ducting	96,000	1.20	115,200
Equipment	312,000	1.10	343,200
Duct and pipe insulation	119,000	1.00	119,000
Temperature controls	53,000	.90	47,700
Plumbing fixtures	43,000	.65	27,950
Water treatment	1,800	.55	990
Test and balance	13,000	.55	7,150
Validation	500	.50	250
	$1,049,300		$1,186,290

$$\text{Estimated gross profit} = \frac{\$1,186,290 - \$1,049,300}{\$1,049,300} = 13\%$$

Billing Breakdown Worksheet

When the billing worksheet has been completed, prepare the billing breakdown as shown on the next page.

The general contractor will usually have a cutoff date each month for progress billings. If the invoice has not been received by that date, it won't be considered until the following month. The cutoff date is usually either the 20th or 25th of the month. The general contractor may permit you to include in each monthly billing work that's not yet complete but should be by the last day of the month.

Column one, Activity List each work item on the billing breakdown worksheet.

Column two, Amount ($) List costs from the Sell ($) column of the billing breakdown worksheet.

Column three, % Complete Show the percentage of completion for each activity listed. Obviously, this figure has to be based on estimates. Be prepared to support your estimate of completion if asked by the general contractor or the lender. If you anticipate a dispute on the figures in this column, call the lender or general contractor to explain the percentage you claim before submitting the bill.

Column four, $ Complete Multiply the percentage in the *% Complete* column by the figure in the *Amount ($)* column. Enter the result in the *$ Complete* column.

Total the *$ Complete* column. Then subtract the retention percentage and progress billings received to date. The new total is the progress billing for the current month.

ACME
Mechanical Contractors

7600 Oak Avenue - Anytown, U.S.A. 12345-6789
Voice 123-456-7890 -- Fax 123-456-0987
Progress Bill No. _____
Date _____

Monthly Progress Billing

(Chancey Bank Building - Job No. 7777)

Activity	Amount ($)	% Complete	$ Complete
Mobilization	3,000		
Underground chilled and heated water piping	162,000		
Underground plumbing & piping	117,450		
Above ground chilled and heating water piping	138,000		
Above ground plumbing piping	104,400		
HVAC ducting	115,200		
Equipment	343,200		
Duct and pipe insulation	119,000		
Temperature controls	47,700		
Plumbing fixtures	27,950		
Water treatment	990		
Test and balance	7,150		
Validation	250		
Change order numbers ___ through ___			
Contract price to date:	$1,186,290		

Total completed to date	$
Less _____% retention	$
Subtotal	$
Less previous billings	$
Amount this billing	$

Monthly Progress Billing

Upon completion of a project, all subcontractors may be required to provide a warranty letter for the work they've completed. These letters are addressed to the general contractor and will probably be forwarded to the owner.

The warranty usually covers defects in labor or material discovered during the first year after the project is turned over to the owner. Occasionally, the owner will request occupancy of a portion of the project before the entire project is completed. This is called a period of *beneficial occupancy*. If there will be a period of beneficial occupancy, a separate warranty letter may be required covering only the areas of the building occupied by the owner and only the portion of the plumbing and HVAC systems used in those areas.

Before preparing a warranty letter, get warranty letters from third-tier subs such as the environmental control, thermal insulation, test and balance subs. Any exclusions or limitations in these warranty letters should also appear as exclusions and limitations in your warranty letter. Otherwise you may be accepting responsibility for repairs on work you never performed.

A sample warranty letter follows.

ACME
Mechanical Contractors

7600 Oak Avenue - Anytown, U.S.A. 12345-6789
Voice 123-456-7890 -- Fax 123-456-0987

Date: _____

To: _____

        _____

        _____

        _____

Project:_____

Gentlemen:

In compliance with the project specifications, we submit our guarantee of equipment, materials and workmanship furnished by Acme Mechanical Contractors for a period of one year. The warranty period is as follows:

From _____ to _____

Equipment and materials furnished by others, but installed by Acme Mechanical Contractors, are not covered by this warranty, except for the installation work performed by Acme Mechanical Contractors.

Ordinary wear is not covered by this warranty. The owner's abuse, neglect or failure to perform recommended maintenance procedures will void this warranty.

Should any problems occur during the specified warranty period, due to faulty equipment, materials or workmanship, Acme Mechanical Contractors will correct the problems, without charge, to the satisfaction of the owner.

Sincerely,

_____
Project Manager

Warranty Letter

National Estimator Quick Start

 Use the Quick Start on the next ten pages to get familiar with National Estimator. In less than an hour you'll be printing labor and material cost estimates for your jobs.

To install National Estimator, put the National Estimator disk in a disk drive (such as A:). Start Windows™.

In Windows 3.1 or 3.11 go to the Program Manager.

1. Click on File.
2. Click on Run . . .
3. Type A:SETUP
4. Press [Enter ↵].

In Windows 95,

1. Click on [Start]
2. Click on Run . . .
3. Type A:SETUP
4. Press [Enter ↵].

Click on File. Click on Run.

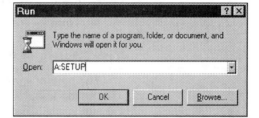

Type A:SETUP in Windows 95.

National Estimator Icon

Then follow instructions on the screen. Installation should take about two minutes. We recommend installing all National Estimator files to the NATIONAL directory. If you have trouble installing National Estimator, call 619-438-7828.

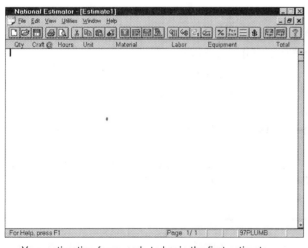

Your estimating form ready to begin the first estimate.

The installation program will create a Construction Estimating group and put the National Estimator Icon in that group. When installation is complete, click on OK to begin using National Estimator. In a few seconds your electronic estimating form will be ready to begin the first estimate.

On the title bar at the top of the screen you see the program name, National Estimator, and [Estimate1], showing that Estimate1 is on the screen. Let's take a closer look at other information at the top of your screen.

The Menu Bar

Below the title bar you see the menu bar. Every option in National Estimator is available on the menu bar. Click with your left mouse button on any item on the menu to open a list of available commands.

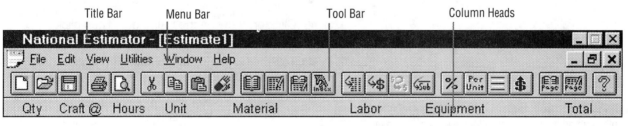

The National Estimator title bar, menu bar, tool bar and column heads.

Buttons on the Tool Bar

Below the menu bar you see 24 buttons that make up the tool bar. The options you use most in National Estimator are only a mouse click away on the tool bar.

Column Headings

Below the tool bar you'll see column headings for your estimate form:

- **Qty** for quantity
- **Craft@Hours** for craft (the crew doing the work) and manhours (to complete the task)
- **Unit** for unit of measure, such as per linear foot or per square foot
- **Material** for material cost
- **Labor** for labor cost
- **Equipment** for equipment cost
- **Total** for the total of all cost columns

The Status Bar

The bottom line on your screen is the status bar. Here you'll find helpful information about the choices available. Notice "Page 1/ 1" near the right end of the status line. That's a clue that you're looking at page 1 of a one-page estimate. When you see PICT at the right end of the status bar, a picture is available in the costbook. Click on PICT to show the picture. Click on PICT again to hide the picture.

| For Help, press F1 | Page 1/ 1 | 97PLUMB | PICT |

Check the status bar occasionally for helpful tips and explanations of what you see on screen.

Beginning an Estimate

Let's start by putting a heading on this estimate.

| The Blinking Cursor (insert point)

1. Press [Enter ←] once to space down one line.

2. Press [Tab] four times (or hold the space bar down) to move the Blinking Cursor (the insert point) near the middle of the line.

Mouse Pointer

3. Type "Estimate One" and press [Enter ←]. That's the title of this estimate, "Estimate One."

Mouse Pointer

4. Press [Enter ←] again to move the cursor down a line. That opens up a little space below the title.

The Costbook

Let's leave your estimating form for a moment to take a look at estimates stored in the costbook. To switch to the costbook, either:

- Click on the [ ] button, -Or-
- Click on View on the menu bar. Then click on Costbook Window, -Or-
- Tap the [Alt] key, tap the letter V (for View) and tap the letter C (for Costbook Window), -Or-
- Press [Esc]. Press the [↓] key to highlight Costbook Window. Then press [Enter ←].

Begin by putting a title on your estimate, such as "Estimate One."

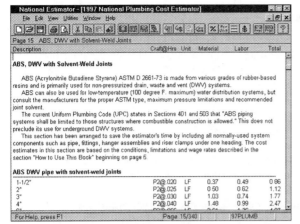

The costbook window has the
entire National Plumbing & HVAC Estimator.

The entire 1997 National Plumbing & HVAC Estimator is available in the Costbook Window. Notice the words *Page 15 ABS, DWV Pipe with Solvent-Weld Joints* at the left side of the screen just below the tool bar. That's your clue that the ABS section of page 15 is on the screen. To turn to the next page, either:

- Press [PgDn] (with Num Lock off), -Or-

- Click on the lower half of the scroll bar at the right edge of the screen.

To move down one line at a time, either:

- Press the [↓] key (with Num Lock off), -Or-

- Click on the arrow on the down scroll bar at the lower right corner of the screen.

Press [PgDn] about 1,400 times and you'll page through the entire National Plumbing & HVAC Estimator. Obviously, there's a better way. To turn quickly to any page, either:

- Click on the [image] button near the right end of the tool bar, -Or-

- Click on View on the menu bar. Then click on Turn to Costbook Page, -Or-

- Press [Esc]. Press the [↓] key to highlight Turn to Costbook Page. Then press [Enter ←].

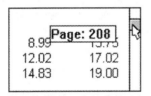

Type the page number you want to see.

Type the number of the page you want to see and press [Enter ←]. National Estimator will turn to the top of the page you requested.

An Even Better Way

Find the small square in the slide bar at the right side of the Costbook Window. Click and hold on that square while rolling the mouse up or down. Keep dragging the square until you see the page you want in the Page: box. Release the mouse button to turn to the top of that page.

Drag the square to see any page.

A Still Better Way: Keyword Search

To find any cost estimate in seconds, search by keyword in the index. To go to the index, either:

- Click on the [image] button near the center of the tool bar, -Or-

- Press [Esc]. Press [Enter ←], -Or-

- Click on View on the menu bar. Then press [Enter ←].

Notice that the cursor is blinking in the Enter Keyword to Locate box at the right of the screen. Obviously, the index is ready to begin a search.

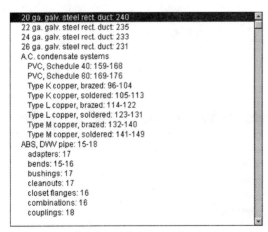

Use the electronic index to find cost estimates for any item.

Your First Estimate

Suppose we're estimating the cost of copper DWV pipe with soldered joints. Let's put the index to work with a search for copper pipe. In the box under Enter Keyword to Locate, type *copper*. The index jumps to the heading *Copper pipe*.

The first item under copper pipe is *drain, waste, vent: 93, 94, 95*. Either:

- Click once on that line and press Enter ↵, -Or-

- Double click on that line, -Or-

- Press Tab and the ↓ key to move the highlight to *drain, waste, vent: 93, 94, 95*. Then press Enter ↵.

Copper pipe
drain, waste, vent: 93, 94, 95
Type K copper, brazed: 96-104
Type K copper, soldered: 105-113
Type L copper, brazed: 114-122
Type L copper, soldered: 123-131
Type M copper, brazed: 133-140
Type M copper, soldered: 141-149

The index jumps to
Copper pipe.

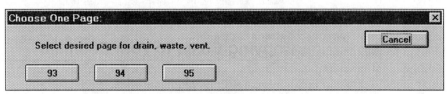

If costs appear on several pages, click on the page you prefer.

To select the page you want to see (page 93 in this case), either:

- Click on number 93, -Or-

- Press Tab to highlight 93. Then press Enter ↵.

National Estimator turns to the top of page 93. Notice that estimates at the top of this page are for copper DWV pipe with soft-soldered joints. Press the ↓ key 34 times until your screen looks like the example at the left below.

Splitting the Screen

Most of the time you'll want to see what's in both the costbook and your estimate. To split the screen into two halves, either:

- Click on the [⊞] button near the center of the tool bar, -Or-

- Press Esc and the ↓ key to move the selection bar to Split Window, -Or-

- Tap the Alt key, tap the letter V (for View), tap the letter S (for Split Window).

Your screen should look like the example at the right below.

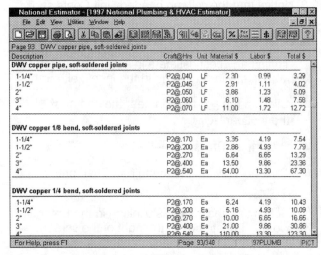

Costs for copper DWV pipe on page 93.

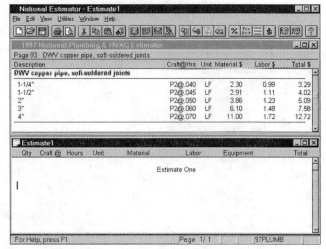

The split window: Costbook above and estimate below.

Notice that eight lines of the costbook are at the top of the screen and your estimate is at the bottom. You should recognize "Estimate One." It's your title for this estimate. Column headings are at the top of the costbook and across the middle of the screen (for your estimate).

To Switch from Window to Window

- Click in the window of your choice, -Or-

- Hold the Ctrl key down and press Tab.

Notice that a window title bar turns dark when that window is selected. The selected window is where keystrokes appear as you type.

Copying Costs to Your Estimate

Next, we'll estimate the cost of 128 feet of 1-1/2" copper DWV pipe. Click on the 🖩 button on the tool bar to be sure you're in the split window. Click anywhere in the costbook (the top half of your screen). Then press the ↓ key until the cursor is on the line:

| 1-1/2" | | | P2@.045 | LF | 2.91 | 1.11 | 4.02 |

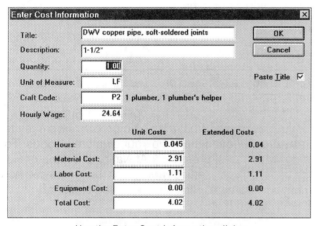

Use the Enter Cost Information dialog box to copy or change costs.

To copy this line to your estimate:

1. Click on the line.

2. Click on the 🗐 button.

3. Click on the 📋 button to open the Enter Cost Information dialog box.

Notice that the blinking cursor is in the Quantity box:

1. Type a quantity of 128 because the job has 128 feet of copper DWV pipe.

2. Press Tab and check the estimate for accuracy.

3. Notice that the column headed Unit Costs shows costs per unit, per "LF" (linear foot) in this case.

4. The column headed Extended Costs shows costs for the 128 feet.

5. The lines opposite Title and Description show what's getting installed. You can change the words in either of these boxes. Just click on what you want to change and start typing or deleting.

6. You can also change any numbers in the Unit Cost column. Just click and start typing.

7. When the words and costs are exactly right, press Enter ↵ or click on OK to copy these figures to the end of your estimate.

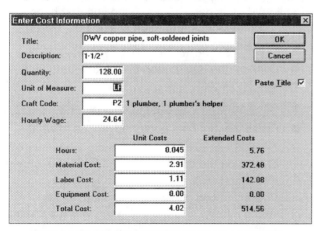

Costs for the 128 feet (extended costs) are on the right.

DWV copper pipe, soft-soldered joints						
1-1/2"						
128.00	P2@5.760	LF	372.48	142.08	0.00	514.56

Extended costs for 128 feet of copper DWV pipe as they appear on your estimate form.

The new line at the bottom of your estimate shows:

128.00 is the quantity of copper DWV pipe

P2 is the recommended crew, a plumber and a helper

@5.760 shows the manhours required for the work

LF is the unit of measure, linear feet

372.48 is the material cost for copper DWV pipe

142.08 is the labor cost for the job

0.00 shows there is no equipment cost

514.56 is the total of material, labor and equipment

Copy Anything to Anywhere in Your Estimate

Anything in the costbook can be copied to your estimate. Just click on the line (or select the words) you want to copy and press the F8 key. It's copied to the last line of your estimating form. If your selection includes costs, you'll have a chance to enter the quantity. To copy to the middle of your estimate:

1. Select what you want to copy.

2. Click on the 🖹 button.

3. Click in the estimate where you want to paste.

4. Click on the 📋 button.

Your Own Wage Rates

The labor cost in the example above is based on a plumber and helper working at an average cost of $24.64 per hour. (See page 7 in the National Plumbing & HVAC Estimator for crew rates used in the costbook.) Suppose $24.64 per hour isn't right for your estimate. What then? No problem! It's easy to use your own wage rate for any crew or even make up your own crew codes. To get more information on setting wage rates, press F1. At National Estimator Help Contents, click on the Search button. Type "setting" and press Enter to see how to set wage rates. To return to your estimate, click on File on the National Estimator Help menu bar. Then click on Exit.

Search for information on setting wage rates.

Changing Cost Estimates

With Num Lock off, use the ↑ or ↓ key to move the cursor to the line you want to change (or click on that line). In this case, move to the line that begins with a quantity of 128. To open the Enter Cost Information Dialog box, either:

■ Press Enter, -Or-

■ Click on the 🔍$ button on the tool bar.

To make a change, either

■ Click on what you want to change, -Or-

■ Press Tab until the cursor advances to what you want to change.

Then type the correct figure. In this case, change the material cost to 2.50.

Press Tab and check the Extended Costs column. If it looks OK, press Enter and the change is made on your estimating form.

Unit Costs	
Hours:	0.045
Material Cost:	2.50
Labor Cost:	1.11
Equipment Cost:	0.00
Total Cost:	3.61

Change the material cost to 2.50.

Changing Text (Descriptions)

Click on the  button on the tool bar to be sure you're in the estimate. With Num Lock off, use the ↑ or ↓ key or click the mouse button to put the cursor where you want to make a change. In this case, we're going to make a change on the line that begins "DWV copper pipe . . ."

To make a change, click where the change is needed. Then either:

■ Press the Del or ←Bksp key to erase what needs deleting, -Or-

■ Select what needs deleting and click on the ✂ button on the tool bar.

■ Type what needs to be added.

In this case, click just after "DWV copper pipe." Then hold the left mouse button down and drag the mouse to the right until you've put a dark background behind ", soft-soldered joints." The dark background shows that these words are selected and ready for editing.

Press the Del key (or click on the ✂ button on the tool bar), and the selection is cut from the estimate. If that's not what you wanted, click on the button and "soft-soldered joints" is back again.

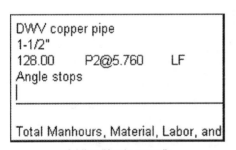

To select, click and hold the mouse
button while dragging the mouse.

Adding Text (Descriptions)

Some of your estimates will require descriptions (text) and costs that can't be found in the National Plumbing & HVAC Estimator. What then? With National Estimator it's easy to add descriptions and costs of your choice anywhere in the estimate. For practice, let's add an estimate for four angle stops to Estimate One.

Click on the button to be sure the estimate window is maximized. We can add lines anywhere on the estimate. But in this case, let's make the addition at the end. Press the ↓ key to move the cursor down until it's just above the horizontal line that separates estimate detail lines from estimate totals. To open a blank line, either:

■ Press Enter↵, -Or-

■ Click on the button on the tool bar.

Type "Angle stops" and press Enter↵.

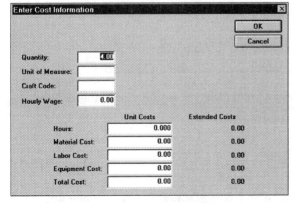

Adding "Angle stops."

Adding a cost line with the
Enter Cost Information dialog box.

Adding a Cost Estimate Line

Now let's add a cost for "Angle stops" to your estimate. Begin by opening the Enter Cost Information dialog box. Either:

■ Click on ⤴$ button on the tool bar, -Or-

■ Click on Edit on the menu bar. Then click on Insert a Cost Line.

1. The cursor is in the Quantity box. Type the number of units (4 in this case) and press Tab.

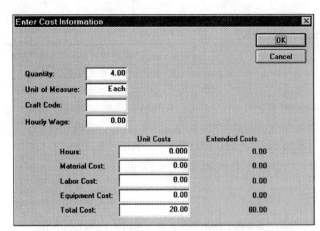

Unit and extended costs for four angle stops.

2. The cursor moves to the next box, Unit of Measure.

3. In the Unit of Measure box, type *Each* and press [Tab].

4. Press [Tab] twice to leave the Craft Code blank and Hourly Wage at zero.

5. Since these angle stops will be installed by the supplier, there's no material, labor or equipment cost. So press [Tab] four times to skip over the Hours, Material Cost, Labor Cost and Equipment Cost boxes.

6. In the Total Cost box, type 20.00. That's the cost per angle stop quoted by your supplier.

7. Press [Tab] once more to advance to OK.

8. Press [Enter ←] and the cost of four panels is written to your estimate.

Note: The sum of material, labor, and equipment costs appears automatically in the Total Cost box. If there's no cost entered in the Material Cost, Labor Cost or Equipment Cost boxes (such as for a subcontracted item), you can enter any figure in the Total Cost box.

Adding Lines to the Costbook

Add lines or make changes in the costbook the same way you add lines or make changes in an estimate. The additions and changes you make become part of the user costbook. For more information on user costbooks, press [F1]. Click on Search. Type "user" and press [Enter ←].

Adding Tax

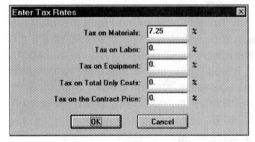

Type the tax rate that applies.

To include sales tax in your estimate:

1. Click on Edit.

2. Click on Tax Rates.

3. Type the tax rate in the appropriate box.

4. Press [Tab] to advance to the next box.

5. Press [Enter ←] or click on OK when done.

In this case, the tax rate is 7.25% on materials only. Tax will appear near the last line of the estimate.

Adding Overhead and Profit

Set markup percentages in the Add for Overhead & Profit dialog box. To open the box, either:

- Click on the [$] button on the tool bar, -Or-
- Click on Edit on the menu bar. Then click on Markup.

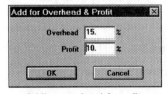

Adding overhead & profit.

Type the percentages you want to add for overhead. For this estimate:

1. Type 15 on the Overhead line.

2. Press [Tab] to advance to Profit.

3. Type 10 on the Profit line.

4. Press [Enter ←].

Markup percentages can be changed at any time. Just reopen the Add for Overhead & Profit Dialog box and type the correct figure.

| File Name: Estimate 1 | | | Construction Estimate | | | Page 1 |
Qty	Craft@ Hours	Unit	Material	Labor	Equipment	Total
			Estimate One			
DWV copper pipe						
1-1/2"						
120.00	P2@ 5.760	LF	320.00	142.08	0.00	462.08
Angle stops						
4.00	--@ .0000	Each	0.00	0.00	0.00	80.00
Total Manhours, Material, Labor, and Equipment:						
5.8			320.00	142.08	0.00	462.08
Total Only (Subcontract) Costs:						80.00
			Subtotal:			542.08
			15.00% Overhead:			81.31
			10.00% Profit:			62.34
			Estimate Total:			685.73
			Tax on Materials:			23.20
			Grand Total:			708.93

A preview of Estimate One.

Preview Your Estimate

You can display an estimate on screen just the way it will look when printed on paper. To preview your estimate, either:

■ Click on the [ ] button on the tool bar, -Or-

■ Click on File on the menu bar. Then click on Print Preview.

In print preview:

■ Click on Next Page or Prev. Page to turn pages.

■ Click on Two Page to see two estimate pages side by side.

■ Click on Zoom In to get a closer look.

■ Click on Close when you've seen enough.

Use buttons on Print Preview to see your estimate as it will look when printed.

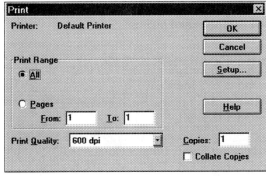

Options available depend on the printer you're using.

Printing Your Estimate

When you're ready to print the estimate, either:

■ Click on the [ ] button on the tool bar, -Or-

■ Click on File on the menu bar. Then click on Print, -Or-

■ Hold the [Ctrl] key down and type the letter P.

■ Press [Enter ◄┘] or click on OK to begin printing.

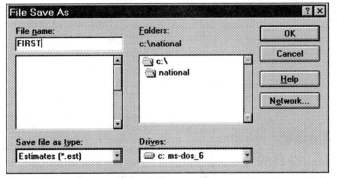

Type the estimate name in the
File Name box to assign a file name.

Save Your Estimate to Disk

To store your estimate on the hard disk where it can be re-opened and changed at any time, either:

■ Click on the [ ] button on the tool bar, -Or-

■ Click on File on the menu bar. Then click on Save, -Or-

■ Hold the [Ctrl] key down and type the letter S.

The cursor is in the File Name box. Type the name you want to give this estimate, such as FIRST. The name can be up to eight letters and numbers, but don't use symbols or spaces. Press [Enter ◄┘] or click on OK and the estimate is written to disk.

Opening Other Costbooks

Many construction cost estimating databases are available for the National Estimator program. For example, a National Estimator costbook comes with each of these estimating manuals:

- *National Construction Estimator*
- *National Repair & Remodeling Estimator*
- *National Electrical Estimator*
- *National Plumbing & HVAC Estimator*
- *National Painting Cost Estimator*
- *National Renovation & Insurance Repair Estimator*

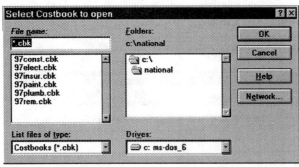

Open the costbook of your choice.

To open any of the other costbooks on your computer:

- Click on File
- Click on Open Costbook
- Be sure the drive and directory are correct, usually c:\national.
- Double click on the costbook of your choice.

To see a list of the costbooks open, click on Window. The name of the current costbook will be checked. Click on any other costbook or estimate name to display that costbook or estimate. Click on Window, then click on Tile to display all open costbooks and estimates.

Select Your Default Costbook

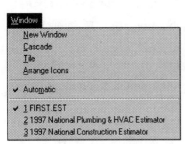

Click to switch costbooks.

Your default costbook opens automatically every time you begin using National Estimator. Save time by making the default costbook the one you use most.

To change your default costbook, click on Utilities on the menu bar. Then click on Options. Next, click on Select Default Costbook. Click on the costbook of your choice. Click on OK. Then click on OK again.

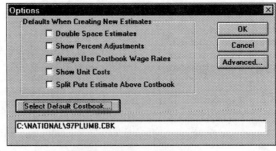

Select the default costbook.

Use National Estimator Help

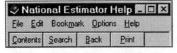

Click on File and then Print Topic.

That completes the basics of National Estimator. You've learned enough to complete most estimates. When you need more information about the fine points, use National Estimator Help. Click on [?] to see Help Contents. Then click on the menu selection of your choice. To print 28 pages of instructions for National Estimator, go to Help Contents. Click on Print All Topics (at the bottom of Help Contents). Click on File on the Help menu bar. Then click on Print Topic. Click on OK.

Index

Other Practical References

Blueprint Reading for the Building Trades

How to read and understand construction documents, blueprints, and schedules. Includes layouts of structural, mechanical, HVAC and electrical drawings. Shows how to interpret sectional views, follow diagrams and schematics, and covers common problems with construction specifications. **192 pages, 5½ x 8½, $14.75**

Rough Framing Carpentry

If you'd like to make good money working outdoors as a framer, this is the book for you. Here you'll find shortcuts to laying out studs; speed cutting blocks, trimmers and plates by eye; quickly building and blocking rake walls; installing ceiling backing, ceiling joists, and truss joists; cutting and assembling hip trusses and California fills; arches and drop ceilings — all with production line procedures that save you time and help you make more money. Over 100 on-the-job photos of how to do it right and what can go wrong.
304 pages, 8½ x 11, $26.50

Roofing Construction & Estimating

Installation, repair and estimating for nearly every type of roof covering available today in residential and commercial structures: asphalt shingles, roll roofing, wood shingles and shakes, clay tile, slate, metal, built-up, and elastometric. Covers sheathing and underlayment techniques, as well as secrets for installing leakproof valleys. Many estimating tips help you minimize waste, as well as insure a profit on every job. Troubleshooting techniques help you identify the true source of most leaks. Over 300 large, clear illustrations help you find the answer to just about all your roofing questions. **432 pages, 8½ x 11, $35.00**

National Painting Cost Estimator

A complete guide to estimating painting costs for just about any type of residential, commercial, or industrial painting, whether by brush, spray, or roller. Shows typical costs and bid prices for fast, medium, and slow work, including material costs per gallon; square feet covered per gallon; square feet covered per manhour; labor, material, overhead, and taxes per 100 square feet; and how much to add for profit. Includes an electronic version of the book on computer disk with a stand-alone *Windows* estimating program FREE on a 3½" high-density (1.44 Mb) disk.
448 pages, 8½ x 11, $38.00. Revised annually

Plumber's Handbook Revised

This new edition shows what will and won't pass inspection in drainage, vent, and waste piping, septic tanks, water supply, fire protection, and gas piping systems. All tables, standards, and specifications completely up-to-date with recent plumbing code changes. Covers common layouts for residential work, how to size piping, selecting and hanging fixtures, practical recommendations, and trade tips. The approved reference for the plumbing contractor's exam in many states. **240 pages, 8½ x 11, $18.00**

National Building Cost Manual

Square foot costs for residential, commercial, industrial, and farm buildings. Quickly work up a reliable budget estimate based on actual materials and design features, area, shape, wall height, number of floors, and support requirements. Includes all the important variables that can make any building unique from a cost standpoint. **240 pages, 8½ x 11, $23.00. Revised annually**

Builder's Guide to Room Additions

How to tackle problems that are unique to additions, such as requirements for basement conversions, reinforcing ceiling joists for second-story conversions, handling problems in attic conversions, what's required for footings, foundations, and slabs, how to design the best bathroom for the space, and much more. Besides actual construction methods, you'll also find help in designing, planning and estimating your room-addition jobs.
352 pages, 8½ x 11, $27.25

Construction Estimating Reference Data

Provides the 300 most useful manhour tables for practically every item of construction. Labor requirements are listed for sitework, concrete work, masonry, steel, carpentry, thermal and moisture protection, door and windows, finishes, mechanical and electrical. Each section details the work being estimated and gives appropriate crew size and equipment needed. Includes an electronic version of the book on computer disk with a stand-alone *Windows* estimating program FREE on a 3½" high-density (1.44 Mb) disk.
432 pages, 8½ x 11, $39.50

Bookkeeping for Builders

Shows simple, practical instructions for setting up and keeping accurate records — with a minimum of effort and frustration. Explains the essentials of a record-keeping system: the payment, income, and general journals, and records for fixed assets, accounts receivable, payables and purchases, petty cash, and job costs. Shows how to keep I.R.S. records, and accurate, organized business records for your own use. **208 pages, 8½ x 11, $19.75**

Construction Forms & Contracts

125 forms you can copy and use — or load into your computer (from the FREE disk enclosed). Then you can customize the forms to fit your company, fill them out, and print. Loads into Word for *Windows*, Lotus 1-2-3, WordPerfect, or Excel programs. You'll find forms covering accounting, estimating, fieldwork, contracts, and general office. Each form comes with complete instructions on when to use it and how to fill it out. These forms were designed, tested and used by contractors, and will help keep your business organized, profitable and out of legal, accounting and collection troubles. Includes a 3½" high-density disk for your PC. For Macintosh disks, add $15. **432 pages, 8½ x 11, $39.75**

National Construction Estimator

Current building costs for residential, commercial, and industrial construction. Estimated prices for every common building material. Manhours, recommended crew, and labor cost for installation. Includes an electronic version of the book on computer disk with a stand-alone *Windows* estimating program FREE on a 3½" high-density (1.44 Mb) disk.
528 pages, 8½ x 11, $37.50. Revised annually

Professional Kitchen Design

Remodeling kitchens requires a "special" touch — one that blends artistic flair with function to create a kitchen with charm and personality as well as one that is easy to work in. Here you'll find how to make the best use of the space available in any kitchen design job, as well as tips and lessons on how to design one-wall, two-wall, L-shaped, U-shaped, peninsula and island kitchens. Also includes what you need to know to run a profitable kitchen design business. **176 pages, 8½ x 11, $24.50**

Builder's Guide to Accounting Revised

Step-by-step, easy-to-follow guidelines for setting up and maintaining records for your building business. This practical, newly-revised guide to all accounting methods shows how to meet state and federal accounting requirements, explains the new depreciation rules, and describes how the Tax Reform Act can affect the way you keep records. Full of charts, diagrams, simple directions and examples, to help you keep track of where your money is going. Recommended reading for many state contractor's exams.
320 pages, 8½ x 11, $26.50

Construction Surveying & Layout

A practical guide to simplified construction surveying. How to divide land, use a transit and tape to find a known point, draw an accurate survey map from your field notes, use topographic surveys, and the right way to level and set grade. You'll learn how to make a survey for any residential or commercial lot, driveway, road, or bridge — including how to figure cuts and fills and calculate excavation quantities. Use this guide to make your own surveys, or just read and verify the accuracy of surveys made by others.
256 pages, 5½ x 8½, $19.25

Drafting House Plans

Here you'll find step-by-step instructions for drawing a complete set of home plans for a one-story house, an addition to an existing house, or a remodeling project. This book shows how to visualize spatial relationships, use architectural scales and symbols, sketch preliminary drawings, develop detailed floor plans and exterior elevations, and prepare a final plot plan. It even includes code-approved joist and rafter spans and how to make sure that drawings meet code requirements.
192 pages, 8½ x 11, $27.50

Building Contractor's Exam Preparation Guide

Passing today's contractor's exams can be a major task. This book shows you how to study, how questions are likely to be worded, and the kinds of choices usually given for answers. Includes sample questions from actual state, county, and city examinations, plus a sample exam to practice on. This book isn't a substitute for the study material that your testing board recommends, but it will help prepare you for the types of questions — and their correct answers — that are likely to appear on the actual exam. Knowing how to answer these questions, as well as what to expect from the exam, can greatly increase your chances of passing.
320 pages, 8½ x 11, $35.00

Contractor's Guide to the Building Code Revised

This completely revised edition explains in plain English exactly what the Uniform Building Code requires. Based on the newly-expanded 1994 code, it explains many of the changes made. Also covers the Uniform Mechanical Code and the Uniform Plumbing Code. Shows how to design and construct residential and light commercial buildings that'll pass inspection the first time. Suggests how to work with an inspector to minimize construction costs, what common building shortcuts are likely to be cited, and where exceptions are granted. **384 pages, 8½ x 11, $39.00**

Plumber's Exam Preparation Guide

Hundreds of questions and answers to help you pass the apprentice, journeyman, or master plumber's exam. Questions are in the style of the actual exam. Gives answers for both the Standard and Uniform plumbing codes. Includes tips on studying for the exam and the best way to prepare yourself for examination day.
320 pages, 8½ x 11, $29.00

The Contractor's Legal Kit

Stop "eating" the costs of bad designs, hidden conditions, and job surprises. Set ground rules that assign those costs to the rightful party ahead of time. And it's all in plain English, not "legalese."
For less than the cost of an hour with a lawyer you'll learn the exclusions to put in your agreements, why your insurance company may pay for your legal defense, how to avoid liability for injuries to your sub and his employees or damages they cause, how to collect on lawsuits you win, and much more. It also includes a FREE computer disk with contracts and forms you can customize for your own use.
352 pages, 8½ x 11, $59.95

Estimating Framing Quantities

Gives you hundreds of time-saving estimating tips. Shows how to make thorough step-by-step estimates of all rough carpentry in residential and light commercial construction: ceilings, walls, floors, and roofs. Lots of illustrations showing lumber requirements, nail quantities, and practical estimating procedures.
285 pages, 5½ x 8½, $34.95

Finish Carpenter's Manual

Everything you need to know to be a finish carpenter: assessing a job before you begin, and tricks of the trade from a master finish carpenter. Easy-to-follow instructions for installing doors and windows, ceiling treatments (including fancy beams, corbels, cornices and moldings), wall treatments (including wainscoting and sheet paneling), and the finishing touches of chair, picture, and plate rails. Specialized interior work includes cabinetry and built-ins, stair finish work, and closets. Also covers exterior trims and porches. Includes manhour tables for finish work, and hundreds of illustrations and photos. **208 pages, 8½ x 11, $22.50**

Handbook of Construction Contracting

Volume 1: Everything you need to know to start and run your construction business; the pros and cons of each type of contracting, the records you'll need to keep, and how to read and understand house plans and specs so you find any problems before the actual work begins. All aspects of construction are covered in detail, including all-weather wood foundations, practical math for the job site, and elementary surveying. **416 pages, 8½ x 11, $28.75**

Volume 2: Everything you need to know to keep your construction business profitable; different methods of estimating, keeping and controlling costs, estimating excavation, concrete, masonry, rough carpentry, roof covering, insulation, doors and windows, exterior finishes, specialty finishes, scheduling work flow, managing workers, advertising and sales, spec building and land development, and selecting the best legal structure for your business.
320 pages, 8½ x 11, $30.75

Contractor's Growth and Profit Guide

Step-by-step instructions for planning growth and prosperity in a construction contracting or subcontracting company. Explains how to prepare a business plan: select reasonable goals, draft a market expansion plan, make income forecasts and expense budgets, and project cash flow. You'll learn everything that most lenders and investors require, as well as the best way to organize your business.
336 pages, 5½ x 8½, $19.00

Basic Plumbing with Illustrations, Revised

This completely-revised edition brings this comprehensive manual fully up-to-date with all the latest plumbing codes. It is the journeyman's and apprentice's guide to installing plumbing, piping, and fixtures in residential and light commercial buildings: how to select the right materials, lay out the job and do professional-quality plumbing work, use essential tools and materials, make repairs, maintain plumbing systems, install fixtures, and add to existing systems. Includes extensive study questions at the end of each chapter, and a section with all the correct answers. **384 pages, 8½ x 11, $33.00**

Estimating Plumbing Costs

Offers a basic procedure for estimating materials, labor, and direct and indirect costs for residential and commercial plumbing jobs. Explains how to read and understand plot plans, design drainage, waste, and vent systems, meet code requirements, and make an accurate take-off for materials and labor. Includes sample cost sheets, manhour production tables, complete illustrations, and all the practical information you need. **224 pages, 8½ x 11, $22.50**

Home Inspection Handbook

Every area you need to check in a home inspection ¾ especially in older homes. Twenty complete inspection checklists: building site, foundation and basement, structural, bathrooms, chimneys and flues, ceilings, interior & exterior finishes, electrical, plumbing, HVAC, insects, vermin and decay, and more. Also includes information on starting and running your own home inspection business. **324 pages, 5½ x 8½, $24.95**

Basic Engineering for Builders

If you've ever been stumped by an engineering problem on the job, yet wanted to avoid the expense of hiring a qualified engineer, you should have this book. Here you'll find engineering principles explained in non-technical language and practical methods for applying them on the job. With the help of this book you'll be able to understand engineering functions in the plans and how to meet the requirements, how to get permits issued without the help of an engineer, and anticipate requirements for concrete, steel, wood and masonry. See why you sometimes have to hire an engineer and what you can undertake yourself: surveying, concrete, lumber loads and stresses, steel, masonry, plumbing, and HVAC systems. This book is designed to help the builder save money by understanding engineering principles that you can incorporate into the jobs you bid. **400 pages, 8½ x 11, $34.00**

Carpentry Estimating

Simple, clear instructions on how to take off quantities and figure costs for all rough and finish carpentry. Shows how to convert piece prices to MBF prices or linear foot prices, use the extensive manhour tables included to quickly estimate labor costs, and how much overhead and profit to add. All carpentry is covered; floor joists, exterior and interior walls and finishes, ceiling joists and rafters, stairs, trim, windows, doors, and much more. Includes *Carpenter's Dream* a material-estimating program, at no extra cost on a 5¼" high-density disk. **336 pages, 8½ x 11, $35.50**

Residential Steel Framing Guide

Steel is stronger and lighter than wood — straight walls are guaranteed — steel framing will not wrap, shrink, split, swell, bow, or rot. Here you'll find full page schematics and details that show how steel is connected in just about all residential framing work. You won't find lengthy explanations here on how to run your business, or even how to do the work. What you will find are over 150 easy-to-ready full-page details on how to construct steel-framed floors, roofs, interior and exterior walls, bridging, blocking, and reinforcing for all residential construction. Also includes recommended fasteners and their applications, and fastening schedules for attaching every type of steel framing member to steel as well as wood. **170 pages, 8½ x 11, $38.80**

How to Succeed With Your Own Construction Business

Everything you need to start your own construction business: setting up the paperwork, finding the work, advertising, using contracts, dealing with lenders, estimating, scheduling, finding and keeping good employees, keeping the books, and coping with success. If you're considering starting your own construction business, all the knowledge, tips, and blank forms you need are here. **336 pages, 8½ x 11, $24.25**

National Repair & Remodeling Estimator

The complete pricing guide for dwelling reconstruction costs. Reliable, specific data you can apply on every repair and remodeling job. Up-to-date material costs and labor figures based on thousands of jobs across the country. Provides recommended crew sizes; average production rates; exact material, equipment, and labor costs; a total unit cost and a total price including overhead and profit. Separate listings for high- and low-volume builders, so prices shown are specific for any size business. Estimating tips specific to repair and remodeling work to make your bids complete, realistic, and profitable. Includes an electronic version of the book on computer disk with a stand-alone *Windows* estimating program FREE on a 3½" high-density (1.44 Mb) disk. **416 pages, 8½ x 11, $38.50. Revised annually**

Planning Drain, Waste & Vent Systems

How to design plumbing systems in residential, commercial, and industrial buildings. Covers designing systems that meet code requirements for homes, commercial buildings, private sewage disposal systems, and even mobile home parks. Includes relevant code sections and many illustrations to guide you though what the code requires in designing drainage, waste, and vent systems. **192 pages, 8½ x 11, $19.25**

Wood-Frame House Construction

Step-by-step construction details, from the layout of the outer walls, excavation and formwork, to finish carpentry and painting. Contains all new, clear illustrations and explanations updated for construction in the '90s. Everything you need to know about framing, roofing, siding, interior finishings, floor covering and stairs — your complete book of wood-frame homebuilding. **320 pages, 8½ x 11, $25.50. Revised edition**

CD Estimator

If your computer has *Windows*™ and a CD-ROM drive, CD Estimator puts at your fingertips 85,000 construction costs for new construction, remodeling, renovation & insurance repair, electrical, plumbing, HVAC and painting. You'll also have the National Estimator program — a stand-alone estimating program for *Windows* that *Remodeling* magazine called a "computer wiz." Quarterly cost updates are available at no charge on the Internet. To help you create professional-looking estimates, the disk includes over 40 construction estimating and bidding forms in a format that's perfect for nearly any word processing or spreadsheet program for *Windows*. And to top it off, a 70-minute interactive video teaches you how to use this CD-ROM to estimate construction costs. **CD Estimator is $59.00**

National Renovation & Insurance Repair Estimator

Current prices in dollars and cents for hard-to-find items needed on most insurance, repair, remodeling, and renovation jobs. All price items include labor, material, and equipment breakouts, plus special charts that tell you exactly how these costs are calculated. Includes an electronic version of the book on computer disk with a stand-alone *Windows* estimating program FREE on a 3½" high density (1.44 Mb) disk. **560 pages, 8½ x 11, $39.50. Revised annually**

Roof Framing

Shows how to frame any type of roof in common use today, even if you've never framed a roof before. Includes using a pocket calculator to figure any common, hip, valley, or jack rafter length in seconds. Over 400 illustrations cover every measurement and every cut on each type of roof: gable, hip, Dutch, Tudor, gambrel, shed, gazebo, and more. **480 pages, 5½ x 8½, $22.00**

Craftsman's Illustrated Dictionary of Construction Terms

Almost everything you could possibly want to know about any word or technique in construction. Hundreds of up-to-date construction terms, materials, drawings and pictures with detailed, illustrated articles describing equipment and methods. Terms and techniques are explained or illustrated in vivid detail. Use this valuable reference to check spelling, find clear, concise definitions of construction terms used on plans and construction documents, or learn about little-known tools, equipment, tests and methods used in the building industry. It's all here. **416 pages, 8½ x 11, $36.00**

Profits in Buying & Renovating Homes

Step-by-step instructions for selecting, repairing, improving, and selling highly profitable "fixer-uppers." Shows which price ranges offer the highest profit-to-investment ratios, which neighborhoods offer the best return, practical directions for repairs, and tips on dealing with buyers, sellers, and real estate agents. Shows you how to determine your profit before you buy, what "bargains" to avoid, and how to make simple, profitable, inexpensive upgrades. **304 pages, 8½ x 11, $19.75**

Vest Pocket Guide To HVAC Electricity

This handy guide will be a constant source of useful information for anyone working with electrical systems for heating, ventilating, refrigeration, and air conditioning. Includes essential tables and diagrams for calculating and installing electrical systems for powering and controlling motors, fans, heating elements, compressors, transformers and every electrical part of an HVAC system. **304 pages, 3½ x 5½, $18.00**

Renovating & Restyling Vintage Homes

Any builder can turn a run-down old house into a showcase of perfection — if the customer has unlimited funds to spend. Unfortunately, most customers are on a tight budget. They usually want more improvements than they can afford — and they expect you to deliver. This book shows how to add economical improvements that can increase the property value by two, five or even ten times the cost of the remodel? Sound impossible? Here you'll find the secrets of a builder who has been putting these techniques to work on Victorian and Craftsman-style houses for twenty years. You'll see what to repair, what to replace and what to leave, so you can remodel or restyle older homes for the least amount of money and the greatest increase in value. **416 pages, 8½ x 11, $33.50**

Contractor's Survival Manual

How to survive hard times and succeed during the up cycles. Shows what to do when the bills can't be paid, finding money and buying time, transferring debt, and all the alternatives to bankruptcy. Explains how to build profits, avoid problems in zoning and permits, taxes, time-keeping, and payroll. Unconventional advice on how to invest in inflation, get high appraisals, trade and postpone income, and stay hip-deep in profitable work. **160 pages, 8½ x 11, $16.75**

Contractor's Year-Round Tax Guide Revised

How to set up and run your construction business to minimize taxes: corporate tax strategy and how to use it to your advantage, and what you should be aware of in contracts with others. Covers tax shelters for builders, write-offs and investments that will reduce your taxes, accounting methods that are best for contractors, and what the I.R.S. allows and what it often questions. **192 pages, 8½ x 11, $26.50**

Craftsman Book Company
6058 Corte del Cedro, P.O. Box 6500
Carlsbad, CA 92018

24 hour order line
1-800-829-8123
Fax (619) 438-0398

Order online
http://www.craftsman-book.com

In A Hurry?
We accept phone orders charged to your
Visa, MasterCard, Discover or American Express

_____
Name

_____
Company

_____
Address

_____
City/State/Zip

Total enclosed_____(In California add 7.25% tax)

We pay shipping when your check covers your order in full.

If you prefer, use your
❑Visa ❑MasterCard ❑Discover or ❑American Express

Card#_____Exp. date_____Initials_____

Tax Deductible: Treasury regulations make these references tax deductible when used in your work. Save the canceled check or charge card statement as your receipt.

10-Day Money Back Guarantee

❑34.00 Basic Engineering for Builders
❑33.00 Basic Plumbing with Illustrations
❑14.75 Blueprint Reading for Building Trades
❑19.75 Bookkeeping for Builders
❑26.50 Builder's Guide to Accounting Revised
❑27.25 Builder's Guide to Room Additions
❑35.00 Building Contractor's Exam Preparation Guide
❑35.50 Carpentry Estimating with FREE *Carpenter's Dream* material-estimating program on a 5¼" HD disk.
❑59.00 CD Estimator
❑39.50 Construction Estimating Reference Data with FREE stand-alone *Windows* estimating program on a 3½" HD disk.
❑39.75 Construction Forms & Contracts with a 3½" HD disk. Add $15.00 if you need ❑Macintosh disks.
❑19.25 Construction Surveying & Layout
❑19.00 Contractor's Growth & Profit Guide
❑39.00 Contractor's Guide to Building Code Revised
❑59.95 Contractor's Legal Kit
❑16.75 Contractor's Survival Manual
❑26.50 Contractor's Year-Round Tax Guide Revised
❑36.00 Craftsman's Illustrated Dictionary of Construction Terms
❑27.50 Drafting House Plans
❑34.95 Estimating Framing Quantities
❑22.50 Estimating Plumbing Costs
❑22.50 Finish Carpenter's Manual
❑28.75 Handbook of Construction Contracting Volume 1
❑30.75 Handbook of Construction Contracting Volume 2

❑24.95 Home Inspection Handbook
❑24.25 How to Succeed w/Your Own Construction Business
❑23.00 National Building Cost Manual
❑37.50 National Construction Estimator with FREE stand-alone *Windows* estimating program on a 3½" HD disk
❑38.00 National Painting Cost Estimator with FREE stand-alone *Windows* estimating program on a 3½" HD disk.
❑39.50 National Renovation & Insurance Repair Estimator with FREE stand-alone *Windows* estimating program on a 3½" HD disk.
❑38.50 National Repair & Remodeling Estimator with FREE stand-alone *Windows* estimating program on a 3½" HD disk.
❑19.25 Planning Drain, Waste & Vent Systems
❑29.00 Plumber's Exam Prep. Guide
❑18.00 Plumber's Handbook Revised
❑24.50 Professional Kitchen Design
❑19.75 Profits in Buying & Renovating Homes
❑33.50 Renovating & Restyling Vintage Homes
❑38.80 Residential Steel Framing
❑22.00 Roof Framing
❑35.00 Roofing Const. & Estimating
❑26.50 Rough Framing Carpentry
❑18.00 Vest Pocket Guide to HVAC Electricity
❑25.50 Wood-Frame House Construction
❑38.25 National Plumbing & HVAC Estimator with FREE stand-alone *Windows* estimating program on a 3½" HD disk.
❑FREE Full Color Catalog

BUSINESS REPLY MAIL
FIRST CLASS MAIL PERMIT NO. 271 CARLSBAD CA

POSTAGE WILL BE PAID BY ADDRESSEE

Craftsman Book Company
6058 Corte del Cedro
P.O. Box 6500
Carlsbad CA 92018-9974

BUSINESS REPLY MAIL
FIRST CLASS MAIL PERMIT NO. 271 CARLSBAD CA

POSTAGE WILL BE PAID BY ADDRESSEE

Craftsman Book Company
6058 Corte del Cedro
P.O. Box 6500
Carlsbad CA 92018-9974

BUSINESS REPLY MAIL
FIRST CLASS MAIL PERMIT NO. 271 CARLSBAD CA

POSTAGE WILL BE PAID BY ADDRESSEE

Craftsman Book Company
6058 Corte del Cedro
P.O. Box 6500
Carlsbad CA 92018-9974